T0207486

Lecture Notes in Computer Science 13526

More information about this series at https://link.springer.com/bookseries/558

Ivana Ljubić · Francisco Barahona ·
Santanu S. Dey · A. Ridha Mahjoub (Eds.)

Combinatorial Optimization

7th International Symposium, ISCO 2022
Virtual Event, May 18–20, 2022
Revised Selected Papers

 Springer

Editors
Ivana Ljubić (iD)
ESSEC Business School of Paris
Cergy Pontoise Cedex, France

Francisco Barahona (iD)
IBM TJ Watson Research Center
Yorktown Heights, NY, USA

Santanu S. Dey (iD)
Georgia Institute of Technology
Atlanta, GA, USA

A. Ridha Mahjoub (iD)
Université Paris-Dauphine
Paris, France

ISSN 0302-9743   ISSN 1611-3349 (electronic)
Lecture Notes in Computer Science
ISBN 978-3-031-18529-8   ISBN 978-3-031-18530-4 (eBook)
https://doi.org/10.1007/978-3-031-18530-4

This Springer imprint is published by the registered company Springer Nature Switzerland AG
The registered company address is: Gewerbestrasse 11, 6330 Cham, Switzerland

Preface

This volume contains 24 regular papers presented at ISCO 2022, the 7th International Symposium on Combinatorial Optimization, during May 18–20, 2022. The conference was held online, attracting more than 250 registered participants from all around the world. This edition of ISCO also included presentations of short papers. ISCO 2022 was preceded during May 16–17 by the doctoral school entitled "An introduction to quantum algorithms for optimization" given by Giacomo Nannicini (IBM, USA). Four eminent invited speakers also gave talks at the symposium: Jon Lee (University of Michigan, USA), Petra Mutzel (Bonn University, Germany), Rekha R. Thomas (University of Washington, USA), and Rico Zenklusen (ETH Zürich, Switzerland).

The ISCO series aims to bring together researchers from all communities related to combinatorial optimization, including algorithms and complexity, mathematical programming, operations research, stochastic optimization, graphs, and polyhedral combinatorics. It is intended to be a forum for presenting original research on all aspects of combinatorial optimization, ranging from mathematical foundations and theory of algorithms to computational studies and practical applications, and especially their intersections. In response to the call for papers, ISCO 2022 received 50 regular submissions. Each submission was reviewed by three to five Program Committee members with the assistance of external reviewers. The submissions were judged on their originality and technical quality. The review process was extremely selective and many good papers could not be accepted. As a result, 24 regular papers were selected, giving an acceptance rate of 48%. The revised versions of these 24 regular papers presented at the symposium are included in this volume. Overall, 52 talks (based on 24 regular papers and 28 short abstracts) of 25–30 minutes each, grouped into 16 sessions, were given at the conference.

We would like to thank all the authors who submitted their work to ISCO 2022, and the Program Committee members and external reviewers for their exceptional work. We would also like to thank our invited speakers as well as the speaker of the doctoral school for their excellent and inspiring lectures. They all contributed to the quality of the symposium. Finally, we would like to thank the members of the Organizing Committee for hosting the conference website, and for their remarkable technical assistance and support.

June 2022

Ivana Ljubić
Francisco Barahona
Santanu S. Dey
A. Ridha Mahjoub

Organization

Conference Chairs

Ivana Ljubić	ESSEC Business School, France
Francisco Barahona	IBM, T. J. Watson Research Center, New York, USA
Santanu Dey	Georgia Institute of Technology, USA
A. Ridha Mahjoub	Paris Dauphine University, France

Organizing Committee

Zacharie Ales	ENSTA Paris, France
Fatiha Bendali	Université Clermont Auvergne, France
Ibrahima Diarrassouba	Université Le Havre Normandie, France
Diego Delle Donne	ESSEC Business School, France
Pierre Fouilhoux	Université Sorbonne Paris Nord, France
Hela Haj Mohamed	University of Monastir, Tunisia
Ivana Ljubić	ESSEC Business School, France
A. Ridha Majhoub	Paris Dauphine University, France

Steering Committee

Mourad Baïou	CNRS, Clermont Auvergne University, France
Pierre Fouilhoux	Université Sorbonne Paris Nord, France
Luis Gouveia	University of Lisbon, Portugal
Nelson Maculan	Federal University of Rio de Janeiro, Brasil
A. Ridha Mahjoub	Paris Dauphine University, France
Vangelis Paschos	Paris Dauphine University, France
Giovanni Rinaldi	IASI-CNR, Italy

Program Committee

Amitabh Basu	Johns Hopkins University, USA
Victor Blanco	University of Granada, Spain
Merve Bodur	University of Toronto, Canada
Flavia Bonomo	University of Buenos Aires, Argentina
Francesco Carrabs	University of Salerno, Italy
Margarida Carvalho	University of Montreal, Canada
Raffaele Cerulli	University of Salerno, Italy
Karthekeyan Chandrasekaran	University of Illinois Urbana-Champaign, USA

Plenary Lectures

Graphical Designs

Rekha R. Thomas

University of Washington, USA
rrthomas@uw.edu

Abstract. A graphical design on an undirected graph is a quadrature rule in the following sense: Given an eigenbasis of the graph Laplacian, a design is a collection of vertices of the graph (with weights) so that the weighted average of a collection of eigenvectors on this subset equals the weighted average on the full set of vertices. Depending on which eigenvectors are to be averaged, and requirements on the weights, one obtains different types of designs. Designs can be computed via linear and integer programming. In this talk I will show that positively weighted designs can be organized on the faces of a polytope, and using this connection, we compute optimal designs in several graph families. Joint work with Catherine Babecki.

Bio: Rekha Thomas received a Ph.D. in Operations Research from Cornell University in 1994 under the supervision of Bernd Sturmfels. This was followed by two postdoctoral positions, the first at the Cowles Foundation for Economics at Yale University and the second at the Konrad-Zuse-Zentrum for Informationstechnik in Berlin. She is currently the Walker Family Endowed Professor in Mathematics at the University of Washington in Seattle. Her research interests are in optimization and applied algebraic geometry.

Advances in Approximation Algorithms for Tree Augmentation

Rico Zenklusen

ETH Zurich, Switzerland
`ricoz@ethz.ch`

Abstract. Augmentation problems are a fundamental class of Network Design problems. The goal is to find a cheapest way to increase the (edge-) connectivity of a graph by adding edges among a given set of options. The Minimum Spanning Tree Problem is one of its most elementary examples, which can be interpreted as determining a cheapest way to increase the edge-connectivity of a graph from 0 to 1. The "next step", to increase from 1 to 2, leads to the heavily studied Tree Augmentation Problem, which is the focus of this talk. This talk has several goals, namely:

1. Providing a brief introduction to Tree Augmentation and some related problems.
2. Discussing relevant algorithmic techniques, including the Relative Greedy method and a new link to local search procedures.
3. Showing how these techniques can be leveraged to address a long-standing open question, namely how to obtain better-than-2 approximations for (Weighted) Tree Augmentation.

Bio: Rico Zenklusen is a Professor in the Mathematics Department at ETH Zurich, heading the Combinatorial Optimization Group. Prior to joining ETH Zurich, Rico was on the faculty of the Johns Hopkins University, before which he worked several years as a postdoc at MIT, and also shortly at EPFL. Rico holds a PhD from ETH Zurich and a master's degree from EPFL. His main research interests lie broadly in Combinatorial Optimization and its applications, ranging from foundational research related to polyhedra, (poly-) matroids, and submodular functions to industrial collaborations.

Algorithmic Data Science

Petra Mutzel

Bonn University, Germany
petra.mutzel@cs.uni-bonn.de

Abstract. The area of algorithmic data science offers new opportunities for researchers in the algorithmic and the optimization community. In this talk we will first survey four fundamental problems for analysing data. The basis for these problems are concepts for distance and similarity. We will discuss similarity concepts for graphs that are relevant for analysis tasks on graph data sets. These approaches are increasingly applied in the context of data analysis tools for systems with a network structure. Applications are, e.g., learning tasks in drug design, social network analysis, and geodesy.

Bio: Petra Mutzel is professor of Computational Analytics at the University of Bonn, where she is also the scientific director of the High Performance Computing and Analytics Lab at the Digital Science Center. Before she was professor at TU Dortmund University and at Vienna University of Technology. She received her Ph.D. in Computer Science at the University of Cologne in 1994, followed by a PostDoc position at the Max Planck Institute for Informatics in Saarbrücken. Her research focuses on algorithm engineering, algorithmic data analysis, and combinatorial optimization for graphs and networks. Currently, the main application areas are in cheminformatics, social and biological network analysis, statistical physics, and geodesy. She is a member of the Steering Committees of ESA, ALENEX, and WALCOM, and Associate Editor of the ACM Journal on Experimental Algorithmics, Journal of Graph Algorithms and Applications (JGAA), and Mathematical Programming Computation (MPC).

Recent Algorithmic Advances for Maximum-Entropy Sampling

Sampling

Jon Lee

University of Michigan, USA
jonxlee@umich.edu

Abstract. The maximum-entropy sampling problem (MESP) is to select a subset, of given size s, from a set of correlated Gaussian random variables, so as to maximize the differential entropy. If C is the covariance matrix, then we are simply seeking to maximize the determinant of an order-s principal submatrix. A key application is for the contraction of an environmental-monitoring network. MESP sits within the intersection of optimization and data science, and so it has attracted a lot of recent attention. The problem is NP-hard, and there have been algorithmic attacks aimed at exact solution of moderate-sized instance for three decades. It is a fascinating problem from the perspective of integer nonlinear optimization, as it does not fit within a framework that is successfully attacked via available general-purpose paradigms. I will give a broad overview of algorithmic work, concentrating on the many useful techniques related to various convex relaxations.

Bio: Jon Lee obtained his Ph.D. at Cornell University. He has held long-term positions at Yale University, University of Kentucky, IBM Research, and New York University. Now at the University of Michigan, Jon is the G. Lawton and Louise G. Johnson Professor of Engineering and Professor of Industrial and Operations Engineering. He is author of "A First Course in Combinatorial Optimization" (Cambridge University Press), "A First Course in Linear Optimization" (Reex Press), and "Maximum-Entropy Sampling: Algorithms and Application" (with M. Fampa). Jon was the founding Managing Editor of the journal Discrete Optimization, and he is currently Editor-in-Chief of Mathematical Programming, Series A. Jon is an INFORMS Fellow, and he has received the INFORMS Computing Society Prize.

Contents

Polyhedra and Algorithms

New Classes of Facets
for Complementarity Knapsack Problems

Alberto Del Pia[1,2] , Jeff Linderoth[1] , and Haoran Zhu[1(✉)]

[1] Department of Industrial and Systems Engineering, University of
Wisconsin-Madison, Madison, WI, USA
{delpia,linderoth,hzhu94}@wisc.edu
[2] Wisconsin Institute for Discovery, University of Wisconsin-Madison,
Madison, WI, USA

Abstract. The *complementarity knapsack problem* (CKP) is a knapsack problem with real-valued variables and complementarity conditions between pairs of its variables. We extend the polyhedral studies of De Farias et al. for CKP, by proposing three new families of cutting-planes that are all obtained from a combinatorial concept known as a *pack*. Sufficient conditions for these inequalities to be facet-defining, based on the concept of a *maximal switching pack*, are also provided. Moreover, we answer positively a conjecture by De Farias et al. about the separation complexity of the inequalities introduced in their work.

Keywords: Complementarity knapsack polytope · Complementarity problem · Pack inequalities · Branch-and-cut

1 Introduction

In this paper, we investigate cutting-planes for the *complementarity knapsack problem* (CKP). Let $M = \{1, \ldots, m\}$, and, for every $i \in M$, let $N_i = \{1, \ldots, n_i\}$, for some $m, n_i \in \mathbb{N}$. Let $b \in \mathbb{R}_+$ and $c_{ij}, a_{ij} \in \mathbb{R}_+$ for every $i \in M, j \in N_i$. Then the *CKP* is

$$\max \quad \sum_{i \in M} \sum_{j \in N_i} c_{ij} x_{ij} \tag{CKP}$$

$$\text{s.t.} \quad \sum_{i \in M} \sum_{j \in N_i} a_{ij} x_{ij} \le b, \tag{1}$$

$$x_{ij} \cdot x_{ij'} = 0, \qquad i \in M, j \ne j' \in N_i, \tag{2}$$

$$0 \le x_{ij} \le 1, \qquad i \in M, j \in N_i.$$

A. Del Pia is partially funded by ONR grant N00014-19–1-2322. Any opinions, findings, and conclusions or recommendations expressed in this material are those of the authors and do not necessarily reflect the views of the Office of Naval Research.

I. Ljubić et al. (Eds.): ISCO 2022, LNCS 13526, pp. 3–21, 2022.
https://doi.org/10.1007/978-3-031-18530-4_1

4   A. Del Pia et al.

The constraints in (2), enforcing that at most one of the variables in the set N_i take a positive value are usually referred to as *complementarity constraints*. Often, in the literature, the set of variables $\{x_{ij}\}_{j \in N_i}$, for any $i \in M$, is called a *special ordered set of type 1* (SOS1). When the variables are binary, the corresponding complementarity constraint is often called a *generalized upper bound* (GUB).

Traditionally, (CKP) is modeled by introducing binary variables y_{ij} for any $i \in M, j \in N_i$, and the complementarity constraints (2) are formulated by $x_{ij} \leq y_{ij}$ and $\sum_{j \in N_i} y_{ij} \leq 1$ for any $i \in M$. Then cutting-planes are added into this mixed-integer programming (MIP) formulation in order to obtain tighter relaxation, specifically cutting-planes from knapsack polytope, see, e.g., [2,9,10,22,23]. Note that all the cutting-planes that are added into such formulation are "general purpose", in the sense that their derivation does not take into consideration the structure of the specific problem at hand. This approach naturally has several computational disadvantages, including the size increase of the problem and the loss of structure, see [7,14]. For these reasons, De Farias et al. [8] conducted a polyhedral study for CKP in the space of the continuous variables. In particular, the authors presented two families of facet-defining inequalities called *fundamental complementarity inequalities*, that can be derived by lifting *cover inequalities* of CKP.[1] In a more recent paper, De Farias et al. [12] first generalized the two families of inequalities introduced in [8], and then numerically tested the performance of these inequalities by using them within a branch-and-cut algorithm. Our paper is motivated by these papers of de Farias et al. [8,12]. In particular, we further investigate the polyhedral study for CKP in the space of the original variables. We remark that, the idea of dispensing with the use of auxiliary binary variables to model complementarity constraints, and enforcing those constraints directly appears also in, e.g., [4,5,13,20].

In a *branch-and-cut* procedure using the inequalities proposed in [12] to solve CKP, a crucial step is *separation*: given an optimal solution of the current linear relaxation, how can we efficiently identify an inequality in a given class to separate the given solution and strengthen the relaxation? de Farias et al. conjectured that the separation of their inequalities is $\mathcal{NP}$-complete. In our first contribution of this paper, we show that it is indeed $\mathcal{NP}$-complete to separate both classes of inequalities proposed in [12]. Our second contribution is the introduction of novel valid inequalities for the convex hull of the feasible set of (CKP), which is coined the *complementarity knapsack polytope* by De Farias et al. We also provide sufficient conditions for these inequalities to be facet-defining. We remark that a key difference between our inequalities and the ones proposed in [8,12], is that the inequalities that we introduce cannot be derived by the same lifting procedure as proposed in [8]. This is further discussed at the end of Sect. 4.

This paper is organized as follows. In Sect. 2, we introduce some notation, assumptions, and previous results in literature. In Sect. 3, we provide the positive answer to the conjecture made by de Farias et al. [12]. In Sect. 4, we introduce the concept of *pack* and *maximal switching pack*, and propose three new families

---

[1] These are not the typical "cover inequalities" in 0–1 programming.

of cutting-planes for (CKP) which are originated from packs. We also show that these inequalities are facet-defining if the pack itself or some particular subset of the pack is a maximal switching pack. In Sect. 5, we discuss directions for future research.

2 Notations, Assumptions, and Previous Work

First, we introduce some commonly used notation. To the best of our knowledge, the papers [8,12] are the only polyhedral studies of the complementarity knapsack polytope. Using the same notation as in [8], we denote by S the set of feasible solutions of (CKP), and by $PS := \text{conv}(S)$ the *complementarity knapsack polytope*. Here PS is a polytope because S can be seen as the union of finitely many polytopes. We denote by d the number of variables in the problem, i.e., $d := \sum_{i \in M} n_i$. For ease of notation, we denote by ij the ordered pair (i, j). We also identify any set with one element with the element itself. We denote by $I := \cup_{i \in M}(i \times N_i)$, which is simply the set of all indices of x. We denote by M_0 the set of indices in M that do not lie in any complementarity constraint, i.e., $M_0 := \{i \in M \mid n_i = 1\}$. For $T \subseteq I$, we define $M_T := \{i \in M \mid ij \in T \text{ for some } j \in N_i\}$. For any $n \in \mathbb{N}$, we let $[n] := \{1, \ldots, n\}$. For any vector $v \in \mathbb{R}^n$, we denote by $\text{supp}(v)$ the support set of vector v, i.e., $\text{supp}(v) := \{i \in [n] \mid v_i \neq 0\}$.

As in [12], we make the following assumptions:

Assumption 1 (Assumption 1 in [12]). $M \neq M_0$.

Assumption 2 (Assumption 2 in [12]). $\sum_{i \in M} \max\{a_{i1}, \ldots, a_{in_i}\} > b$.

Assumption 3 (Assumption 3 in [12]). $b > 0, a_{ij} \geq 0$, for all $ij \in I$.

Assumption 4 (Assumption 6 in [12]). $a_{i1} \geq \ldots \geq a_{in_i}$, for all $i \in M - M_0$.

All these assumptions can be made without loss of generality (w.l.o.g.): If the first two assumptions do not hold, then problem (CKP) is trivial. For Assumption 3, if $a_{ij} \leq 0$ for some index ij, then x_{ij} can be fixed to zero. Assumption 4 is made because we are dealing with a single knapsack inequality, and we make this assumption to simplify the presentation.

Next, we introduce a few basic properties of PS, and review the concepts and results introduced in De Farias et al. [8,12]. Throughout this section, known results are not proved, but referenced. The results that are proved are, to the best of our knowledge, new.

Lemma 1 (Proposition 2 in [8]). *PS is full-dimensional.*

Lemma 2. *If $\tilde{x}$ is a non-integral vertex of PS, then it has exactly one fractional component, and we have $\sum_{i \in M} \sum_{j \in N_i} a_{ij} \tilde{x}_{ij} = b$.*

6   A. Del Pia et al.

Proof. If $\tilde{x}$ has at least two fractional components, say, $\tilde{x}_{i'j'}, \tilde{x}_{i''j''} \in (0,1)$, for some $i' \neq i'' \in M$, $j' \in N_{i'}$, $j'' \in N_{i''}$. Then the 2-dimensional point $(\tilde{x}_{i'j'}, \tilde{x}_{i''j''})$ is contained in the 2-dimensional polyhedron

$$\left\{ (x_{i'j'}, x_{i''j''}) \in [0,1]^2 \mid a_{i'j'}x_{i'j'} + a_{i''j''}x_{i''j''} \leq b - \sum_{i \in M}\sum_{j \in N_i} a_{ij}\tilde{x}_{ij} + a_{i'j'}\tilde{x}_{i'j'} + a_{i''j''}\tilde{x}_{i''j''} \right\}.$$

It is simple to check that the extreme points of this 2-dimensional polyhedron all have at most one fractional component, therefore $(\tilde{x}_{i'j'}, \tilde{x}_{i''j''})$ can be written as the convex combination of some other 2-dimensional points in the above polyhedron. Note that replacing the $x_{i'j'}$ and $x_{i''j''}$ components of $\tilde{x}$ with the components of these 2-dimensional points will lead to vectors in S, and $\tilde{x}$ can be written as the convex combination of these resulting points. This gives us a contradiction, since $\tilde{x}$ is an extreme point of PS by assumption.

We now assume, without loss of generality, that $\tilde{x}_{11}$ is the only fractional component of vertex $\tilde{x}$. If $\sum_{i \in M}\sum_{j \in N_i} a_{ij}\tilde{x}_{ij} < b$, then replacing the x_{11} component of $\tilde{x}$ by $\tilde{x}_{11} + \epsilon$ or $\tilde{x}_{11} - \epsilon$, for some small enough ϵ, will still satisfy the inequality $\sum_{i \in M}\sum_{j \in N_i} a_{ij}x_{ij} \leq b$ and the complementarity constraints. But this also implies that $\tilde{x}$ is not a vertex of PS, which gives the contradiction. $\square$

Lifting is an important concept in integer programming that is used to strengthen valid inequalities [2,3,6,16,17]. In the case of 0–1 knapsack problems, the *cover inequalities* are one of the most well-known family of cuts. Applying the lifting process to cover inequalities produces the widely used *lifted cover inequalities*, discovered independently by Balas [2] and Wolsey [24]. (See also the survey [1]). In [8], the authors extended the concept of a cover and cover inequalities to CKPs as follows:

Definition 1 (Definition 1 in [8]). *Let* $C = \{i_1j_1, \ldots, i_kj_k\} \subset I$, *where* $i_1, \ldots, i_k$ *are all distinct. The set* C *is called a* cover *if* $\sum_{ij \in C} a_{ij} > b$. *Given a cover* C, *the inequality* $\sum_{ij \in C} a_{ij}x_{ij} \leq b$ *is called a* cover inequality.

Next, we present two families of cutting-planes for PS proposed in [12], which are obtained by sequentially lifting the cover inequalities with respect to covers which satisfy different conditions. The obtained inequalities generalize the two families of inequalities given in [8].

Theorem 1 (Theorem 3 in [12]). *Let* C *be a cover. Denote the elements of* C *as* ir_i, $\forall i \in M_C$. *Assume that* $\sum_{i \in M_C - i'} a_{ir_i} + a_{i'j'} < b$ *for some* $i' \in M_C$ *and* $j' \in \{j \in N_{i'} \mid r_{i'} < j' \leq n_{i'}\}$. *Then,*

$$\sum_{i \in M_C}\sum_{j=1}^{r_i-1} a_{ir_i}x_{ij} + \sum_{i \in M_C}\sum_{j=r_i}^{n_i} \max\left\{ a_{ij}, b - \sum_{k \in M_C - i} a_{kr_k} \right\} x_{ij} \leq b \quad (3)$$

is valid for PS. *Inequality* (3) *is facet-defining when* $r_i = 1$, $\forall i \in M_C$.

Theorem 2 (Theorem 4 in [12]). *Let C be a cover and $i'j' \in C$ with $j' < n_{i'}$. Denote the elements of C other than $i'j'$ as it_i, $\forall i \in M_C - i'$. Suppose that $\sum_{i \in M_C - i'} a_{it_i} + a_{i'n_{i'}} < b$. Then,*

$$
\sum_{j \in N_{i'}} \max \left\{ a_{i'j}, b - \sum_{k \in M_C - i'} a_{kt_k} \right\} x_{i'j}
$$
$$
+ \sum_{i \in M_C - i'} \left(\sum_{j=1}^{t_i} a_{it_i} \max \left\{ 1, \frac{a_{ij}}{b - \sum_{k \in M_C - i, i'} a_{kt_k} - a_{i'n_{i'}}} \right\} x_{ij} + \sum_{j=t_i+1}^{n_i} a_{ij} x_{ij} \right) \le b
$$
$$
(4)
$$

is valid for PS. Inequality (4) is facet-defining when $t_i = n_i$, $\forall i \in M_C - i'$.

3 Separation Complexity of Lifted Cover Inequalities for CKP

A fundamental component of cutting plane algorithms is separation—Does there exist a cutting plane within a given family that is violated by a given infeasible solution. For many classic classes of inequalities for integer programming, separation is known to be $\mathcal{NP}$-complete. In knapsack problems, [21] showed that the separation for cover inequalities is $\mathcal{NP}$-complete, and [15] showed the same result for lifted cover inequalities. Moreover, recently Del Pia et al. [11] further extended the complexity result for 0–1 knapsack problems to extended cover inequalities [2], $(1, k)$-configuration inequalities [22] and weight inequalities [23]. In [12], De Farias et al. also conjectured that the separation problems for both (3) and (4) are $\mathcal{NP}$-complete. In this section, we show that this conjecture is true.

First, we formally define the separation problems for (3) and (4). Here the LP relaxation of (CKP) is simply the optimization problem without the complementarity constraints (2).

Problem SP1
Input: $(c, a, b) \in (\mathbb{Z}_+^d, \mathbb{Z}_+^d, \mathbb{Z}_+)$ and an optimal solution x^* to the LP relaxation of (CKP).
Question: Is there a cover $C = \{ir_i \mid i \in M_C\}$ of (CKP), such that

$$
\sum_{i \in M_C} \sum_{j=1}^{r_i - 1} a_{ir_i} x_{ij}^* + \sum_{i \in M_C} \sum_{j=r_i}^{n_i} \max \left\{ a_{ij}, b - \sum_{k \in M_C - i} a_{kr_k} \right\} x_{ij}^* > b? \quad (5)
$$

Problem SP2
Input: $(c, a, b) \in (\mathbb{Z}_+^d, \mathbb{Z}_+^d, \mathbb{Z}_+)$ and an optimal solution x^* to the LP relaxation of (CKP).
Question: Is there a cover $C = \{it_i \mid i \in M_C - i'\} \cup \{i'j'\}$ of (CKP), with $\sum_{i \in M_C - i'} a_{it_i} + a_{i'n_{i'}} < b$, such that

$$\sum_{j \in N_{i'}} \max \left\{ a_{i'j}, b - \sum_{k \in M_C - i'} a_{kt_k} \right\} x_{i'j}$$

$$+ \sum_{i \in M_C - i'} \left(\sum_{j=1}^{t_i} a_{it_i} \max \left\{ 1, \frac{a_{ij}}{b - \sum_{k \in M_C - i,i'} a_{kt_k} - a_{i'n_{i'}}} \right\} x_{ij} + \sum_{j=t_i+1}^{n_i} a_{ij} x_{ij} \right) > b?$$

$$(6)$$

Now, we show that Problem SP1 is $\mathcal{NP}$-complete. The reduction is from the *partition problem*: Given $(\alpha_1, \ldots, \alpha_k; \beta) \in (\mathbb{Z}_+^k, \mathbb{Z}_+)$ with $\sum_{i=1}^k \alpha_i = 2\beta$, does there exist a subset $S \subseteq [k]$ such that $\sum_{i \in S} \alpha_i = \beta$? The partition problem is one of the original 21 problems that Karp demonstrated to be $\mathcal{NP}$-complete [19].

Theorem 3. *Problem SP1 is $\mathcal{NP}$-complete.*

Proof. Since verifying if a given point violates a given inequality can be done in polynomial time with respect to the input size of the point and the inequality, the separation problem SP1 is clearly in the class $\mathcal{NP}$. Therefore, we only have to show that SP1 is $\mathcal{NP}$-hard.

Given an instance $(\alpha_1, \ldots, \alpha_k; \beta) \in (\mathbb{Z}_+^k, \mathbb{Z}_+)$ of the partition problem, we construct the following instance of (CKP), where the data (a, b, c, x^*) is defined as follows:

$$M := [k+1], n_1 = \ldots = n_k := 1, n_{k+1} := \beta + 1,$$
$$a_{i1} := \alpha_i \ \forall i \in [k], a_{k+1,1} := 3, a_{k+1,2} = \ldots = a_{k+1,\beta+1} := 1,$$
$$b := \beta + 2, \quad c := a, \tag{7}$$
$$x_{11}^* = \ldots = x_{k1}^* := \frac{2\beta - 3}{6\beta}, x_{k+1,1}^* := 1, x_{k+1,2}^* = \ldots = x_{k+1,\beta+1}^* := \frac{1}{3}.$$

Since $\sum_{i=1}^n \alpha_i = 2\beta$, here $\sum_{i \in M} \sum_{j \in N_i} a_{ij} x_{ij}^* = b$ and $c = a$, so x^* is an optimal solution to the LP relaxation of our constructed CKP instance. Hence, (a, b, c, x^*) is a correct input to problem SP1, and the encoding size of (a, b, c, x^*) is polynomial in the encoding size of $(\alpha_1, \ldots, \alpha_k; \beta)$.

First, assume $S \subseteq [k]$ is a yes-certificate to the above partition problem, with $\sum_{i \in S} \alpha_i = \beta$. Then, we consider $C = \{i1 \mid i \in S \cup \{k+1\}\}$. We have $\sum_{ij \in C} a_{ij} = \sum_{i \in S} \alpha_i + 3 = \beta + 3 = b + 1$, so C is a cover. Now we verify (5). By plugging C and x^* into the left-hand side of (5), we obtain: $\sum_{i \in S} \alpha_i \cdot \frac{2\beta - 3}{6\beta} + (3 \cdot 1 + \sum_{j=2}^{\beta+1} 2 \cdot \frac{1}{3}) = \beta + \frac{5}{2}$, which is larger than b. Hence, from a yes-certificate to the partition problem, we can obtain a yes-certificate to problem SP1 with the above constructed input data.

Second, we assume C is a cover to our constructed CKP, such that the corresponding inequality (3) is not satisfied by x^*. The assumption in Theorem 1 implies that there exists some $i' \in M_C$ and j' such that $\sum_{i \in M_C - i'} a_{ir_i} + a_{i'j'} < b$, since otherwise inequality (3) with be dominated by original knapsack constraint (1). In our constructed CKP instance, we know that i' must be $k+1$ and $(k+1, 1) \in C$. In other words, $\sum_{i \in M_C \cap [k]} \alpha_i + 1 < b$. Since C is a cover, we also have $\sum_{i\in M_C\cap[k]}\alpha_i + 3 > b$. Here $\alpha_i \in \mathbb{Z}_+$ for all $i \in [k]$. Therefore, $\sum_{i\in M_C\cap[k]} = b - 2 = \beta$, which means $M_C \cap [k]$ is a yes-certificate for the above partition problem.

We have shown that there is a yes-certificate to SP1 with input (a,b,c,x^*) in (7) if and only if there is a yes-certificate to the partition problem with input $(\alpha_1,\ldots,\alpha_k;\beta)$. Since the partition problem is $\mathcal{NP}$-hard, we obtain that SP1 is $\mathcal{NP}$-hard as well. $\square$

The proof of the next theorem is almost identical to the proof of Theorem 3, for the completeness of this paper, we include it here.

Theorem 4. *Problem SP2 is $\mathcal{NP}$-complete.*

Proof. Same as the proof for Theorem 3, here we only have to prove the $\mathcal{NP}$-hardness of SP2. Given an instance $(\alpha_1,\ldots,\alpha_k;\beta)$ of the partition problem, we construct the instance of (CKP) with data (a,b,c,x^*) according to (7). As we have shown in the proof for Theorem 3, (a,b,c,x^*) is a correct input to SP2 with polynomial encoding size with respect to that of $(\alpha_1,\ldots,\alpha_k;\beta)$.

First, assume $S \subseteq [k]$ is a yes-certificate to the above partition problem, with $\sum_{i\in S}\alpha_i = \beta$. Then, we consider $C = \{i1 \mid i \in S \cup \{k+1\}\}$, and pick $(i',j') = (k+1,1)$. Here $\sum_{ij\in C}a_{ij} = \sum_{i\in S}\alpha_i + 3 = \beta + 3 = b + 1$, so C is a cover. Also, $\sum_{i\in M_C-\{k+1\}}a_{i1} + a_{k+1,\beta+1} = \sum_{i\in S}\alpha_i + 1 = \beta + 1 < b$. Now we verify (6). By plugging C and x^* into the left-hand side of (6), we obtain: $(3\cdot 1 + \sum_{j=2}^{\beta+1}2\cdot\frac{1}{3}) + \sum_{i\in S}\alpha_i\cdot\frac{2\beta-3}{6\beta} = \beta + \frac{5}{2}$, which is larger than b. Hence, from a yes-certificate to the partition problem, we can obtain a yes-certificate to problem SP2 with input data (a,b,c,x^*) constructed in (7).

Second, we assume C is a cover to our constructed CKP and $i'j' \in C$, such that $\sum_{i\in M_C-i'}a_{it_i} + a_{i'n_{i'}} < b$, and the corresponding inequality (4) is not satisfied by x^*. In our constructed CKP instance, we know that (i',j') must be $(k+1,1)$. In other words, $\sum_{i\in M_C\cap[k]}\alpha_i + 1 < b$. Since C is a cover, we also have $\sum_{i\in M_C\cap[k]}\alpha_i + 3 > b$. Here $\alpha_i \in \mathbb{Z}_+$ for all $i \in [k]$. Therefore, $\sum_{i\in M_C\cap[k]} = b - 2 = \beta$, which means $M_C \cap [k]$ is a yes-certificate for the above partition problem.

We have shown that there is a yes-certificate to SP2 with input (a,b,c,x^*) in (7) if and only if there is a yes-certificate to the partition problem with input $(\alpha_1,\ldots,\alpha_k;\beta)$. Since the partition problem is $\mathcal{NP}$-hard, we obtain that SP2 is $\mathcal{NP}$-hard as well. $\square$

4 New Families of Facet-Defining Inequalities

In this section, we propose three new families of inequalities valid for PS, that are fundamentally different from (3) and (4). We also give sufficient conditions for the inequalities to be facet-defining. The inequalities are derived using the concept of a *pack*, a complementary concept to a cover, and we call the inequalities *complementarity pack inequalities*. We refer the reader to [1] for discussions on how packs are used to obtain strong valid inequalities in 0–1 programming.

Definition 2. *Let* $P = \{i_1 j_1, \ldots, i_k j_k\} \subset I$, *where* $i_1, \ldots, i_k$ *are all distinct. The set* P *is called a* pack *if* $\sum_{ij \in P} a_{ij} < b$. *A pack* P *is further called a* maximal switching pack, *if* $j = n_i$ *for any* $ij \in P$, *and* $\sum_{ij \in P} a_{ij} - a_{i'n_{i'}} + a_{i', n_{i'}-1} > b$, *for any* $i' \in M_P - M_0$.

Now we present the first family of complementarity pack inequalities.

Theorem 5. *Let* P *be a pack for* PS, *and* $s := \sum_{ij \in P} a_{ij}$. *Then*

$$\sum_{i \in M_P} \sum_{j \in N_i} a_{ij} x_{ij} + \sum_{\substack{ij \in P, \\ i \in M_P - M_0}} (b-s) x_{ij} \leq b + (|M_P - M_0| - 1)(b-s) \tag{8}$$

is a valid inequality for PS. *Furthermore, when* P *is a maximal switching pack, we have:*

1. *(8) induces a face of* PS *of dimension at least* $d - |M_P - M_0|$;
2. *When* $M_P \cap M_0 \neq \emptyset$, *(8) is facet-defining.*

Proof. First, we show that any feasible point $\tilde{x}$ in S satisfies (8). We denote $\bar{M} := M_P - M_0$, and $\bar{P} := \{ij \in P \mid i \in \bar{M}\}$. We have $|\bar{P}| = |\bar{M}| = |M_P - M_0|$.
Case 1: $\bar{P} \subseteq \text{supp}(\tilde{x})$. We know that, for any $i_1 j_1, i_2 j_2 \in P$, there is $i_1 \neq i_2$, and $\tilde{x}$ satisfies the complementarity constraint $\tilde{x}_{ij} \tilde{x}_{i,j'} = 0$ for any $j' \neq j \in N_i$, $i \in \bar{M}$. Therefore, in this case, $\sum_{i \in M_P} \sum_{j \in N_i} a_{ij} \tilde{x}_{ij} \leq \sum_{ij \in P} a_{ij} = s$. Hence we have

$$\sum_{i \in M_P} \sum_{j \in N_i} a_{ij} \tilde{x}_{ij} + (b-s) \sum_{ij \in \bar{P}} \tilde{x}_{ij} \leq \sum_{i \in M_P} \sum_{j \in N_i} a_{ij} \tilde{x}_{ij} + (b-s)|\bar{P}|$$

$$\leq s + (b-s)|\bar{P}|$$
$$= b + (|\bar{P}| - 1)(b-s)$$
$$= b + (|M_P - M_0| - 1)(b-s).$$

Case 2: $\bar{P} \not\subseteq \text{supp}(\tilde{x})$. In this case, we have $|\{ij \mid ij \in \bar{P}, ij \in \text{supp}(\tilde{x})\}| \leq |\bar{P}| - 1 = |M_P - M_0| - 1$. Hence

$$\sum_{i \in M_P} \sum_{j \in N_i} a_{ij} \tilde{x}_{ij} + (b-s) \sum_{ij \in \bar{P}} \tilde{x}_{ij} \leq b + (b-s)|\{ij \mid ij \in \bar{P}, ij \in \text{supp}(\tilde{x})\}|$$

$$\leq b + (|M_P - M_0| - 1)(b-s).$$

From the discussion in the above two cases, we have shown that (8) is satisfied by any feasible point $\tilde{x} \in S$, which means (8) is valid for PS.

Now, we assume P to be a maximal switching pack. A point x with $x_{ij} = 1 \ \forall ij \in P$, and 0 elsewhere, satisfies (8) at equality, and is in S, since by definition of pack we have $\sum_{ij \in P} a_{ij} < b$. Also, for any $i' \notin M_P$, $j' \in N_{i'}$, the point x with $x_{i'j'} = \min\{1, \frac{b-s}{a_{i'j'}}\}$, $x_{ij} = 1 \ \forall ij \in P$, and 0 elsewhere, is in S and satisfies (8) at equality. Hence so far we have found $\sum_{i \notin M_P} n_i + 1$ affinely independent feasible points of S that satisfy (8) at equality.

Recall that $j = n_i$ for any $ij \in P$ when P is a maximal switching pack. In the following, we discuss the dimension of the face induced by (8), and we consider separately two cases:

Case 1: $M_P \cap M_0 = \emptyset$. For any $i' \in M_P$, $j' < n_{i'}$, consider the point x with $x_{in_i} = 1 \; \forall i \in M_P - i'$, $x_{i'j'} = \frac{a_{i'n_{i'}} + b - s}{a_{i'j'}}$, and 0 elsewhere. By definition of maximal switching pack and Assumption 4, we know $x_{i'j'} < 1$. It is easy to verify that x is in S, and satisfies (8) at equality. Therefore, we find another $\sum_{i \in M_P} (n_i - 1)$ points in S that satisfy (8) at equality, together with the previous $\sum_{i \notin M_P} n_i + 1$ points in S, we find $\sum_{i \in M} n_i - |M_P| + 1 = d + 1 - |M_P|$ points of S in total which satisfy (8) at equality. From our characterization of these points, we know that they are affinely independent. Therefore, the face given by (8) has dimension at least $d - |M_P| = d - |M_P - M_0|$.

Case 2: $M_P \cap M_0 \neq \emptyset$. In this case we simply have $|M_P \cap M_0| \geq 1$. For any $i' \in \bar{M}$, we consider the following set

$$\Big\{ x \in [0,1]^d \mid x_{in_i} = 1 \; \forall i \in \bar{M} - i', x_{ij} = 0 \; \forall i \in \bar{M} - i' \text{ and } j \in N_i - n_i,$$

$$x_{i'n_{i'}} = 0, x_{ij} = 0 \; \forall i \notin M_P \text{ and } j \in N_i,$$

$$\sum_{i \in M_P \cap M_0} a_{i1} x_{i1} + \sum_{j \in N_{i'} - n_{i'}} a_{i'j} x_{i'j} = b - \sum_{i \in \bar{M} - i'} a_{in_i}, \qquad (9)$$

$$\text{set } \{x_{i'1}, \ldots, x_{i',n_{i'}-1}\} \text{ is SOS1} \Big\}.$$

By definition of maximal switching pack, we have $\sum_{i \in M_P \cap M_0} a_{i1} + a_{i'j} > b - \sum_{i \in \bar{M} - i'} a_{in_i}$ for any $j \in N_{i'} - n_{i'}$, so this set (9) has $|M_P \cap M_0| + n_{i'} - 1$ affinely independent points, which are all in S and satisfy (8) at equality. Hence in total we obtain $\sum_{i \in \bar{M}} (|M_P \cap M_0| + n_i - 1) = \sum_{i \in \bar{M}} n_i + |\bar{M}|(|M_P \cap M_0| - 1)$ affinely independent points in S that satisfy (8) at equality. Notice that these points all lie in the affine space

$$\Big\{ x \in \mathbb{R}^d \mid x_{ij} = 0 \; \forall i \notin M_P \text{ and } j \in N_i, \sum_{i \in \bar{M}} x_{in_i} = |\bar{M}| - 1,$$

$$\sum_{i \in M_P \cap M_0} a_{i1} x_{i1} + \sum_{j \in N_{i'} - n_{i'}} a_{i'j} x_{i'j} = b - \sum_{i \in \bar{M} - i'} a_{in_i} \Big\}, \qquad (10)$$

whose dimension is $d - \sum_{i \notin M_P} n_i - 2 = \sum_{i \in M_P} n_i - 2$. Since $|M_P \cap M_0| \geq 1$, we know that the number of previous points $\sum_{i \in \bar{M}} n_i + |\bar{M}|(|M_P \cap M_0| - 1)$ is at least $\sum_{i \in \bar{M}} n_i + |M_P \cap M_0| - 1$, which equals $\sum_{i \in M_P} n_i - 1$. Hence, we obtain another $\sum_{i \in M_P} n_i - 1$ affinely independent points in S that satisfy (8) at equality. Together with the previous $\sum_{i \notin M_P} n_i + 1$ points in S, in total we have found d points in S which satisfy (8) at equality, and this means that inequality (8) is facet-defining when $M_P \cap M_0 \neq \emptyset$.   □

Next, we provide two examples where inequality (8) is either facet-defining, or it defines a face of high dimension.

Example 1. Let $M = [5]$, $n_1 = n_2 = n_3 = 1, n_4 = n_5 = 2$, and let the knapsack inequality in (CKP) be

$$2x_{11} + 4x_{21} + 8x_{31} + (10x_{41} + 6x_{42}) + (8x_{51} + 4x_{52}) \leq 21.$$

Take $P_1 = \{(1,1), (3,1), (4,2), (5,2)\}$. Since $2 + 8 + 6 + 4 = 20 < 21$, and $2 + 8 + 10 + 4 > 21, 2 + 8 + 6 + 8 > 21$, we know P_1 is a maximal switching pack. Furthermore, $M_{P_1} \cap M_0 = \{1, 3\} \neq \emptyset$, so Theorem 5 gives facet-defining inequality $2x_{11} + 8x_{31} + (10x_{41} + 7x_{42}) + (8x_{51} + 5x_{52}) \leq 22$.

Next, consider the pack $P_2 = \{(3,1), (4,2), (5,2)\}$. Also P_2 is a maximal switching pack, with $M_{P_2} \cap M_0 \neq \emptyset$. P_2 gives us another facet $8x_{31} + (10x_{41} + 9x_{42}) + (8x_{51} + 7x_{52}) \leq 24$. $\diamond$

Example 2. Let $M = [4]$, $n_1 = 1, n_2 = n_3 = n_4 = 2$, and let the knapsack inequality in (CKP) be

$$2x_{11} + (14x_{21} + 10x_{22}) + (13x_{31} + 9x_{32}) + (9x_{41} + 6x_{42}) \leq 22.$$

Take $P_1 = \{(1,1), (2,2), (3,2)\}$, which is a maximal switching pack, with $M_{P_1} \cap M_0 \neq \emptyset$. Hence, Theorem 5 gives facet-defining inequality $2x_{11} + (14x_{21} + 11x_{22}) + (13x_{31} + 10x_{32}) \leq 23$.

Now pick $P_2 = \{(2,2), (3,2)\}$, which is also a maximal switching pack, but with $M_{P_2} \cap M_0 = \emptyset$. So the corresponding inequality (8): $(14x_{21} + 13x_{22}) + (13x_{31} + 12x_{32}) \leq 25$ is not necessarily facet-defining. Theorem 5 states that it induces a face of PS with dimension at least $d - |M_{P_2} - M_0| = 7 - 2 = 5$, which is exactly the dimension of the corresponding face of PS. $\diamond$

The next theorem provides us with the second class of complementarity pack inequalities for PS and provides a sufficient condition for them to be facet-defining.

Theorem 6. *Let P be a pack for PS with $|M_P - M_0| \geq 2$, and $s := \sum_{ij \in P} a_{ij}$. For any $i^*j^* \in P$ with $i^* \notin M_0$ and $j^* = n_{i^*}$, the inequality*

$$\sum_{i \in M_P - i^*} \sum_{j \in N_i} a_{ij} x_{ij} + \sum_{\substack{ij \in P, \\ i \in M_P - M_0 - i^*}} (b - s) x_{ij} + a_{i^*j^*} x_{i^*j^*}$$

$$+ \sum_{j \in N_{i^*} - j^*} a_{i^*j^*} \max\left\{1, \frac{a_{i^*j}}{a_{i^*j^*} + b - s}\right\} x_{i^*j} \leq b + (|M_P - M_0| - 2)(b - s)$$

(11)

is valid for PS. Furthermore, when P is a maximal switching pack, (11) is facet-defining.

Proof. First, let $\tilde{x}$ in S. We show that inequality (11) is valid for $\tilde{x}$. Here we denote by $\bar{M} := \{i \in M_P \mid i \notin M_0, i \neq i^*\}$, $\bar{P} := \{ij \in P \mid i \in \bar{M}\}$. We have $|\bar{P}| = |\bar{M}| = |M_P - M_0| - 1$.

Case 1: $\bar{P} \subseteq \mathrm{supp}(\tilde{x})$. In this case, we have $\sum_{i \in M_P, i \neq i^*} \sum_{j \in N_i} a_{ij} \tilde{x}_{ij} + a_{i^*j^*} \leq \sum_{ij \in P} a_{ij} = s$.

Case 1a: $\tilde{x}_{i^*j^*} > 0$. Because $\tilde{x}$ satisfies the complementarity constraint, we know that $\tilde{x}_{i^*j} = 0$ for all $j \in N_{i^*} - j^*$. So the left-hand side of (11) evaluated in $\tilde{x}$ is

$$\sum_{\substack{i \in M_P, j \in N_i \\ i \neq i^*}} a_{ij}\tilde{x}_{ij} + \sum_{\substack{ij \in P, \\ i \notin M_0, i \neq i^*}} (b-s)\tilde{x}_{ij} + a_{i^*j^*}\tilde{x}_{i^*j^*} \leq s + (b-s)|\bar{P}|$$

$$= s + (|M_P - M_0| - 1)(b-s)$$
$$= b + (|M_P - M_0| - 2)(b-s).$$

Case 1b: There exists $j' \in N_{i^*} - j^*$ such that $\tilde{x}_{i^*j'} \in (0,1]$. Since $\tilde{x} \in S$, we have $a_{i^*j'}\tilde{x}_{i^*j'} \leq b - \sum_{i \in M_P, i \neq i^*}\sum_{j \in N_i} a_{ij}\tilde{x}_{ij}$. Let $K := \sum_{i \in M_P, i \neq i^*}\sum_{j \in N_i} a_{ij}\tilde{x}_{ij}$. We then have $K \leq s - a_{i^*j^*}$. Thus, the left-hand side of (11) evaluated in $\tilde{x}$ is

$$\sum_{\substack{i \in M_P, j \in N_i \\ i \neq i^*}} a_{ij}\tilde{x}_{ij} + \sum_{\substack{ij \in P, \\ i \notin M_0, i \neq i^*}} (b-s)\tilde{x}_{ij} + a_{i^*j^*}\max\left\{1, \frac{a_{i^*j'}}{a_{i^*j^*}+b-s}\tilde{x}_{i^*j'}\right\}$$

$$\leq K + (|M_P - M_0| - 1)(b-s) + a_{i^*j^*}\max\left\{1, \frac{b-K}{a_{i^*j^*}+b-s}\right\}$$

$$= \max\left\{K + a_{i^*j^*}, K + a_{i^*j^*}\frac{b-K}{a_{i^*j^*}+b-s}\right\} + (|M_P - M_0| - 1)(b-s)$$

$$\leq \max\left\{s, \frac{K(b-s)+a_{i^*j^*}b}{a_{i^*j^*}+b-s}\right\} + (|M_P - M_0| - 1)(b-s)$$

$$\leq \max\left\{s, \frac{(s-a_{i^*j^*})(b-s)+a_{i^*j^*}b}{a_{i^*j^*}+b-s}\right\} + (|M_P - M_0| - 1)(b-s)$$

$$= \max\{s, s\} + (|M_P - M_0| - 1)(b-s)$$

$$= b + (|M_P - M_0| - 2)(b-s).$$

Case 1c: $\tilde{x}_{i^*j} = 0$ for all $j \in N_{i^*}$. The left-hand side of (11) evaluated in $\tilde{x}$ is

$$\sum_{\substack{i \in M_P, j \in N_i \\ i \neq i^*}} a_{ij}\tilde{x}_{ij} + \sum_{\substack{ij \in P, \\ i \notin M_0, i \neq i^*}} (b-s)\tilde{x}_{ij} \leq s + (|M_P - M_0| - 1)(b-s)$$

$$= b + (|M_P - M_0| - 2)(b-s).$$

Case 2: $\bar{P} \not\subseteq \text{supp}(\tilde{x})$. In this case, we have $|\{ij \mid ij \in \bar{P}, ij \in \text{supp}(\tilde{x})\}| \leq |\bar{P}| - 1 = |M_P - M_0| - 2$. Note that $j^* = n_{i^*}$, and $a_{i^*j^*}\max\{1, \frac{a_{i^*j}}{a_{i^*j^*}+b-s}\} \leq a_{i^*j}$ for all $j \in N_{i^*} - j^*$, so the left-hand side of (11) evaluated in $\tilde{x}$ is at most

$$b + (b-s)|\{ij \mid ij \in \bar{P}, ij \in \text{supp}(\tilde{x})\}| \leq b + (|M_P - M_0| - 2)(b-s).$$

So far we have concluded that (11) is valid for PS. Next, we show that, if P is a maximal switching pack, then (11) is facet-defining.

Consider the point x with $x_{ij} = 1 \; \forall ij \in P$, and 0 elsewhere. It is easy to check that x is in S and satisfies (11) at equality. Also, for any $i' \in M_P$, $j' \in N_{i'}$, consider the point x with $x_{ij} = 1 \; \forall ij \in P$, $x_{i'j'} = \min\{1, \frac{b-s}{a_{i'j'}}\}$, and 0 elsewhere.

Also this point is in S, and satisfies (11) at equality. So in total, we obtain $\sum_{i \notin M_P} n_i + 1$ affinely independent points in S, which all satisfy (11) at equality.

Next, for any $j' \in N_{i^*} - n_{i^*}$, consider the point x with $x_{ij} = 1 \; \forall ij \in P$, $x_{i^* j'} = \frac{a_{i^* n_{i^*}} + b - s}{a_{i^* j'}}$, and 0 elsewhere. Since P is a maximal switching pack and $i^* \notin M_0$, we have $s - a_{i^* n_{i^*}} + a_{i^* j'} > b$, so $x_{i^* j'} = \frac{a_{i^* n_{i^*}} + b - s}{a_{i^* j'}} < 1$, and $x \in PS$. The left-hand side of (11) evaluated in $\tilde{x}$ is

$$s - a_{i^* n_{i^*}} + (|M_P - M_0| - 1)(b - s) + a_{i^* n_{i^*}} \cdot \frac{a_{i^* j'}}{a_{i^* n_{i^*}} + b - s} \cdot \frac{a_{i^* n_{i^*}} + b - s}{a_{i^* j'}}$$

$$= b + (|M_P - M_0| - 2)(b - s).$$

Therefore, we have also found $n_{i^*} - 1$ points in S that satisfy (11) at equality.

Recall that when P is maximal switching pack, then for any $ij \in P$, we have $j = n_i$. Arbitrarily pick $i' \in \bar{M}$, and consider the following set:

$$\Big\{ x \in [0,1]^d \mid x_{ij} = 0 \; \forall i \notin M_P \text{ and } j \in N_i, x_{i' n_{i'}} = 0,$$

$$x_{i n_i} = 1 \text{ and } x_{ij} = 0 \; \forall i \in \bar{M} - i' \text{ and } j < n_i, x_{i^* j} = 0 \; \forall j < n_{i^*},$$

$$\sum_{i \in M_P \cap M_0} a_{i1} x_{i1} + a_{i^* n_{i^*}} x_{i^* n_{i^*}} + \sum_{j < n_{i'}} a_{i'j} x_{i'j} = b - \sum_{i \in \bar{M} - i'} a_{i n_i},$$

$$\text{set } \{x_{i'1}, \ldots, x_{i', n_{i'} - 1}\} \text{ is SOS1} \Big\}.$$

$$(12)$$

It is simple to verify that any point in the set (12) satisfies the complementarity constraint, as well as the knapsack constraint (1) (in fact, it is satisfied at equality), and it satisfies (11) at equality. Since P is a maximal switching pack, we have $\sum_{i \in M_P} a_{i n_i} - a_{i' n_{i'}} + a_{i'j} > b$ for any $j < n_{i'}$, so $\sum_{i \in M_P \cap M_0} a_{i1} + a_{i^* n_{i^*}} + a_{i'j} > b - \sum_{i \in \bar{M} - i'} a_{i n_i}$ for any $j < n_{i'}$. Therefore, the set (12) contains $|M_P \cap M_0| + n_{i'}$ affinely independent points which all satisfy (11) at equality. Lastly, for any $i'' \in \bar{M} - i'$, we consider the set

$$\Big\{ x \in [0,1]^d \mid x_{ij} = 0 \; \forall i \notin M_P \text{ and } j \in N_i, x_{i'' n_{i''}} = 0, x_{i1} = 1 \; \forall i \in M_P \cap M_0,$$

$$x_{i n_i} = 1 \text{ and } x_{ij} = 0 \; \forall i \in \bar{M} - i'' \text{ and } j < n_i, x_{i^* j} = 0 \; \forall j < n_{i^*},$$

$$a_{i^* n_{i^*}} x_{i^* n_{i^*}} + \sum_{j < n_{i''}} a_{i''j} x_{i''j} = b - s + a_{i^* n_{i^*}} + a_{i'' n_{i''}},$$

$$\text{set } \{x_{i''1}, \ldots, x_{i'', n_{i''} - 1}\} \text{ is SOS1} \Big\}.$$

$$(13)$$

The points in the set (13) are in S, and satisfy (11) at equality. Furthermore, we can find $n_{i''}$ affinely independent points in the set (13).

In total, we have found $\sum_{i \notin M_P} n_i + 1 + n_{i^*} - 1 + |M_P \cap M_0| + n_{i'} + \sum_{i'' \in \bar{M} - i'} n_{i''} = d$ points in S that satisfy (11) at equality. Furthermore, according to their construction, all these points are affinely independent. This concludes the proof that (11) is facet-defining when P is a maximal switching pack. $\quad \square$

Next, we present some examples which illustrate that, with the same knapsack constraint, by picking different maximal switching packs and different indices i^*, we obtain several different facet-defining inequalities.

Example 3. Let $M = [5]$, $n_1 = n_2 = 1$, $n_3 = n_4 = n_5 = 2$, and let the knapsack inequality in (CKP) be

$$x_{11} + 6x_{21} + (14x_{31} + 10x_{32}) + (13x_{41} + 9x_{42}) + (12x_{51} + 8x_{52}) \leq 36. \quad (14)$$

For the maximal switching pack $P_1 = \{(1,1), (2,1), (3,2), (4,2), (5,2)\}$, we have $s = \sum_{ij \in P_1} a_{ij} = 1 + 6 + 10 + 9 + 8 = 34$. When $i^* = 3$, (11) gives the facet-defining inequality $x_{11} + 6x_{21} + (13x_{41} + 9x_{42}) + (12x_{51} + 8x_{52}) + (36 - 34) \cdot (x_{42} + x_{52}) + 10x_{32} + 10 \cdot \max\{1, \frac{14}{10+36-34}\}x_{31} \leq 36 + (36 - 34) \cdot (3 - 2)$, which can be simplified as

$$x_{11} + 6x_{21} + \left(\frac{35}{3}x_{31} + 10x_{32}\right) + (13x_{41} + 11x_{42}) + (12x_{51} + 10x_{52}) \leq 38.$$

For the same maximal switching pack P_1, when $i^* = 4$ and 5, inequality (11) gives two other facet-defining inequalities:

$$x_{11} + 6x_{21} + (14x_{31} + 12x_{32}) + \left(\frac{117}{11}x_{41} + 9x_{42}\right) + (12x_{51} + 10x_{52}) \leq 38,$$

$$x_{11} + 6x_{21} + (14x_{31} + 12x_{32}) + (13x_{41} + 11x_{42}) + \left(\frac{48}{5}x_{51} + 8x_{52}\right) \leq 38.$$

Consider now the maximal switching pack $P_2 = \{(2,1), (3,2), (4,2), (5,2)\}$. Setting $i^* = 3$, 4, and 5 gives us the following three facet-defining inequalities, respectively:

$$6x_{21} + \left(\frac{140}{13}x_{31} + 10x_{32}\right) + (13x_{41} + 12x_{42}) + (12x_{51} + 11x_{52}) \leq 39,$$

$$6x_{21} + (14x_{31} + 13x_{32}) + \left(\frac{39}{4}x_{41} + 9x_{42}\right) + (12x_{51} + 11x_{52}) \leq 39,$$

$$6x_{21} + (14x_{31} + 13x_{32}) + (13x_{41} + 12x_{42}) + \left(\frac{96}{11}x_{51} + 8x_{52}\right) \leq 39.$$

◇

Our final family of facet-defining complementarity pack inequalities is defined in Theorem 7.

Theorem 7. *Let P be a pack for PS with $|M_P - M_0| \geq 2$, and $s := \sum_{ij \in P} a_{ij}$. For any $i^*j^* \in P$ with $i^* \notin M_0, j^* = n_{i^*}$ and any $i' \in M_P \cap M_0$, the inequality*

$$\frac{a_{i^*j^*} \cdot a_{i'1}}{a_{i^*j^*} + b - s} x_{i'1} + \sum_{i \in M_P - \{i', i^*\}} \sum_{j \in N_i} a_{ij} x_{ij} + \sum_{\substack{ij \in P, \\ i \in M_P - M_0 - i^*}} \left(b - s + \frac{(b-s)a_{i'1}}{a_{i^*j^*} + b - s} \right) x_{ij}$$

$$+ a_{i^*j^*} x_{i^*j^*} + \sum_{j \in N_{i^*} - j^*} a_{i^*j^*} \max \left\{ 1, \frac{a_{i^*j}}{a_{i^*j^*} + b - s} \right\} x_{i^*j}$$

$$\leq b + (|M_P - M_0| - 2)(b - s) \left(1 + \frac{a_{i'1}}{a_{i^*j^*} + b - s} \right)$$

(15)

is valid for PS. Furthermore, when $P - i'1$ is a maximal switching pack, (15) is facet-defining.

We should remark that, the inequality (15) is simply obtained by tilting the previous inequality (11): for a fixed index $i' \in M_0$, we: (i) subtract from the original coefficient $a_{i'1}$ in (11) the quantity $(b - s)\frac{a_{i'1}}{a_{i^*j^*}+b-s}$, (ii) add the same amount $(b - s)\frac{a_{i'1}}{a_{i^*j^*}+b-s}$ to the coefficients of x_{ij}, for $ij \in P$ and $i \notin i^*$, and (iii) multiply the right-hand side of the inequality by $(1 + \frac{a_{i'1}}{a_{i^*j^*}+b-s})$.

Proof. Let $\tilde{x}$ be a vector in S. We show that the inequality (15) is satisfied by $\tilde{x}$. For ease of exposition, we define $\bar{M} := M_P - M_0 - i^*$ and $\bar{P} := \{ij \in P \mid i \in \bar{M}\}$.
Case 1: $\bar{P} \subseteq \text{supp}(\tilde{x})$. In this case, since $\tilde{x} \in S$, we have $\sum_{i \in M_P - i'} \sum_{j \in N_i} a_{ij} \tilde{x}_{ij} + a_{i'1} \leq \sum_{ij \in P} a_{ij} = s$.
Case 1a: $\tilde{x}_{i^*j} = 0$ for any $j \in N_{i^*} - j^*$. In this case, the left-hand side of inequality (15) evaluated in $\tilde{x}$ is

$$\sum_{i \in M_P - i'} \sum_{j \in N_i} a_{ij} \tilde{x}_{ij} + \frac{a_{i^*j^*} a_{i'1}}{a_{i^*j^*} + b - s} \tilde{x}_{i'1} + \sum_{ij \in \bar{P}} \left(b - s + \frac{(b-s)a_{i'1}}{a_{i^*j^*} + b - s} \right) \tilde{x}_{ij}$$

$$\leq s - a_{i'1} + \frac{a_{i^*j^*} a_{i'1}}{a_{i^*j^*} + b - s} + \left(b - s + \frac{(b-s)a_{i'1}}{a_{i^*j^*} + b - s} \right) (|M_P - M_0| - 1)$$ (16)

$$= b + (|M_P - M_0| - 2)(b - s) \left(1 + \frac{a_{i'1}}{a_{i^*j^*} + b - s} \right).$$

So (15) is satisfied by point $\tilde{x}$.
Case 1b: There exists $j' \in N_{i^*} - j^*$ such that $\tilde{x}_{i^*j'} \in (0, 1]$. Let $K := \sum_{i \in M_P - \{i^*, i'\}} \sum_{j \in N_i} a_{ij} \tilde{x}_{ij}$. Since $\bar{P} \subseteq \text{supp}(\tilde{x})$, we obtain $K \leq s - a_{i'1} - a_{i^*j^*}$, as well as $a_{i'1}\tilde{x}_{i'1} + a_{i^*j'}\tilde{x}_{i^*j'} + K \leq b$. Thus, the left-hand side of inequality (15) evaluated in $\tilde{x}$ is

$$K + \frac{a_{i'1}a_{i^*j^*}\tilde{x}_{i'1}}{a_{i^*j^*}+b-s} + \max\left\{a_{i^*j^*}\tilde{x}_{i^*j'}, \frac{a_{i^*j^*}a_{i^*j'}\tilde{x}_{i^*j'}}{a_{i^*j^*}+b-s}\right\}$$

$$+ \sum_{ij\in\bar{P}}\left(b-s+\frac{(b-s)a_{i'1}}{a_{i^*j^*}+b-s}\right)\tilde{x}_{ij}$$

$$\leq \max\left\{K+\frac{a_{i'1}a_{i^*j^*}}{a_{i^*j^*}+b-s}+a_{i^*j^*}, K+\frac{a_{i^*j^*}(a_{i'1}\tilde{x}_{i'1}+a_{i^*j'}\tilde{x}_{i^*j'})}{a_{i^*j^*}+b-s}\right\}$$

$$+\left(b-s+\frac{(b-s)a_{i'1}}{a_{i^*j^*}+b-s}\right)(|M_P-M_0|-1)$$

$$\leq \max\left\{K+\frac{a_{i'1}a_{i^*j^*}}{a_{i^*j^*}+b-s}+a_{i^*j^*}, K+\frac{a_{i^*j^*}(b-K)}{a_{i^*j^*}+b-s}\right\} \tag{17}$$

$$+(|M_P-M_0|-1)(b-s)\left(1+\frac{a_{i'1}}{a_{i^*j^*}+b-s}\right)$$

$$\leq \max\left\{s-a_{i'1}+\frac{a_{i'1}a_{i^*j^*}}{a_{i^*j^*}+b-s}, \frac{(b-s)K+a_{i^*j^*}b}{a_{i^*j^*}+b-s}\right\}$$

$$+(|M_P-M_0|-1)(b-s)\left(1+\frac{a_{i'1}}{a_{i^*j^*}+b-s}\right)$$

$$= s-a_{i'1}+\frac{a_{i'1}a_{i^*j^*}}{a_{i^*j^*}+b-s}+(|M_P-M_0|-1)(b-s)(1+\frac{a_{i'1}}{a_{i^*j^*}+b-s})$$

$$= b+(|M_P-M_0|-2)(b-s)\left(1+\frac{a_{i'1}}{a_{i^*j^*}+b-s}\right).$$

Case 2: $\bar{P}\not\subseteq\text{supp}(\tilde{x})$. In this case, we have $|\{ij \mid ij\in\bar{P}, ij\in\text{supp}(\tilde{x})\}|\leq|\bar{P}|-1 = |M_P-M_0|-2$. Note that $\frac{a_{i^*j^*}\cdot a_{i'1}}{a_{i^*j^*}+b-s}\leq a_{i'1}$ and $a_{i^*j^*}\max\{1,\frac{a_{i^*j}}{a_{i^*j^*}+b-s}\}\leq a_{i^*j}$ for all $j\in N_{i^*}-j^*$, so the left-hand side of (15) evaluated in $\tilde{x}$ is at most

$$b+(b-s)(1+\frac{a_{i'1}}{a_{i^*j^*}+b-s})|\{ij\mid ij\in\bar{P}, ij\in\text{supp}(\tilde{x})\}|$$

$$\leq b+(|M_P-M_0|-2)(b-s)\left(1+\frac{a_{i'1}}{a_{i^*j^*}+b-s}\right).$$

So far we have shown that inequality (15) is valid for PS. Next, we assume that $P-i'1$ is a maximal switching pack, and we show that (15) is facet-defining. We note that, since $P-i'1$ is a maximal switching pack and P is a pack, we have that P is also a maximal switching pack. Hence, for any $ij\in P$, $j=n_i$.

Consider the point x with $x_{ij}=1 \ \forall ij\in P$, and 0 elsewhere. It is easy to check that $x\in S$ and it satisfies (15) at equality. For any $i'\notin M_P$ and $j'\in N_{i'}$, consider the point x with $x_{ij}=1 \ \forall ij\in P$, $x_{i'j'}=\min\{1,\frac{b-s}{a_{i'j'}}\}$, and 0 elsewhere. Also this point is in S and it satisfies (15) at equality. Thus we have found $\sum_{i\notin M_P}n_i+1$ points in S that satisfy inequality (15) at equality.

Next, we consider the following set

$$\Big\{ x \in [0,1]^d \mid x_{i n_i} = 1, x_{ij} = 0 \ \forall i \in M_P - i' - i^*, j < n_i,$$

$$x_{i^* n_{i^*}} = 0, x_{ij} = 0 \ \forall i \notin M_P, j \in N_i,$$

$$a_{i'1} x_{i'1} + \sum_{j < n_{i^*}} a_{i^* j} x_{i^* j} = b - s + a_{i'1} + a_{i^* n_{i^*}}, \tag{18}$$

$$\text{set } \{x_{i^* 1}, \dots, x_{i^*, n_{i^*} - 1}\} \text{ is SOS1} \Big\}.$$

Since P is a maximal switching pack, we have $s - a_{i^* n_{i^*}} + a_{i^* j} > b$ for any $j < n_{i^*}$. Hence, the set (18) has n_{i^*} affinely independent points. For any point x in this set (18), the left-hand side of (15) evaluated in x is

$$\frac{a_{i^* n_{i^*}} (b - s + a_{i'1} + a_{i^* n_{i^*}})}{a_{i^* n_{i^*}} + b - s} + s - a_{i'1} - a_{i^* n_{i^*}}$$

$$+ (|M_P - M_0| - 1)(b - s) \left(1 + \frac{a_{i'1}}{a_{i^* j^*} + b - s} \right)$$

$$= s - a_{i'1} + \frac{a_{i'1} a_{i^* n_{i^*}}}{a_{i^* n_{i^*}} + b - s} + (|M_P - M_0| - 1)(b - s) \left(1 + \frac{a_{i'1}}{a_{i^* j^*} + b - s} \right)$$

$$= b + (|M_P - M_0| - 2)(b - s) \left(1 + \frac{a_{i'1}}{a_{i^* j^*} + b - s} \right).$$

It is easy to observe that any point in (18) is in S. Hence, in total, we have found another n_{i^*} affinely independent points in S that satisfy (15) at equality.

Now we arbitrarily pick an index $i'' \in \bar{M}$. Consider the following set:

$$\Big\{ x \in [0,1]^d \mid x_{ij} = 0 \ \forall i \notin M_P \text{ and } j \in N_i, x_{i'1} = 0, x_{i'' n_{i''}} = 0,$$

$$x_{i n_i} = 1 \text{ and } x_{ij} = 0 \ \forall i \in \bar{M} - i'' \text{ and } j < n_i, x_{i^* j} = 0 \ \forall j < n_{i^*},$$

$$\sum_{i \in M_P \cap M_0 - i'} a_{i1} x_{i1} + a_{i^* n_{i^*}} x_{i^* n_{i^*}} + \sum_{j < n_{i''}} a_{i'' j} x_{i'' j} = b - \sum_{i \in \bar{M} - i''} a_{i n_i},$$

$$\text{set } \{x_{i'' 1}, \dots, x_{i'', n_{i''} - 1}\} \text{ is SOS1} \Big\}.$$

$$\tag{19}$$

It is easy to verify that any point in the set (19) satisfies the complementarity constraints in (2), as well as the knapsack constraint (1) (in fact, it is satisfied at equality), and it satisfies (15) at equality. Since $P - i'1$ is a maximal switching pack, we have $\sum_{i \in M_P - i'} a_{i n_i} - a_{i'' n_{i''}} + a_{i'' j} > b$ for any $j < n_{i''}$, therefore $\sum_{i \in M_P \cap M_0 - i'} a_{i1} + a_{i^* n_{i^*}} + a_{i'' j} > b - \sum_{i \in \bar{M} - i''} a_{i n_i}$ for any $j < n_{i'}$. Hence, the set (12) contains $|M_P \cap M_0| + n_{i''} - 1$ affinely independent points which all satisfy (11) at equality.

Lastly, for any $\hat{i} \in \bar{M} - i''$, we consider the set

$$\Big\{ x \in [0,1]^d \mid x_{ij} = 0 \; \forall i \notin M_P \text{ and } j \in N_i, x_{i'1} = 0, x_{\hat{i}n_{\hat{i}}} = 0,$$

$$x_{in_i} = 1 \text{ and } x_{ij} = 0 \; \forall i \in \bar{M} - \hat{i} \text{ and } j < n_i, x_{i^*j} = 0 \; \forall j < n_{i^*},$$

$$a_{i^*n_{i^*}} x_{i^*n_{i^*}} + \sum_{j<n_{\hat{i}}} a_{\hat{i}j} x_{\hat{i}j} = b - s + a_{i^*n_{i^*}} + a_{\hat{i}n_{\hat{i}}} + a_{i'1},$$

$$x_{i1} = 1 \; \forall i \in M_P \cap M_0 - i', \text{ set } \{x_{\hat{i}1}, \ldots, x_{\hat{i},n_{\hat{i}}-1}\} \text{ is SOS1} \Big\}.$$
(20)

Again, the set (20) contains $n_{\hat{i}}$ affinely independent points in S, and they all satisfy (15) at equality.

In total, we have found $\sum_{i \notin M_P} n_i + 1 + n_{i^*} + |M_P \cap M_0| + n_{i''} - 1 + \sum_{\hat{i} \in \bar{M} - i''} n_{\hat{i}} = d$ points in S that satisfy (15) at equality. Furthermore, according to their construction, all these points are affinely independent. This concludes the proof that (15) is facet-defining when $P - i'1$ is maximal switching pack. $\square$

Example 4. Consider the same knapsack constraint (14) studied in Example 3.

Consider $P = \{(1,1), (2,1), (3,2), (4,2), (5,2)\}$, and $i' = 1 \in M_P \cap M_0$. Here $s = \sum_{ij \in P} a_{ij} = 34$. Then both P and $P - i'1$ are maximal switching packs, satisfying the condition in Theorem 7. For $i^* = 3$, (15) gives the facet-defining inequality

$$\frac{5}{6}x_{11} + 6x_{21} + \left(\frac{35}{3}x_{31} + 10x_{32}\right) + \left(13x_{41} + \frac{67}{6}x_{42}\right) + \left(12x_{51} + \frac{61}{6}x_{52}\right) \leq 38 + \frac{1}{6}.$$

Similarly, picking $i^* = 4$ and 5, (15) gives another two facet-defining inequalities:

$$\frac{9}{11}x_{11} + 6x_{21} + \left(14x_{31} + \frac{134}{11}x_{32}\right) + \left(\frac{117}{11}x_{41} + 9x_{42}\right) + \left(12x_{51} + \frac{112}{11}x_{52}\right) \leq 38 + \frac{2}{11},$$

$$\frac{4}{5}x_{11} + 6x_{21} + \left(14x_{31} + \frac{61}{5}x_{32}\right) + \left(13x_{41} + \frac{56}{5}x_{42}\right) + \left(\frac{48}{5}x_{51} + 8x_{52}\right) \leq 38 + \frac{1}{5}.$$

$\diamond$

Here we briefly mention the main difference between our complementarity pack inequalities and those lifted cover inequalities (3) and (4). As introduced in [12], (3) and (4) are both obtained through sequential lifting of the original cover inequalities $\sum_{ij \in C} a_{ij} x_{ij} \leq b$. For a complete survey about lifting procedure, we refer the readers to Sect. 3 in [18]. One fundamental property of all lifting procedures is, the difference between the original inequality and the lifted version of the inequality only occurs on the coefficients of variables that do not appear in the original inequality. In the cases of lifted cover inequalities (3) and (4), one can easily verify that the coefficients of those variables x_{ij} for $ij \in C$ remain to be a_{ij}. However, for the complementarity pack inequalities (8), (11) and (15), the coefficients of variables x_{ij} for $ij \in P$ all increase. This implies that our complementarity pack inequalities cannot be easily obtained through some lifting process of the original inequality $\sum_{ij \in P} a_{ij} x_{ij} \leq b$.

5   Future Direction

In order for the inequalities proposed in this paper to be of practical use, it is necessary to develop efficient exact or heuristic separation methods. Therefore, future work on this topic includes further investigation of separation algorithms and conducting numerical experiments. Another interesting direction of research spawns from the observation that the embedding conflict graph of the complementarity constraints in (CKP) is given by the union of some non-overlapping cliques. An interesting question is then, whether we can gain a deep understanding of the valid inequalities for the following set, with respect to a general graph $G = ([n], E)$:

$$\text{conv}\left(\left\{x \in [0,1]^n \mid a^T x \le b, x_i \cdot x_j = 0 \ \forall \{i,j\} \in E\right\}\right).$$

References

1. Atamtürk, A.: Cover and pack inequalities for (mixed) integer programming. Ann. Oper. Res. **139**(1), 21–38 (2005)
2. Balas, E.: Facets of the knapsack polytope. Math. Program. **8**(1), 146–164 (1975)
3. Balas, E., Zemel, E.: Facets of the knapsack polytope from minimal covers. SIAM J. Appl. Math. **34**(1), 119–148 (1978)
4. Beale, E.M.L., Tomlin, J.A.: Special facilities in a general mathematical programming system for non-convex problems using ordered sets of variables. OR **69**(447–454), 99 (1970)
5. Beaumont, N.: An algorithm for disjunctive programs. Eur. J. Oper. Res. **48**(3), 362–371 (1990)
6. Crowder, H., Johnson, E.L., Padberg, M.: Solving large-scale zero-one linear programming problems. Oper. Res. **31**(5), 803–834 (1983)
7. De Farias, I., Johnson, E.L., Nemhauser, G.L.: Branch-and-cut for combinatorial optimization problems without auxiliary binary variables. Knowl. Eng. Rev. **16**(1), 25–39 (2001)
8. De Farias Jr, I.R., Johnson, E.L., Nemhauser, G.L.: Facets of the complementarity knapsack polytope. Math. Oper. Res. **27**(1), 210–226 (2002)
9. Del Pia, A., Linderoth, J., Zhu, H.: Multi-cover inequalities for totally-ordered multiple knapsack sets. In: International Conference on Integer Programming and Combinatorial Optimization, pp. 193–207. Springer, Heidelberg (2021). https://doi.org/10.1007/s10107-022-01817-4
10. Del Pia, A., Linderoth, J., Zhu, H.: Multi-cover inequalities for totally-ordered multiple knapsack sets: theory and computation. arXiv preprint arXiv:2106.00301 (2021)
11. Del Pia, A., Linderoth, J., Zhu, H.: On the complexity of separation from the Knapsack Polytope. In: International Conference on Integer Programming and Combinatorial Optimization, pp. 168–180. Springer, Cham
12. de Farias, I.R., Kozyreff, E., Zhao, M.: Branch-and-cut for complementarity-constrained optimization. Math. Program. Comput. **6**(4), 365–403 (2014). https://doi.org/10.1007/s12532-014-0070-2
13. de Farias, I.R., Zhao, M.: A polyhedral study of the semi-continuous knapsack problem. Math. Program. **142**(1–2), 169–203 (2013)

14. Fischer, T., Pfetsch, M.E.: Branch-and-cut for linear programs with overlapping SOS1 constraints. Math. Program. Comput. **10**(1), 33–68 (2018)
15. Gu, Z., Nemhauser, G.L., Savelsbergh, M.W.P.: Lifted cover inequalities for 0-1 integer programs: complexity. INFORMS J. Comput. **11**(1), 117–123 (1999)
16. Gu, Z., Nemhauser, G.L., Savelsbergh, M.W.: Lifted cover inequalities for 0–1 integer programs: computation. INFORMS J. Comput. **10**(4), 427–437 (1998)
17. Gu, Z., Nemhauser, G.L., Savelsbergh, M.W.: Sequence independent lifting in mixed integer programming. J. Comb. Optim. **4**(1), 109–129 (2000)
18. Hojny, C., Gally, T., Habeck, O., Lüthen, H., Matter, F., Pfetsch, M.E., Schmitt, A.: Knapsack polytopes: a survey. Ann. Oper. Res. **292**, 1–49 (2019)
19. Karp, R.M.: Reducibility among combinatorial problems. In: Complexity of Computer Computations, pp. 85–103 (1972)
20. Keha, A.B., de Farias Jr, I.R., Nemhauser, G.L.: A branch-and-cut algorithm without binary variables for nonconvex piecewise linear optimization. Oper. Res. **54**(5), 847–858 (2006)
21. Klabjan, D., Nemhauser, G.L., Tovey, C.: The complexity of cover inequality separation. Oper. Res. Lett. **23**(1–2), 35–40 (1998)
22. Padberg, M.W.: (1, k)-configurations and facets for packing problems. Math. Program. **18**(1), 94–99 (1980)
23. Weismantel, R.: On the 0/1 knapsack polytope. Math. Program. **77**(3), 49–68 (1997)
24. Wolsey, L.A.: Faces for a linear inequality in 0–1 variables. Math. Program. **8**(1), 165–178 (1975)

Branch-and-Cut for a 2-Commodity Flow Relocation Model with Time Constraints

José Luis Figueroa González[(✉)], Mourad Baïou, Alain Quilliot,
Hélène Toussaint , and Annegret Wagler

LIMOS UMR 6158 CNRS/Université Clermont-Auvergne, Aubière, France
{jose_luis.figueroa_gonzalez,mourad.baiou,alain.quilliot,
helene.toussaint,annegret.wagler}@uca.fr

Abstract. We deal here with a general 2-commodity flow model designed for the management of shared mobility systems which operate on a given transit network. The model involves an integral flow which represents carriers together with an integral flow which represents the objects transported by those carriers. It may be viewed as the projection on the transit network of a flow model formulated on a time expanded network which simultaneously copes with temporal and resource issues, but does not fit practical computation. In order to make this projected model compatible with the time expanded network model, we introduce specific constraints whose handling involves a separation process. We prove that this separation process can be performed in polynomial time, discuss the experimental behaviour of the related Branch-and-Cut algorithm and briefly address the lift issue to turn an optimal solution of our projected model into a solution of the original problem.

Keywords: Relocation problem · Network flows · Branch-and-cut · Time expanded networks

1 Introduction

Emerging mobility systems, based on vehicle sharing, aim at finding a flexible compromise between individual mobility and rigid public transportation systems (see e.g. [12,16,17] for surveys). Such systems include collective taxis, transport on demand, car pooling, shared riding, etc. and share some key features: they rely on emerging mobility technologies (e.g. electric vehicles, autonomous vehicles), require a responsive day-to-day operational management through intensive use of internet platforms, and aim at answering environmental concerns and urban congestion, while keeping part of the flexibility of individual transportation.

Managing such emerging mobility systems requires decisions at a strategic level (for pricing, infrastructure dimensioning, demand/cost analysis, integration into multimodality, see [3,12,19]), at an operational level (for real time handling of the demands, relocation of free vehicles, synchronization, and unexpected event management, see [12,15]), and also at a tactical level (for in advance management of recurrent demands or maintenance scheduling, see [4,7,10,18]).

I. Ljubić et al. (Eds.): ISCO 2022, LNCS 13526, pp. 22–34, 2022.
https://doi.org/10.1007/978-3-031-18530-4_2

The resulting problems are difficult and must be usually addressed in a dynamic way, taking uncertainty into account. To overcome these difficulties, the problems are often studied in a static paradigm using aggregated representations of the circulation of vehicles and passengers to make decisions about the dimensions of the system and to pre-compute routes and schedules for its operational management. A popular approach is the use of multi-commodity flow models, where the different commodities correspond to the different kinds of objects whose circulation is supported by the network (see e.g. [1,5,6,8,14]).

Hereby, it is crucial to take time constraints into account that are imposed by the synchronization of vehicle routes and users demands. In order to deal with this specific temporal dimension while taking profit from the powerful network flow machinery, mobility models are often cast in time expanded networks (TENs), containing one copy of the original transit network for each considered time unit (see e.g. [2,5,6,9]). This powerful conceptual tool is well fitted to the modelling of systems and decision problems involving the circulation of resources (vehicles, objects, passengers, ...) over time. The disadvantage is that such networks grow with the size of the time space so that solving multi-commodity flow problems in TENs typically requires very long computation times.

One possible approach to overcome this difficulty consists in restricting the TEN to an active part of the network of limited size (see [2,14]) which typically results in heuristics without optimality guarantee. Another approach consists in projecting the TEN model to the original network or to some auxiliary simpler network (see [7,11]), and so making the temporal dimension become implicit: the algorithmic process consists in first solving the projected model, and next turning the resulting solution into a feasible or even optimal solution of the original TEN model.

In this work, we adopt the latter approach. For the sake of simplicity, we work on a generic 2-commodity flow model, related to the relocation problem (see [4, 7,11,16]), which arises when carriers are required to periodically exchange items between stations in order to rebalance the access to those items for potential users. According to such an interpretation, one flow is related to carriers and the other to the (identical) items which are transported by the carriers. In most cases, this relocation process has to be performed within a given time horizon, and carriers are allowed to meet in order to exchange part of their load if necessary. Our model is generic in the sense that it embraces the whole spectrum of possible quality criteria: the number of involved carriers, carrier operational costs, and item riding time.

A major difficulty for deriving a projected model from a TEN is to set the constraints in such a way that neither the time dimension of the problem is lost nor the structure of the performance criteria. Another difficulty is imposed by the algorithmic handling of those constraints if their number grows exponentially with the size of the network. In the present paper, we focus on a specific class of such constraints, called *Extended Subtour Constraints*, which link the duration of the relocation process to the number of involved carriers. We show how those constraints can be separated in polynomial time, giving rise to the implementation of an efficient Branch-and-Cut algorithm. Finally, we briefly address the problem how the solutions of the projected model may be lifted, i.e., turned into a feasible or even optimal solution of the original TEN model, taking the temporal dimension into account.

Our contribution is organized as follows: we first introduce in Sect. 2 the studied relocation problem and its formulation through the TEN framework. In Section 3, we describe the projected model and the way this model may be enhanced through the use of Extended Subtour Constraints. We propose two versions of those constraints, which we prove to be both separable in polynomial time. Section 4 is devoted to the description of the resulting Branch-and-Cut algorithm and to numerical experiments. Finally, we propose in Sect. 5 a brief discussion about the Lift Problem and conclude with some perspectives of future research.

2   A TEN Model for the Item Relocation Problem

We consider a mobility system where a set of identical items is distributed in a transit network and a fleet of carriers is available to rebalance the distribution of the items in the network.

The Item Relocation Problem. The considered *transit network* $N = (X, A)$ is composed of vertices and arcs. There exists in X a distinguished *depot vertex* s, where carriers are located at the beginning and at the end of the relocation process, and every arc $a = (x, y)$ is provided with a discrete *time cost value* $T_{x,y}$ and an *economic cost value* $C_{x,y}$ which respectively encode the time and the economical cost required by a carrier in order to move from x to y. We define the set $T = \{T_{x,y} : (x, y) \in A\}$ and the set $C = \{C_{x,y} : (x, y) \in A\}$.

Items are situated in the vertices of the network, rebalancing their distribution may be required to guarantee the operationability of the mobility system. The relocation demand of items in the network is expressed by an integral vector **b** of *balance coefficients* b_x for all $x \in X$ satisfying $\sum_{x \in X} b_x = 0$, where $b_x > 0$ indicates that x is an *excess vertex* and carriers must remove b_x items from x, $b_x < 0$ indicates that x is a *deficit vertex* and carriers must bring $-b_x$ items to x, whereas $b_x = 0$ indicates that x is *neutral*.

For the relocation process, a fleet Q of carriers is available. Each carrier has *capacity* κ, specifying the maximal number of items which may be transported by a carrier at the same time. Finally, a global *time horizon* $[0, \Omega]$ is given to perform the relocation.

The *Item Relocation Problem* (IRP) consists in organizing the transfer by carriers of items from excess vertices to deficit vertices, while meeting time horizon and carrier capacity requirements.

For the relocation process, it is in order to construct routes for the carriers $q \in Q$. A *carrier route* is a circuit which starts and ends in the depot vertex s. We assume that no more than one arc $a = (x, y)$ may connect a vertex x to another vertex y so that a carrier route may be represented as a sequence of vertices, with both first and last elements equal to s.

That is, all carriers start and end their route in the depot vertex s; their loads cannot exceed the capacity κ at any time during the process; the process must take place within the time horizon $[0, \Omega]$. Furthermore, *preemption* is allowed, which means that carriers may exchange items. If such an exchange occurs between two carriers at some vertex x, then *weak synchronization* is required: the *receiving carrier* q^r cannot leave x before the *emitting carrier* q^e could reach x.

A solution of the IRP with the help of carriers $q \in Q$ can be represented as a collection $\Gamma = \{\Gamma^q : q \in Q\}$ of carrier routes $\Gamma^q = \{x_0^q = s, x_1^q, \ldots, x_{\nu(q)}^q = s\}$ with $\nu(q) + 1$ vertices (where $\nu : Q \rightarrow \mathbb{Z}_+$), a *schedule* for Γ^q is defined by two *time sequences* $t^q = (t_0^q, \ldots, t_{\nu(q)}^q)$, $\bar{t}^q = (\bar{t}_0^q, \ldots, \bar{t}_{\nu(q)}^q)$, and a *load sequence* $\ell^q = (\ell_0^q, \ldots, \ell_{\nu(q)-1}^q)$. Here, the value t_i^q is the time when the carrier q arrives at vertex x_i^q, the value $\bar{t}_i^q$ is the time when the carrier q leaves vertex x_i^q), and the value ℓ_i^q is the load of carrier q when it leaves vertex x_i^q.

The following canonical constraints have to be satisfied: loads $\ell_i^q, q \in Q, i = 1, \ldots, \nu(q)$, must never exceed the capacity κ; time values $t_i^q, q \in Q, i = 1, \ldots, \nu(q)$, must belong to the interval $[0, \Omega]$ and fit with the lengths $\{T_{x,y} : (x, y) \in A\}$; for any vertex x, we must have $\sum_{(q,i) \in In(x)} \ell_i^q - \sum_{(q,i) \in Out(x)} \ell_i^q = b_x$ where $In(x) = \{(q, i) \in Q \times \mathbb{Z}_+ : x_{i+1}^q = x\}$ and $Out(x) = \{(q, i) \in Q \times \mathbb{Z}_+ : x_i^q = x\}$.

Considering a mobility system where carriers must periodically transfer vehicles from excess vertices to deficit vertices, we see that the main components of the costs should be the *number of carriers* $c_1 = |Q|$ corresponding, e.g. to human resources; the economic *carrier ride costs* $c_2 = \sum_{q \in Q} \sum_{i=0}^{\nu(q)} C_{x_i^q, x_{i+1}^q}$ related to the routes $\Gamma^q, q \in Q$, involving all the costs related to the carriers; and the *items ride time* c_3 which is the time during which items are not available to the users, expressed as $\sum_{q \in Q} (\sum_{i=0}^{\nu(q)-1} \ell_{i+1}^q (t_{x_{i+1}}^q - \bar{t}_{x_i}^q))$. In order to avoid a multiobjective formulation, we consider the minimization of a hybrid cost $\alpha \cdot c_1 + \beta \cdot c_2 + \gamma \cdot c_3$, where α, β, γ are scaling coefficients.

Example 1. Consider the network $N = (X, A)$ depicted in Fig. 1. It shows two carrier routes $\Gamma^1 = (s, v, x, y, s)$ and $\Gamma^2 = (s, w, x, z, s)$, whose related schedules are $t^1 = (0, 1, 3, 4, 5)$, $\bar{t}^1 = (0, 1, 3, 4, 5)$, $\ell^1 = (0, 5, 3, 0)$, $t^2 = (0, 1, 2, 5, 6)$, $\bar{t}^2 = (0, 1, 3, 5, 6)$, and $\ell^2 = (0, 5, 7, 0)$. Note that there is a synchronization involved at vertex x: two items are transferred from the carrier in Γ^1 to the carrier in Γ^2, and we have that $c_1 = 2$, $c_2 = 10$, and $c_3 = 32$.

A TEN 2-Commodity Flow Formulation of the IRP.

In order to cast the IRP into the TEN framework (see [2,5,9,14]), we first derive from the transit network $N = (X, A)$ its time expansion $N^\Omega = (X^\Omega, A^\Omega)$ according to Ω. The vertex set X^Ω is the set of all pairs $x_t = (x, t)$, $x \in X$, $t \in \{0, 1, \ldots, \Omega\}$, augmented with two distinguished vertices, a *source* $\hat{s}$ and a *sink* $\hat{t}$. Every arc $a \in A^\Omega$ has associated an *item cost* CI_a and a *carrier cost* CC_a and belongs to one of the following types of arcs:

- *input-arcs* $a = (\hat{s}, x_0)$, with $x \in X$, $CI_a = 0$, and $CC_a = 0$;
- *output-arcs* $a = (x_\Omega, \hat{t})$, with $x \in X$, and $CI_a = CC_a = 0$;

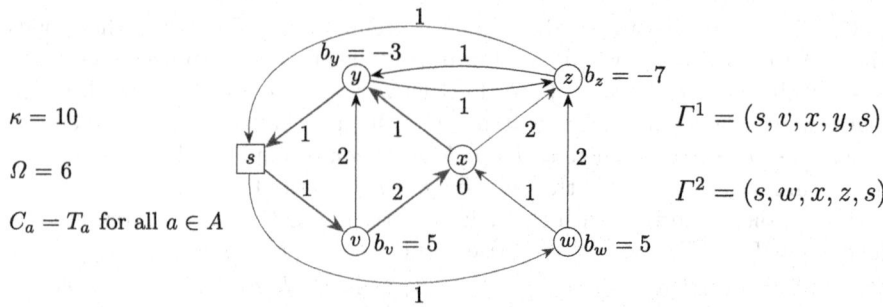

Fig. 1. The transit network $N = (X, A)$ used in Example 1.

- *waiting-arcs* $a = (x_t, x_{t+1})$, $x \in X$, $t \in \{0, \ldots, \Omega - 1\}$, and $CI_a = CC_a = 0$;
- *active-arcs* $a = (x_t, y_{t+T_{x,y}})$, $(x, y) \in A$, $t \in \{0, \ldots, \Omega - T_{x,y}\}$, $CI_a = \gamma \cdot T_{x,y}$, and $CC_a = \beta \cdot C_{x,y}$;
- *backward-arc* $a = (\hat{t}, \hat{s})$ with $CI_a = 0$ and $CC_a = \alpha$.

Now, we formalize the IRP as a 2-commodity flow model on $N^\Omega = (X^\Omega, A^\Omega)$.

TEN IRP Formulation: *Compute functions* $H : A^\Omega \to \mathbb{Z}_+$ *and* $h : A^\Omega \to \mathbb{Z}_+$ *(for carriers and items, respectively) such that:*

- *H and h satisfy flow conservation at any vertex of X^Ω;*   (E1)
- *for any active-arc* $a = \big((x, t), (y, t + T_{x,y})\big)$, $h(a) \le \kappa \cdot H(a)$;   (E2)
- *for any input-arc* $a = \big(\hat{s}, (x, 0)\big)$, $x \ne s$: $H(a) = 0$; $h(a) = \min(b_x, 0)$;   (E3)
- *for any output-arc* $a = \big((y, \Omega), \hat{t}\big)$, $y \ne s$: $H(a) = 0$; $h(a) = \min(-b_y, 0)$;   (E4)
- *the global cost* $Cost(H, h) = \sum_{a \in A^\Omega} (H(a) \cdot CC_a + h(a) \cdot CI_a)$ *is minimized.*

Thereby, the flow conservation constraints (E1) express the circulation of carriers and items inside the transit network N, condition (E2) ensures that any item moving between two vertices x and y must be contained into some carrier. Conditions (E3) and (E4) provide us with initial and final constraints: carriers must start and end their route in s, while the flow h means that for any excess vertex x, b_x items must leave x, and that for any deficit vertex y, b_y items must arrive into y.

Example 2. Figure 2 shows the construction of the Time Expanded Network $N^\Omega = (X^\Omega, A^\Omega)$ associated to the network $N = (X, A)$ of Fig. 1. It also illustrates how to turn the carrier routes and schedules of Example 1 into a 2-commodity flow (H, h).

Theorem 1. *Solving the TEN IRP also solves the IRP.*

Proof. Any scheduled route collection Γ of IRP can be turned into a 2-commodity flow (H, h) which meets (E1)-(E4), with the same global cost value.

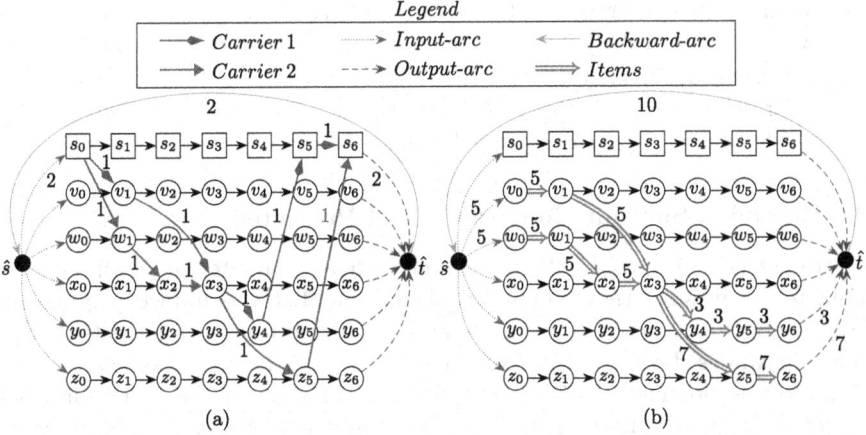

Fig. 2. The routes and schedules from Example 1 viewed as a 2-commodity flow on the Time Expanded Network $N^{\Omega} = (X^{\Omega}, A^{\Omega})$. (a) Carriers flow H. (b) Items flow h.

Conversely, if (H, h) is a feasible solution of the above TEN IRP formulation, then we may decompose H into a sum of $c_1 = |Q| = H\big((\hat{t}, \hat{s})\big)$ flows $(0, 1)$-valued, $H_1, \ldots, H_{|Q|}$, which are the supports of circuits containing the arc $(\hat{t}, \hat{s})$ and so connect vertex $(s, 0)$ with the vertex (s, Ω). Those $(0, 1)$-valued flows may be turned into a carrier route collection $\Gamma = \{\Gamma^1, \ldots, \Gamma^{|Q|}\}$, together with a time sequence $t^q = (t_0^q, \ldots, t_{\nu(q)}^q)$, which fit with IRP requirements. What remains to be done is to deduce from the coupling constraint (E2) that we may distribute values $h(a), a \in A^{\Omega}$ among the routes $\Gamma^q, q = 1, \ldots, |Q|$, in such a way that we get a load sequence $\ell^q = (\ell_0^q, \ldots, \ell_{\nu(q)}^q - 1)$ which meets IRP requirements. $\qquad \square$

3 The Projected IRP Model

This section is devoted to the major results of the present contribution. We are going to deal with the studied IRP while projecting the TEN IRP model on the original transit network N.

Given a network $N = (X, A)$ and a set $U \subseteq X$, we will use the notation $\partial_N^-(U) = \{(x, y) \in A : x \notin U, y \in U\}$, $\partial_N^+(U) = \{(x, y) \in A : x \in U, y \notin U\}$, $\partial_N(U) = \partial^-(U) \cup \partial^+(U)$, and $A(U, U) = \{(x, y) \in A : x \in U, y \in U\}$. For a singleton $\{x\}$ we will write $\partial_N^-(x)$, $\partial_N^+(x)$, and $\partial_N(x)$, instead of $\partial_N^-(\{x\})$, $\partial_N^+(\{x\})$, and $\partial_N(\{x\})$, respectively. Also, we define for any $U \subseteq X \setminus \{s\}$, a value $d_U^+ = \inf_{\{x \in X : (x, y) \in \partial_N^-(U)\}} T^{*+}(x)$, where $T^{*+}(x)$ means the shortest path distance from s to vertex x in N, according to the lengths $\{T_{x, y} : (x, y) \in A\}$; and a value $d_U^- = \inf_{\{y \in X : (x, y) \in \partial_N^+(U)\}} T^{*-}(y)$, where $T^{*-}(y)$ means the shortest path distance from y to s in N, according to the lengths $\{T_{x, y} : (x, y) \in A\}$.

Projecting the TEN IRP on the network N means to deal with the projections F and f, of flows H and h on the arcs of N. If we restrict ourselves to N (i.e. we do not take source $\hat{s}$ and sink $\hat{t}$ into account), then we see that:

- F satisfies flow conservation at any vertex of X;   (E5.1)
- for any vertex x of N, $\sum_{a\in\partial_N^+(x)} f(a) - \sum_{a\in\partial_N^-(x)} f(a) = b_x$;   (E5.2)
- for any arc a of N, $f(a) \leq \kappa \cdot F(a)$;   (E6)
- carrier ride cost $c_2 = \beta \cdot (\sum_{a\in A} C_a \cdot F(a))$;   (E7.1)
- item ride time $c_3 = \gamma \cdot (\sum_{a\in A} T_a \cdot f(a))$.   (E7.2)

3.1   Extended Subtour Constraints and Projected Cost

Constraints (E5.1) and (E5.2) are not enough to characterize F and f in a satisfactory way, since they do not forbid subtours, likely to induce a significant distortion between carrier ride cost c_2 and items ride time c_3 related to H and h, and the values $\beta \cdot (\sum_{a\in A} C_a \cdot F(a))$ and $\gamma \cdot (\sum_{a\in A} T_a \cdot f(a))$. Besides, (E5.1)–(E7.2) do not provide us with a well-fitted estimation of the carrier number $c_1 = |Q| = H\big((\hat{t}, \hat{s})\big)$. However, we can check that the following statements hold.

Lemma 1. *Let (H, h) be a feasible solution of TEN IRP and let (F, f) be its projection on the network N, then $\frac{(\sum_{a\in A} T_a \cdot F(a))}{\Omega}$ is a lower bound for the carrier number.*

Proof. The quantity $\sum_{a\in A} T_a \cdot F(a)$ provide us with the global time carriers spend running inside N, waiting times being excluded. Since the whole process must be performed in no more than Ω time units, we see that we need at least $\frac{(\sum_{a\in A} T_a \cdot F(a))}{\Omega}$ carriers in order to achieve it.   □

As a consequence, we should search for (F, f) that minimizes the projected cost: $PCost(F, f) = \alpha \cdot \frac{(\sum_{a\in A} T_a \cdot F(a))}{\Omega} + \beta \cdot (\sum_{a\in A} C_a \cdot F(a)) + \gamma \cdot (\sum_{a\in A} T_a \cdot f(a))$.

Lemma 2. *For all $U \subseteq X \setminus \{s\}$, the Weak Extended Subtour Inequality holds:*

$$\Omega \cdot \left(\sum_{a\in\partial_N^-(U)} F(a) \right) \geq \sum_{a\in\partial_N(U)\cup A(U,U)} T_a \cdot F(a). \qquad (E8.1)$$

Proof. Let us suppose that Q is the set of carriers involved in a TEN IRP solution (H, h). Some of these carriers can enter into U, one or several times, get out of U, and move inside U. Hence the time they globally spend while doing it is equal to $\sum_{a\in\partial_N(U)\cup A(U,U)} T_a \cdot F(a)$. For each of those $|Q|$ carriers, this time cannot exceed Ω. We deduce that $|Q| \cdot \Omega \geq \sum_{a\in\partial_N(U)\cup A(U,U)} T_a \cdot F(a)$. Since $\sum_{a\in\partial_N^-(U)} F(a) \geq |Q|$, we conclude.   □

Lemma 2 may be strengthened as follows.

Lemma 3. *For all $U \subseteq X \setminus \{s\}$, the Strong Extended Subtour Inequality holds:*

$$(\Omega - d_U^+ - d_U^-) \cdot \left(\sum_{a\in\partial_N^-(U)} F(a) \right) \geq \sum_{a\in\partial_N(U)\cup A(U,U)} T_a \cdot F(a). \qquad (E8.2)$$

Proof. We adapt the proof of Lemma 2, while noticing that for everyone among the $|Q|$ carriers involved into U, the time it spends while entering into U, getting out of U and moving inside U, cannot exceed $\Omega - d_U^+ - d_U^-$, since it moves first from s until some vertex v such that $(v, x) \in \partial_N^-(U)$ and next go back the same way from some vertex y, such that $(y, z) \in \partial_N^+(U)$ into s.   □

Remark 1. Though Lemma 3 implies Lemma 2, we distinguish both, because they are going to induce very different separation algorithms, and because in practice, only constraints (E8.1) are going to be efficient.

We deduce that we should search for (F, f) as a solution of the following projected model.

Projected Item Relocation Problem (PIRP). *Compute on the network $N = (X, A)$ two functions $F : A \to \mathbb{Z}_+$ and $f : A \to \mathbb{Z}_+$ such that:*

- *F satisfies flow conservation at any vertex of X;*   (E5.1)
- *for any vertex $x \in X$, $\sum_{a \in \partial_N^-(x)} f(a) - \sum_{a \in \partial_N^+(x)} f(a) = b_x$;*   (E5.2)
- *for any arc $a \in A$, $f(a) \leq \kappa \cdot F(a)$;*   (E6)
- *for any $U \subseteq X \setminus \{s\}$,*
$$\left(\Omega - d_U^+ - d_U^-\right) \cdot \left(\sum_{a \in \partial_N^-(U)} F(a)\right) \geq \sum_{a \in \partial_N(U) \cup A(U,U)} T_a \cdot F(a);$$ (E8.2)
- *Minimize $PCost(F, f) =$*

$$\alpha \cdot \frac{\left(\sum_{a \in A} T_a \cdot F(a)\right)}{\Omega} + \beta \cdot \left(\sum_{a \in A} C_a \cdot F(a)\right) + \gamma \cdot \left(\sum_{a \in A} T_a \cdot f(a)\right).$$ (E9)

Remark 2 (The Lift Issue). The example depicted in Fig. 3 shows that a solution (F, f) of the above PIRP cannot always be viewed as the projection of a feasible solution (H, h) of IRP TEN. In Fig. 3, we see that the carrier follows the route (s, y, x, z, y, s), but cannot transport this way any item from z to x. In fact, it is known that computing (H, h) from (F, f) in such a way that (F, f) is the projection of (H, h) with identical cost value, is NP-Hard [7].

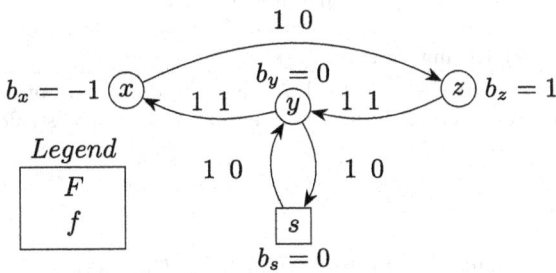

Fig. 3. A solution of PIRP which cannot be lifted.

In the following, we are going to focus on the handling of the PIRP through a Branch-and-Cut algorithm. This includes the design of polynomial time algorithms which separate the Extended Subtour Constraints, while Sect. 4 will describe the way how the Branch-and-Cut is implemented, together with some numerical experiments. We will return to the lift issue only in Sect. 5.

3.2 Separating the Extended Subtour Constraints

We have the two following results.

Theorem 2. *The Weak Extended Subtour Constraints* (E8.1) *can be separated in polynomial time, through a classical Min-Cut algorithm.*

Proof. Let F be a flow on a network $N = (X, A)$, and $\Delta = \frac{\sum_{a \in A} T_a \cdot F(a)}{\Omega}$. Let t be a new vertex. For any $a = (x, y) \in A$, we define $w_a = \frac{T_a \cdot F(a)}{\Omega}$ and we create two arcs: one arc $a' = (x, y)$ with weight $F(a) - w_a$, and one arc $a'' = (x, t)$ with weight w_a. It can be shown that separating the (E8.1) constraints on N is equivalent to search for a Min-Cut on $N' = (X \cup \{t\}, A' = \{a', a'' : a \in A\})$. □

Theorem 3. *The Strong Extended Subtour Constraints* (E8.2) *can be separated in polynomial time, by a sequence of* $|X|^2$ *applications of a Min-Cut algorithm.*

4   Algorithmic Handling and Numerical Experiments

We handle the PIRP model through a Branch-and-Cut algorithm. For that, we used CPLEX12 MILP libraries, and could restrict ourselves to the implementation of the callback procedure which commands the cut generation, while letting the library decide for the choice of a branching strategy.

4.1 Separation Algorithm

We deal with both versions of Extended Subtour Constraints in Algorithm 1.

Algorithm 1: Separation of constraints (E8.1)

---

input : Network N, flow F, and time horizon $[0, \Omega]$.
output: A succes signal if the constraints (E8.1) are satisfied by F, otherwise
a (E8.2) constraint violated by F.

1 Derive from F, N, and Ω the network N' described in the proof of Theorem 2;
2 Apply the Ford-Fulkerson algorithm to compute a maximum s-t-flow φ in N', and let val(φ) be the resulting flow value;
3 **if** val(φ) $\geq \Delta$ **then**
4 | **return succes**;
5 **else**
6 | Retrieve an s-t-cut $\partial^+(U)$ with $s \in U$, $t \in X \setminus U$, and capacity val(φ);
7 | **return** constraint (E8.2) related to U;
8 **end**

Remark 3. Notice that we proceed here in a hybrid way, since we apply a separation approach designed for constraints (E8.1), but send back related constraints (E8.2) to the main Branch-and-Cut process.

4.2   Numerical Experiments

We only present here results involving the Weak Extended Subtour Constraints. The reason is that, as one could expect, the above strong separation algorithm is significantly more time consuming than its weak counterpart, for a marginal cutting power which remains relatively small: in most cases, subsets U which are going to be involved in the separation process are such that the values d_U^+ and d_U^- are small, so that replacing (E8.1) by (E8.2) does not significanlty increase the optimal value of the rational relaxation of PIRP. We proceed in a hybrid way, according to the above description of the weak separation algorithm: we search for subsets $U \subseteq X \setminus \{s\}$ for which the related constraint (E8.1) is violated by the current flow F, and, in case we find such a subset U, we send back the related constraint (E8.2) to the Branch-and-Cut resolution process.

Since we solved PIRP in an exact way and since we do not deal here with the Lift issue, we are not interested in error estimation. We focus here on the way the Branch-and-Cut algorithm behaves (running time, number of nodes visited during the tree search, number of cuts generated) depending on the characteristics of the network N and the ability of the cutting process to enhance the quality of the solution (F, f).

Experiments were performed on a computer with a 2.3 GHz Intel Core i5 processor and 16 GB 1333 MHz RAM. The implementations use the CPLEX12.10 MILP libraries, and they were built in C++ 11.

No standardized benchmarks exist for the generic IRP. So we built instances as follows: the station set X is a set of n points inside a 100×100 grid, the arcs are generated randomly, the set T corresponds to the rounded Euclidean Distance and the set C to the Manhattan Distance, each vertex x but the depot s has assigned a b_x value in $\{-10, \ldots, 10\}$, the capacity κ is chosen from $\{2, 5, 10, 20\}$, the time horizon limit Ω is a product $\lambda \cdot (\max_{x, y \in X} T_{x, y})$ taking $\lambda \in \{4, 8\}$, and the scaling coefficients α, β, γ are chosen in such a way that when they are non-zero, the values of cost components $\alpha \cdot numberofcarriers$, $\beta \cdot carrierridecost$ and $\gamma \cdot itemsridetime$ become comparable.

For any instance with n vertices and m arcs, we compute the optimal value G1 of the PIRP, without considering the term to approximate the number of carriers. T1 is the running time in seconds for the computation of G1. G2 is the optimal value of the PIRP, without the Extended Subtour Constraints. T2 is the running time in seconds for the computation of G2. V2 is the estimated number of vehicles used in the solution with cost G2. G3 is the optimal value of the PIRP, including the Extended Subtour Constraints from PIRP. T3 is the running time in seconds for the computation of G3. V3 is the estimated number of vehicles used in the solution with cost G3. Nd is the number of visited nodes throughout the tree search performed by the Branch-and-Cut process. Ct is the number of cuts generated for the computation of G3.

Table 1 provides us with the output values for the 20 instances analyzed. We see that the Extended Subtour Constraints improve the cost of solutions (F, f), so they are closer to the projections of optimal TEN IRP solutions (H, h).

Table 1. Numerical results.

Id	n	m	κ	Ω	λ	α	β	γ	G1	T1	G2	T2	V2	G3	T3	V3	LB3	Nd	Ct
1	20	78	2	324	4	304	1.0	1.000	999.00	0.01	1621.11	0.03	2	2109.12	0.47	3	1323.04	642	12
2	20	65	5	320	4	150	0.4	0.500	765.90	0.05	1127.71	0.05	3	1290.09	0.08	3	971.52	36	2
3	20	77	10	440	4	328	0.2	0.250	441.85	0.02	819.47	0.07	2	854.83	0.12	2	546.78	34	2
4	20	75	2	680	8	328	1.0	1.000	2925.00	0.02	3708.34	0.02	3	3805.81	0.07	3	3599.04	52	2
5	20	50	5	536	8	392	0.4	0.250	1226.20	0.02	2323.21	0.03	3	2432.86	0.03	3	2064.80	0	1
6	20	57	10	840	8	376	0.2	0.250	1016.20	0.02	1532.38	0.03	2	1532.38	0.03	2	1276.16	0	1
7	20	62	5	420	6	300	0.4	0.500	1600.10	0.01	2445.81	0.02	3	2727.30	0.07	4	2448.12	167	2
8	50	163	2	460	4	170	1.0	1.000	11606.00	0.04	13951.30	0.08	14	15561.30	0.27	17	15163.21	39	3
9	50	155	5	260	4	196	0.4	0.500	2783.70	0.13	4292.61	0.27	8	5612.59	6.71	12	4927.96	7035	33
10	50	149	10	440	4	164	1.0	0.500	6910.50	0.23	7829.03	0.39	6	7966.03	0.78	6	7443.35	481	8
11	50	146	20	436	4	312	0.1	0.125	687.08	0.25	1590.26	0.45	3	1840.17	2.49	4	933.32	5173	3
12	50	175	2	728	8	268	1.0	1.000	5640.00	0.06	6878.19	0.10	5	6976.11	0.14	5	6531.15	225	1
13	50	217	5	912	8	672	0.4	0.250	777.25	0.45	1525.93	0.96	2	1643.76	0.98	2	1044.28	131	5
14	50	154	10	1040	8	416	0.2	0.125	1627.28	0.77	2547.22	0.78	3	2643.62	3.20	3	2230.72	2850	4
15	100	363	2	336	4	252	1.0	1.000	9587.00	1.39	13507.00	4.61	16	17131.00	44.97	22	16286.10	33878	9
16	100	236	5	516	4	188	0.4	0.250	3113.05	0.40	4391.96	1.38	7	4826.24	1.05	8	4257.80	117	7
17	100	289	10	432	4	360	0.2	0.250	1743.65	0.21	3018.55	0.78	4	3227.92	0.60	4	2696.90	59	8
18	100	419	2	1032	8	412	1.0	1.000	16164.00	0.57	19882.20	0.65	9	20219.30	1.00	10	19556.76	161	1
19	100	327	5	552	8	392	0.5	0.200	3146.00	1.44	5748.40	7.51	7	5944.23	5.99	7	5175.84	1829	2
20	100	313	10	712	8	312	0.5	0.500	4706.50	0.63	5883.62	2.46	4	6091.24	4.68	4	4885.78	1026	3

5   Conclusion: A Brief Discussion of the Lift Issue

While starting from a problem simultaneously involving routing and scheduling in a transit network N, we focused here on a 2-commodity flow model which can be viewed as a projection of a formulation of the original problem on a Time Expanded Network on the original transit network N. This 2-commodity flow model is complex by itself, and we mainly focused here on the way to efficiently solve it through Branch-and-Cut and the use of the above proposed Extended Subtour Constraints. Hereby, we neglegted so far the Lift issue, i.e. the way how a solution (F, f) of the PIRP may be turned into a feasible (optimal) solution of the original problem. We observe that this question may be understood in several ways.

First, we are intereseted in finding collections of valid constraints that should be inserted into the PIRP in order to increase the probability that an optimal solution (F, f) becomes liftable, that means to be the projection of some feasible solution (H, h) of the original TEN IRP model. The main contribution of this work is to provide a partial answer to this question.

Starting from a solution (F, f), a second question is how to design an exact algorithm which will test the existence of a feasible (optimal) solution (H, h) of

TEN IRP, whose projection is (F, f)? A further question is which kind of efficient heuristics can we design, specially for the case of large transit networks, which will derive feasible flows (H, h) from (F, f) with the best possible cost value?

Those questions will be at the core of our future research. Simultaneously, we shall study those among the variants presented in Sect. 1, which are related to the case when items or carriers cannot be considered as identical, and so must be represented as multicommodity flow, possibly constrained by time windows.

Acknowledgement. This work has been carried out in the context of the H2020 Marie Skłodowska-Curie Research and Innovation Staff Exchange European project 691161 "GEO-SAFE" [13].

On behalf of all authors, the corresponding author states that there is no conflict of interest.

References

1. Ahuja, R.V., Magnanti, T.L., Orlin, J.B.: Network Flows: Theory, Algorithms and Applications. Prentice hall, Englewood Cliffs (1993)
2. Aronson, J.: A survey of dynamic network flows. Ann. Oper. Res. **20**, 1–66 (1989)
3. Ashok, K., Ben-Akiva, M.: Estimation and prediction of time-dependent origin-destination flows with a stochastic mapping to path flows and link flows. Transp. Sci. **36**(2), 184–198 (2002)
4. Benchimol, M., et al.: Balancing the stations of a self service "bike hire" system. RAIRO - Oper. Res. **45**, 37–61 (2011)
5. Bsaybes, S., Quilliot, A., Wagler, A.K.: Fleet management for autonomous vehicles using flows in time-expanded networks. TOP **27**(2), 288–311 (2019). https://doi.org/10.1007/s11750-019-00506-4
6. Bsaybes, S., Quilliot, A., Wagler A.: Fleet management for autonomous vehicles using multicommodity coupled flows in time-expanded networks. In: 17th International Symposium on Experimental Algorithms (SEA 2018), Leibniz International Proceedings in Informatics (LIPIcs) (2018)
7. Chemla, D., Meunier, F., Wolfler-Calvo, R.: Bike sharing systems: solving the static rebalancing problem. Disc. Optim. **10**(2), 120–146 (2013)
8. Chifflet, J., Mahey, P., Reynier, V.: Proximal decomposition for multicommodity flow problems with convex costs. Telecommun. Syst. **3**, 1–10 (1994)
9. Contardo, C., Morency, C., Rousseau, L.: Balancing a Dynamic Public Bike-Sharing System. Univ. MONTREAL, Rapport CIRRELT (2012)
10. Dell'Amico, M., Hadjicostantinou, E., Iori, M., Novellani, S.: The bike sharing rebalancing problem: mathematical formulations and benchmark instances. Omega **45**, 7–19 (2014)
11. Erdoğan, G., Battarra, M., Wolfler-Calvo, R.: An exact algorithm for the static rebalancing problem arising in bicycle sharing systems. Eur. J. Oper. Res. **245**(3), 667–679 (2015)
12. Gavalas, D., Konstantopoulos, C., Pantziou, G.: Design and management of vehicle-sharing systems: a survey of algorithmic approaches. Smart Cities Homes, 261–289 (2016)
13. GEO-SAFE. Geospatial based Environment for Optimisation Systems Addressing Fire Emergencies (2020). https://cordis.europa.eu/project/id/691161. Accessed 21 Dec 2020

14. Krumke, S., Quilliot, A., Wagler, A., Wegener, J.: Relocation in carsharing systems using flows in time-expanded networks. Lect. Notes Comput. Sci. **8504**, 87–98 (2014)
15. Nourinejad, M., Roorda, M.: A dynamic carsharing decision support system. Transp. Res. Part E: Logist. Transp. Rev. **66**, 36–50 (2014)
16. Raviv, T., Tzur, M., Forma, I.A.: Static repositioning in a bike-sharing system: models and solution approaches. EURO J. Transp. Logist. **2**, 187–229 (2013)
17. Shaheen, S., Guzman, S., Zhang, H.: Bikesharing in Europe, the Americas, and Asia: Past, Present, and Future. Institute of Transportation Studies, UC Davis, Institute of Transportation Studies, Working Paper Series. 2143 (2010)
18. Schuijbroek, J., Hampshire, R.C., van Hoeve, W.: Inventory rebalancing and vehicle routing in bike sharing systems. Eur. J. Oper. Res. **257**(3), 992–1004 (2017)
19. Waserhole, A., Jost, V.: Vehicle sharing system pricing regulation: a fluid approximation (2012)

The Constrained-Routing and Spectrum Assignment Problem: Valid Inequalities and Branch-and-Cut Algorithm

Ibrahima Diarrassouba[1]([✉]) and Youssouf Hadhbi[2]

[1] Le Havre Normandie University, LMAH, FR CNRS 3335, 76600 Le Havre, France
`diarrasi@univ-lehavre.fr`
[2] Clermont Auvergne University, LIMOS (UMR 6158 CNRS), 1 Rue de la Chebarde, Aubiere, 63178 Clermont Ferrand, France
`youssouf.hadhbi@uca.fr`

Abstract. We consider the Constrained-Routing and Spectrum Assignment (C-RSA) problem. Given an undirected, loopless, and connected graph G, an optical spectrum $\mathbb{S}$ of available contiguous frequency slots, and a multiset of traffic demands K, the C-RSA consists in assigning for each traffic demand $k \in K$ a path in G between its origin and destination, and an interval of contiguous frequency slots in $\mathbb{S}$ while satisfying some technological constraints, and minimizing the total cost of the paths used for routing the demands. In this paper, we give an integer linear programming formulation for the C-RSA problem. Moreover, we investigate the related polyhedron and describe several valid inequalities. We also devise separation routines for these inequalities. Based on this, we propose a Branch-and-Cut algorithm for the problem along with an extensive computational study showing the effectiveness of our approach.

Keywords: Optical network design · Routing · Spectrum assignment · Integer programming · Polyhedron · Dimension · Valid inequality · Facet · Separation · Branch-and-cut

1 Introduction

The global Internet Protocol (IP) traffic is expected to reach 396 exabytes per month by 2022, up from 194.4 Exabytes per month in 2020 [4]. Optical networks are then facing a serious challenge related to continuous growth in bandwidth capacity due to the growth of global communication services and networking: mobile internet network (e.g., 5th generation mobile network), cloud computing (e.g., data centers), Full High-definition (HD) interactive video (e.g., TV channel, social networks) [3]. To deal with this, a new generation of optical network architecture called Spectrally Flexible Optical Networks (SFONs) has been introduced as promising technology because of its flexibility, scalability, efficiency, reliability, and survivability [3] compared with the traditional Fixed Grid

This work was supported by the French National Research Agency grant ANR-17-CE25-0006.

I. Ljubić et al. (Eds.): ISCO 2022, LNCS 13526, pp. 35–47, 2022.
https://doi.org/10.1007/978-3-031-18530-4_3

Optical Wavelength Division Multiplexing (WDM) [11]. In this context, SFONs introduce a new technology called Orthogonal Frequency Division Multiplexing (OFDM) to divide the optical spectrum into spectral units, called frequency slots. They have the same frequency of 12.5 GHz where WDM uses 50 GHz as recommended by ITU-T [3].

The Routing and Spectrum Assignment (RSA) problem is a key issue when dimensioning and designing SFONs. It consists of assigning for each traffic demand, a physical optical path in the network between its origin and destination, and an interval of contiguous slots in an optical spectrum while optimizing some linear objective(s) (e.g., minimizing the total cost or length of the paths used to route the demands) and satisfying the following technological constraints [9]:

1. *contiguity*: an interval of contiguous slots should be allocated to each demand k with a width equal to the number of slots requested by demand k;
2. *continuity*: the interval of contiguous slots allocated to each traffic demand stills the same along the chosen path;
3. *non-overlapping*: the intervals of contiguous slots of demands whose paths are not edge-disjoints in the network cannot share any slot over the shared edges.

The RSA is known to be NP-hard [13]. Various mathematical programming formulations and algorithms have been proposed to solve it. However, and to the best of our knowledge, those formulation didn't take into account some additional real technological constraints required by network operators. In this context, the route for each traffic demand should not exceed a certain length called *transmission-reach*. We refer the reader to [7] for more details.

In this paper, we focus on a constrained version of the RSA problem, called the Constrained-Routing and Spectrum Assignment (C-RSA) problem. Recently, Hadhbi et al. [9] developed an exact Branch-and-Cut algorithm for the C-RSA problem. This has then been used in the study of Colares et al. [5].

So far the exact algorithms proposed in the literature could not solve large-scale instances of the RSA and C-RSA problems. We believe that a cutting-plane-based approach could be powerful for the problem. To the best of our knowledge, such an approach has not been yet considered before except in the works done by Bianchetti et al. [2] for the RSA problem. For this, we first introduce an integer linear programming formulation, called cut formulation. We investigate its associated polytope and identify several classes of valid inequalities for the polytope. We then devise separation routines for these inequalities. Using the polyhedral results and the separation routines, we develop an exact Branch-and-Cut (B&C) algorithm for solving the C-RSA problem.

The rest of the paper is organized as follows. In Sect. 2, we present the C-RSA problem. We then introduce the so-called cut formulation in Sect. 3. Section 4 is devoted to the polyhedral investigation of the C-RSA and description of several valid inequalities for the polytope. We then present the B&C algorithm in Sect. 5. The computational results are presented in Sect. 6. Finally, we give some concluding remarks and future outlook in Sect. 7.

2   The Constrained-Routing and Spectrum Assignment Problem

In terms of graphs, the C-RSA problem can be stated as follows. Consider an optical spectrum $\mathbb{S} = \{1, \ldots, \bar{s}\}$ of $\bar{s} \in \mathbb{Z}_+$ available frequency slots, and a spectrally flexible optical network as an undirected, loopless, and connected graph $G = (V, E)$, which is given by a set of nodes V (data centers, users, stations,...), and a multiset[1] E of links (optical-fibers). Each link $e = ij \in E$ is specified by a length $\ell_e \in \mathbb{R}_+$ (in kms), and a cost $c_e \in \mathbb{R}_+$ such that each fiber-link $e \in E$ is divided into $\bar{s}$ slots. Let K be a multiset[2] of demands such that each demand $k \in K$ is specified by an origin node $o_k \in V$, a destination node $d_k \in V \setminus \{o_k\}$, a slot-width $w_k \in \mathbb{Z}_+$, and a transmission-reach $\bar{\ell}_k \in \mathbb{R}_+$ (in kms). The C-RSA problem consists in determining for each demand $k \in K$, a (o_k, d_k)-path p_k in G such that $\sum_{e \in E(p_k)} \ell_e \leq \bar{\ell}_k$ (*transmission-reach constraint*), where $E(p_k)$ denotes the set of edges in the path p_k, and a subset of contiguous frequency slots $S_k \subset \mathbb{S}$ of width w_k (*continuity and contiguity constraints*) such that $S_k \cap S_{k'} = \emptyset$ for each pair of demands $k, k' \in K$ ($k \neq k'$) with $E(p_k) \cap E(p_{k'}) \neq \emptyset$ (*non-overlapping constraint*) so that the total cost of the paths used for routing the demands (i.e., $\sum_{k \in K} \sum_{e \in E(p_k)} c_e$) is minimum. Figure 1 shows the set of paths and spectrums assigned for the set of demands in a graph G of 7 nodes and 10 edges such that each edge e is characterized by a triplet $[\ell_e, c_e, \bar{s}]$ with $\bar{s} = 9$.

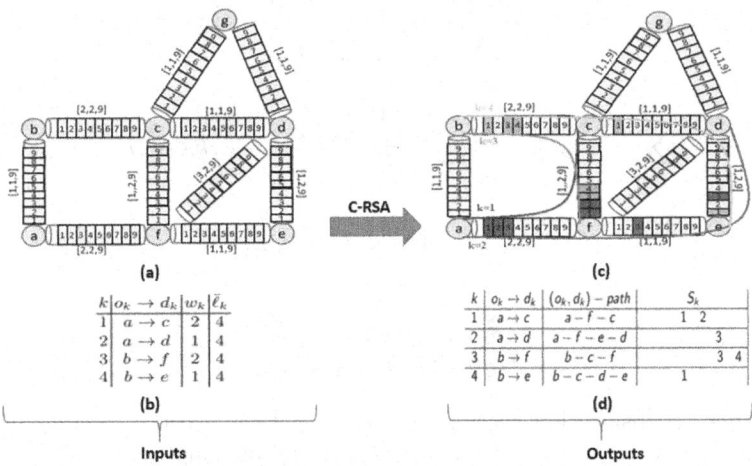

Fig. 1. Set of established paths and spectrums in graph G (Fig. 1(a)) for the set of demands $\{k_1, k_2, k_3, k_4\}$ defined in Fig. 1(b).

---

[1] We take into account the presence of parallel fibers such that two edges e, e' which have the same extremities i and j are independent.

[2] We take into account that we can have several demands between the same origin-node and destination-node.

3 Integer Linear Programming Formulation

Here we introduce an integer linear programming based on the so-called cut formulation. For $k \in K$ and $e \in E$, let x_e^k be a variable which takes 1 if demand k goes through edge e and 0 if not. For $k \in K$ and $s \in \mathbb{S}$, let z_s^k be a variable which takes 1 if slot s is the last slot allocated for the routing of demand k and 0 if not. The contiguous slots $s' \in \{s - w_k + 1, ..., s\}$ should be assigned to demand k whenever $z_s^k = 1$. Note that a slot $s \in \mathbb{S}$ is assigned to a demand $k \in K$ iff $\sum_{s'=s}^{min(\bar{s}, s+w_k-1)} z_{s'}^k = 1$. For $X \subset V$ with $X \neq \emptyset$, let $\delta(X)$ denote cut induced by X, that is the set of edges having one extremity in X and the other in $\bar{X} = V \setminus X$.

The C-RSA is equivalent to the following integer linear program

$$\min \sum_{k \in K} \sum_{e \in E} c_e x_e^k, \tag{1}$$

subject to

$$\sum_{e \in \delta(X)} x_e^k \geq 1, \forall k \in K, \forall X \subseteq V \text{ such that } |X \cap \{o_k, d_k\}| = 1, \tag{2}$$

$$\sum_{e \in E} \ell_e x_e^k \leq \bar{\ell}_k, \forall k \in K, \tag{3}$$

$$z_s^k = 0, \forall k \in K, \forall s \in \{1, ..., w_k - 1\}, \tag{4}$$

$$\sum_{s=w_k}^{\bar{s}} z_s^k = 1, \forall k \in K, \tag{5}$$

$$x_e^k + x_e^{k'} + \sum_{s'=s}^{min(s+w_k-1,\bar{s})} z_{s'}^k + \sum_{s'=s}^{min(s+w_{k'}-1,\bar{s})} z_{s'}^{k'} \leq 3, \forall e \in E, k \in K, k' \in K \setminus \{k\}, s \in \mathbb{S}, \tag{6}$$

$$0 \leq x_e^k \leq 1, \forall k \in K, \forall e \in E, \tag{7}$$

$$z_s^k \geq 0, \forall k \in K, \forall s \in \mathbb{S}, \tag{8}$$

$$x_e^k \in \{0, 1\}, \forall k \in K, \forall e \in E, \tag{9}$$

$$z_s^k \in \{0, 1\}, \forall k \in K, \forall s \in \mathbb{S}. \tag{10}$$

Inequalities (2) ensure that there is an (o_k, d_k)-path between o_k and d_k for each demand k. They are called cut inequalities. By minimizing the total cost of the routing paths (1), and given that the cost of all edges are positive, this ensures that there is exactly one (o_k, d_k)-path as optimal path for each demand k. We suppose that we have sufficient capacity $\bar{s}$ in the network so that all the demands can be routed. This ensures the existence of a feasible solution for the problem. Inequalities (3) express the length limit on the routing paths which are called the transmission-reach inequalities. Equations (4) express the fact that a demand k

cannot use slot $s \leq w_k - 1$ as the last-slot. Inequalities (5) ensure that exactly one slot $s \in \{w_k, \ldots, \bar{s}\}$ can be assigned to demand k as last-slot. Inequalities (6) express the contiguity and non-overlapping constraints such that they ensure that a slot s over the edge e can be assigned to at most by one demand $k \in K$. Inequalities (7)-(8) are the trivial inequalities, and constraints (9)-(10) are the integrality constraints. We call this formulation the *Cut Formulation*.

Note that the linear relaxation of the C-RSA can be solved in polynomial time given that cut inequalities (2) are separable in polynomial time using network flows.

In the next section, we investigate the related polytope, that is the convex hull of all the solutions of (1)-(10).

4   Valid Inequalities and Facets

An instance of the C-RSA is defined by a triplet $(G, K, \mathbb{S})$. Let $P(G, K, \mathbb{S})$ be the convex hull of all the solutions of the cut formulation (1)-(10). For a demand $k \in K$, let E_0^k denote the set of all *forbidden* edges of demand k that is those edges e such that the length of each (o_k, d_k)-path in G going through e exceeds $\bar{l}_k$, and let E_1^k be the set of *essential* edges of demand k that is those edges which belong to every feasible (o_k, d_k)-path in G. Note that sets E_0^k and E_1^k can be determined in polynomial time [7]. Hence, for each demand $k \in K$, we have $x_e^k = 0$ for each $e \in E_0^k$, and $x_e^k = 1$ for each $e \in E_1^k$. For $e \in E$, let K_e denote the subset of demands K having e as essential edge, and $\bar{K}_e$ denote the subset of demands K having e as forbidden edge.

In what follows, we present several valid inequalities for the polytope $P(G, K, \mathbb{S})$. We refer the reader to [6] for more details about each proof.

4.1   Edge-Capacity-Cover Inequalities

Consider an edge $e \in E$, and a subset of demands $C \subseteq K$ with $e \notin E_0^k \cup E_1^k$ for each demand $k \in C$. C is called a cover for the edge e if $\sum_{k \in C} w_k > \bar{s} - \sum_{k' \in K_e} w_{k'}$. Moreover, it is said to be minimal if $C \setminus \{k\}$ is not a cover for all $k \in C$, i.e., $\sum_{k' \in C \setminus \{k\}} w_{k'} \leq \bar{s} - \sum_{k'' \in K_e} w_{k''}$.

Theorem 1. *Let C be a minimal cover for an edge $e \in E$. Then, the inequality*

$$\sum_{k \in C} x_e^k \leq |C| - 1, \tag{11}$$

is valid for $P(G, K, \mathbb{S})$.

Proof. If C is a minimal cover for edge $e \in E$, this means that there is at most $|C| - 1$ demands from the set C that can be routed through the edge e.

Theorem 2. *Let C be a minimal cover for an edge $e \in E$. Then, inequality (11) is facet defining for the polytope*

$$P(G, K, \mathbb{S}, e, C) = conv\{(x, z) \in P(G, K, \mathbb{S}) \mid \sum_{k' \in K \setminus (K_e \cup C \cup \bar{K}_e)} x_e^{k'} = 0\}.$$

One can use the so-called sequential lifting procedure [1] to sequentially lift inequality (11) and generate lifted facets for the polytope $P(G, K, \mathbb{S})$ as follows.

Theorem 3. *Let C be a minimal cover for an edge $e \in E$. Let $K \setminus (K_e \cup C \cup \bar{K}_e)$ = $\{k_1, ..., k_r\}$ be arbitrarily ordered with $r = |K \setminus (K_e \cup C \cup \bar{K}_e)|$. Consider a sequence of knapsack problems defined as*

$$
\begin{cases}
z_i = \max \sum_{j \in C} u_j + \sum_{j=1}^{i-1} \alpha_j u_j, \\
\sum_{j \in C} w_j u_j + \sum_{j=1}^{i-1} w_{k_j} u_j \leq \bar{s} - \sum_{k' \in K_e} w_{k'} - w_{k_i}, \\
u_j \in \{0, 1\}, \forall j \in C \cup \{1, ..., i-1\},
\end{cases}
\tag{12}
$$

for all $i \in \{1, ..., r\}$ with $\alpha_j = |C| - 1 - z_j$ for all $j \in \{1, ..., i - 1\}$. Then, the inequality

$$\sum_{k \in C} x_e^k + \sum_{j=1}^{r} \alpha_j x_e^{k_j} \leq |C| - 1, \tag{13}$$

is valid for $P(G, K, \mathbb{S})$. Moreover, inequality (13) defines facet of $P(G, K, \mathbb{S})$.

4.2   Edge-Interval-Capacity-Cover Inequalities

An interval $I = [s_i, s_j]$ represents a set of contiguous slots situated between the two slots s_i and s_j with $j \geq i + 1$ and $s_j \leq \bar{s}$.

For an interval of contiguous slots $I = [s_i, s_j]$, a subset of demands $\tilde{K} \subseteq K$ is called a cover for the interval I iff $\sum_{k \in \tilde{K}} w_k > |I|$ and $w_k < |I|$ for each $k \in \tilde{K}$. Moreover, it is said to be minimal if $\sum_{k' \in \tilde{K} \setminus \{k\}} w_{k'} \leq |I|$ for each demand $k \in \tilde{K}$.

Theorem 4. *Let $I = [s_i, s_j]$ be an interval of contiguous slots in $[1, \bar{s}]$. Let $\tilde{K}$ be a minimal cover for the interval I such that $\tilde{K}$ does not define a minimal cover for an edge e, where $e \notin E_0^k$ for each demand $k \in \tilde{K}$. Then, the inequality*

$$\sum_{k \in \tilde{K}} x_e^k + \sum_{k \in \tilde{K}} \sum_{s=s_i+w_k-1}^{s_j} z_s^k \leq 2|\tilde{K}| - 1, \tag{14}$$

is valid for $P(G, K, \mathbb{S})$.

Proof. The interval $I = [s_i, s_j]$ can at most cover $|\tilde{K}| - 1$ demands if they pass together through the edge e (i.e., $\sum_{k \in \tilde{K}} x_e^k = |\tilde{K}|$) given that $\tilde{K}$ is a minimal cover for interval I. We ensure that the inequalities (14) are verified by any feasible solution of the problem taking into account that $\sum_{k \in \tilde{K}} x_e^k \leq |\tilde{K}|$ and $\sum_{k \in \tilde{K}} \sum_{s=s_i+w_k-1}^{s_j} z_s^k \leq |\tilde{K}|$. Otherwise, the contiguity and non-overlapping constraints are violated.

Theorem 5. *Let $I = [s_i, s_j]$ be an interval of contiguous slots in $[1, \bar{s}]$. Let $\tilde{K}$ be a minimal cover for the interval I such that $\tilde{K}$ does not define a minimal cover for an edge e, where $e \notin E_0^k$ for each demand $k \in \tilde{K}$. Then, inequality (14) is facet defining for the polytope*

$$P(G, K, \mathbb{S}, e, I, \tilde{K}) = conv\{(x, z) \in P(G, K, \mathbb{S})| \sum_{k' \in K_e \backslash \tilde{K}} \sum_{s'=s_i+w_{k'}-1}^{s_j} z_{s'}^{k'} = 0\}$$

if and only if there does not exist an interval of contiguous slots $I' = [s_i', s_j']$ in $[1, \bar{s}]$ with $I \subset I'$ (i.e., $s_i' \leq s_i$ and $s_j' \geq s_j$) such that $\tilde{K}$ defines a minimal cover for the interval I'.

Inequality (14) can also be lifted using a sequential lifting procedure [1] to be facet defining and generate lifted facets for the polytope $P(G, K, \mathbb{S})$.

Theorem 6. *Let $I = [s_i, s_j]$ be an interval of contiguous slots in $[1, \bar{s}]$. Let $\tilde{K}$ be a minimal cover for the interval I such that $\tilde{K}$ does not define a minimal cover for an edge e, where $e \notin E_0^k$ for each demand $k \in \tilde{K}$. Let $K_e \backslash \tilde{K} = \{k_1, ..., k_r\}$ be arbitrarily ordred with $r = |K_e \backslash \tilde{K}|$. Consider the following sequence of knapsack problems defined as*

$$\begin{cases} z_i = \max \sum_{j \in \tilde{K}} a_j + \sum_{j=1}^{i-1} \beta_j a_j, \\ \sum_{j \in \tilde{K}} w_j a_j + \sum_{j=1}^{i-1} w_{k_j} a_j \leq |I| - w_{k_i}, \\ a_j \in \{0, 1\}, \forall j \in \tilde{K} \cup \{1, ..., i-1\}, \end{cases} \quad (15)$$

for all $i \in \{1, ..., r\}$ with $\beta_j = |\tilde{K}| - 1 - z_j$ for all $j \in \{1, ..., i-1\}$. Then, the inequality

$$\sum_{k \in \tilde{K}} x_e^k + \sum_{k \in \tilde{K}} \sum_{s=s_i+w_k-1}^{s_j} z_s^k + \sum_{j=1}^{r} \sum_{s'=s_i+w_{k_j}-1}^{s_j} \beta_j z_{s'}^{k_j} \leq 2|\tilde{K}| - 1, \quad (16)$$

is valid for $P(G, K, \mathbb{S})$. Moreover, inequality (16) defines facet of $P(G, K, \mathbb{S})$ if and only if there does not exist an interval of contiguous slots $I' = [s_i', s_j']$ in $[1, \bar{s}]$ with $I \subset I'$ such that $\tilde{K}$ defines a minimal cover for the interval I'.

In what follows, we will describe a set of valid inequalities that can be obtained using some conflict graphs related to the problem.

4.3 Edge-Interval-Clique Inequalities

Consider an edge $e \in E$. Let $I = [s_i, s_j]$ be an interval of contiguous slots in $[1, \bar{s}]$. Consider the conflict graph $\tilde{G}_I^e$ defined as follows. For each demand $k \in K$ with $w_k \leq |I|$ such that $e \notin E_0^k$, consider a node v_k in $\tilde{G}_I^e$. Two nodes v_k and $v_{k'}$ are linked by an edge in $\tilde{G}_I^e$ if $w_k + w_{k'} > |I|$.

Theorem 7. *Let $e \in E$, and $I = [s_i, s_j]$ be an interval of contiguous slots in $[1, \bar{s}]$, and C be a clique in $\tilde{G}_I^e$. Then, the inequality*

$$\sum_{v_k \in C} x_e^k + \sum_{s=s_i+w_k-1}^{s_j} z_s^k \leq |C| + 1, \tag{17}$$

is valid for $P(G, K, \mathbb{S})$.

Proof. Inequality (17) can also be obtained using the so-called Chvátal-Gomory procedure based the inequality $x_e^k + x_e^{k'} + \sum_{s=s_i+w_k-1}^{s_j} z_s^k + \sum_{s'=s_i+w_{k'}-1}^{s_j} z_{s'}^{k'} \leq 3$, for each $(v_k, v_{k'}) \in C$, which is valid for $P(G, K, \mathbb{S})$. This ensures that the two demands (k, k') cannot share the interval I if they pass together through edge e. Otherwise, the contiguity and non-overlapping constraints are violated.

Theorem 8. *Let $e \in E$, and $I = [s_i, s_j]$ be an interval of contiguous slots in $[1, \bar{s}]$, and C be a clique in $\tilde{G}_I^e$. Then, inequality (17) is facet defining for $P(G, K, \mathbb{S})$ if and only if the following conditions hold*

1. *there does not exist a demand $k' \in K_e \setminus C$ with $w_{k'} \leq |I|$ such that $w_k + w_{k'} > |I|$ for each $v_k \in C$,*
2. *and $|\{s_i + w_k - 1, ..., s_j\}| \geq w_k$ for at least one demand k with $v_k \in C$,*
3. *and there does not exist an interval I' of contiguous slots with $I \subset I'$ such that C defines also a clique in the associated conflict graph $\tilde{G}_{I'}^e$.*

4.4 Edge-Slot-Assignment-Clique Inequalities

For $e \in E$, let $\tilde{G}_S^e$ be the conflict graph defined as follows. For each slot $s \in \{w_k, ..., \bar{s}\}$ and demand $k \in K$ with $e \in E \setminus E_0^k$, consider a node $v_{k,s}$ in $\tilde{G}_S^e$. Two nodes $v_{k,s}$ and $v_{k',s'}$ are linked by an edge in $\tilde{G}_S^e$ if $k = k'$, or $\{s - w_k + 1, ..., s\} \cap \{s' - w_{k'} + 1, ..., s'\} \neq \emptyset$ if $k \neq k'$.

Theorem 9. *Consider an edge $e \in E$. Let C be a clique in the conflict graph $\tilde{G}_S^e$ with $|C| \geq 3$, and $\sum_{k \in C} w_k \leq \bar{s} - \sum_{k' \in K_e \setminus C} w_{k'}$. Then, the inequality*

$$\sum_{v_{k,s} \in C} (x_e^k + z_s^k) \leq |C| + 1, \tag{18}$$

is valid for $P(G, K, \mathbb{S})$.

Proof. Inequality (18) can also be obtained using the Chvátal-Gomory procedure based on the inequalities (5) and (6).

Theorem 10. *Consider an edge $e \in E$. Let C be a clique in the conflict graph $\tilde{G}_S^e$ with $|C| \geq 3$, and $\sum_{k \in C} w_k \leq \bar{s} - \sum_{k' \in K_e \setminus C} w_{k'}$. Then, inequality (18) is facet defining for $P(G, K, \mathbb{S})$ if and only if the following conditions hold*

- *there does not exist a nodes $v_{k',s'}$ with $k' \in K_e \setminus \{k \in K | \exists s \in \mathbb{S} \text{ with } v_{k,s} \in C\}$ such that $v_{k',s'}$ is linked with all nodes $v_{k,s} \in C$ in the conflict graph $\tilde{G}_S^e$,*
- *and there does not exist an interval of contiguous slots $I = [s_i, s_j] \subset [1, \bar{s}]$ with $\left[\min_{v_{k,s} \in C} (s - w_k + 1), \max_{v_{k,s} \in C} s \right] \subset I$, $w_k + w_{k'} \geq |I| + 1$ for each $(v_k, v_{k'}) \in C$, and $w_k \leq |I|$ for each $v_k \in C$.*

4.5   Slot-Assignment-Clique Inequalities

Let $\tilde{G}_S^E$ be the conflict graph defined as follows. For all slot $s \in \{w_k, ..., \bar{s}\}$ and demand $k \in K$, consider a node $v_{k,s}$ in $\tilde{G}_S^E$ with $|E_1^k| \geq 1$. Two nodes $v_{k,s}$ and $v_{k',s'}$ are linked by an edge in $\tilde{G}_S^E$ if $k = k'$, or $E_1^k \cap E_1^{k'} \neq \emptyset$ and $\{s - w_k + 1, ..., s\} \cap \{s' - w_{k'} + 1, ..., s'\} \neq \emptyset$ if $k \neq k'$.

Theorem 11. *Let C be a clique in conflict graph $\tilde{G}_S^E$. Then, the inequality*

$$\sum_{v_{k,s} \in C} z_s^k \leq 1, \tag{19}$$

is valid for $P(G, K, \mathbb{S})$.

Theorem 12. *Let C be a clique in conflict graph $\tilde{G}_S^E$. Then, inequality (19) is facet defining for $P(G, K, \mathbb{S})$ if and only if the following conditions hold*

- *C is a maximal clique in the conflict graph $\tilde{G}_S^E$,*
- *and there does not exist an interval of contiguous slots $I = [s_i, s_j] \subset [1, \bar{s}]$ with $\left[\min_{v_{k,s} \in C} (s - w_k + 1), \max_{v_{k,s} \in C} s \right] \subset I$, $w_k + w_{k'} \geq |I| + 1$ for each $(v_k, v_{k'}) \in C$, and $w_k \leq |I|$ for each $v_k \in C$.*

Further valid inequalities are introduced in [7]. Necessary and sufficient conditions for these inequalities to be facet for $P(G, K, \mathbb{S})$ are discussed in [6].

5   Branch-and-Cut Algorithm

Our algorithm aims at solving the cut formulation using a classical B&C framework. Since inequalities (2) are exponential in number, we need to address the so-called separation problem associated with these inequalities. Recall that the separation problem associated with a family $\mathcal{F}$ of inequalities consists in saying if a given solution $(\bar{x}, \bar{z}) \in \mathbb{R}^{|E| \times |K|} \times \mathbb{R}^{|K| \times |\mathbb{S}|}$ satisfies all the inequalities of $\mathcal{F}$ or not, and if not, exhibit at least one violated inequality.

It is easy to see that the separation problem of inequalities (2) for any $k \in K$, reduces to solving a maximum flow problem in G. Thus the separation of inequalities (2) can be solved in polynomial time. For our purpose, we use Goldberg and

Tarjan [8] algorithm to compute the maximum flow in G. Thus our separation algorithm runs in $O(|K| \cdot |V|^3)$.

In addition, when all the inequalities (2) are satisfied and the solution $(\bar{x}, \bar{z})$ is still fractional, the B&C algorithm looks for inequalities (11), (14), (17), (18), (19) that may be violated by $(\bar{x}, \bar{z})$. We believe that the separation problem of these families of inequalities is NP-Hard. Thus we devise heuristics to separate them. Notice that our heuristics are implemented to run in polynomial time. We refer the reader to [7] for more details.

6   Computational Study

Our B&C algorithm has been implemented in C++ under Linux and using the Solving Constraint Integer Programs (SCIP 7.0) framework [12] together with CPLEX 12.9 as LP-solver. We setup the maximum CPU time to 5 h (18000 s). At each iteration of the B&C algorithm, we limit to 200 the number of inequalities (11), (14), (17),(18), (19) added during the separation phase.

In the test instances, the demands are randomly generated having $|K|$ up to 200 and $\bar{s}$ up to 320, while the graphs are taken either from real data or from SND-Lib [10]. Table 1 presents the different graphs we use in our tests.

Table 1. Characteristics of different graphs used in the experiments.

Graph		Number of nodes	Number of links	Max node degree	Min node degree	Average node degree
Real	German	17	25	5	2	2.94
	Nsfnet	14	21	4	2	3
	Coronet100	100	136	5	2	2.72
Realistic SND-LIB	India35	35	80	9	2	4.57
	Pioro40	40	89	5	4	4.45
	Giul39	39	86	8	3	4.41

Our objective in this study is to show the efficiency of the inequalities we have introduced for solving the C-RSA problem. For this, we first run our B&C algorithm with SCIP in which we deactivate all the internal cuts automatically added by SCIP. We call this run B&C_OWN. Then, we run the B&C algorithm with SCIP using only the cut-set inequalities (2), and activating all the internal cuts we had deactivated prior in run 1. We call this run B&C_SCIP. Table 2 below reports the results obtained for the two runs. For each run and each instance, we report the number of nodes in B&C tree (Nbr_Nd), the optimality gap (Gap), and the total CPU Time (TT) in seconds. Finally, notice that each line of Table 2 corresponds to the average results of 4 instances.

We can see from Table 2 that our B&C algorithm (B&C_OWN) is able to solve to optimality more instances than B&C_SCIP. Indeed, 137 instances are solved to optimality when our inequalities are used (B&C_OWN) while 101 instances are solved to optimality in run B&C_SCIP. Also, when our inequalities are used, the number of nodes in the B&C tree is decreased in most cases

Table 2. The impact of adding all classes of valid inequalities in the B&C_OWN performance.

Instances			B&C_SCIP			B&C_OWN		
Graph	$|K|$	$\bar{s}$	Nbr_Nd	Gap	TT	Nbr_Nd	Gap	TT
German	10	15	1310,25	0,00	14,35	59	0,00	0,83
	20	45	185956	0,27	3895,5	141,00	0,00	3,89
	30	45	401335,75	1,60	11740,04	160376,50	1,46	8334,95
	40	45	315993,66	8,33	16206,36	383058,66	3,70	16624,87
	50	55	246146,50	9,62	16675,88	251152,50	13,73	17074,95
	100	140	1158,50	0,00	340,10	3014,25	0,00	617,80
	150	210	12759	0,01	7329,06	3609,00	0,00	3057,79
	200	260	5099,33	0,78	10095,88	3067,00	0,00	6770,75
Coronet100	10	320	1	0,00	10,37	1	0,00	462,15
	20	320	10,50	0,00	19,21	15	0,00	832,22
	30	40	66534,75	16,40	17677,13	11304,25	6,08	5006,31
	40	40	81051	3,96	17701,62	2127,00	0,00	707,54
	50	80	11385,25	0,01	4496,92	19,75	0,00	139,55
	100	120	12787,50	13,36	14228,34	8390,25	7,66	10920,70
	150	200	4454,50	27,12	13692,63	3165,75	29,13	15527,10
	200	280	3579,25	33,35	18000	1	38,97	18000
Nsfnet	10	15	13462	0,00	113,64	1	0,00	0,15
	20	20	699646	9,51	14242,82	21586	0,00	192,27
	30	30	272065	40,99	16558,66	281569,66	3,29	11048,71
	40	35	225696,67	46,74	16813,88	119841,66	1,17	5673,46
	50	50	247873,25	43,09	16914,53	148476,50	5,91	17405,09
	100	120	56598,50	57,19	17779,07	1	0,00	40,87
	150	160	12663	58,50	18000	1	0,00	136,02
	200	210	7726,50	54,85	18000	710,00	0,28	9121,79
India35	10	40	1907,25	0,00	87,60	1	0,00	1,80
	20	40	9	0,00	4,00	7	0,00	5,92
	30	40	91798	0,00	7821,5	32156,75	0,00	2309,66
	40	40	161514	2,42	17486,08	191812	0,18	17333,53
	50	80	34	0,00	22,13	69,25	0,00	112,19
	100	120	24797	0,32	9137,26	23403,75	0,44	9494,52
	150	200	16809	0,21	13739,65	1026,00	0,00	4101,80
	200	280	11197	0,37	13930,35	2027,75	3,69	14516,65
Pioro40	10	40	1	0,00	1,49	1	0,00	1,69
	20	40	1,50	0,00	3,44	1	0,00	4,88
	30	40	1,50	0,00	5,72	6,25	0,00	10,54
	40	40	83597	0,20	8692,5	67151	0,12	8711,30
	50	80	14	0,00	15,93	4	0,00	54,39
	100	80	21281,75	0,04	9087,52	23785,75	0,04	9916,63
	150	160	823,50	0,00	816,89	124,50	0,00	1509,87
	200	280	1503,75	0,00	3772,9	423,50	0,00	7424,98
Giul39	10	40	1	0,00	1,58	1	0,00	1,83
	20	40	1,50	0,00	2,92	1	0,00	3,71
	30	40	4	0,00	4,50	1	0,00	6,10
	40	40	4,50	0,00	7,17	1	0,00	10,15
	50	40	54420	0,00	4376,98	52156,75	0,00	4361,26
	100	40	55472,50	6,88	17781,71	54675,50	8,38	17802,83
	150	120	836,00	0,00	1050,13	11655,50	0,00	9411,30
	200	120	10191,25	0,24	13794,32	6518,00	0,01	9914,02

compared to the case where they are not used. Moreover, the CPU time is, in general, smaller when our inequalities are used. Finally, when comparing the instances which are not solved to optimality, we can see that the optimality gap is smaller, for most of the instances, when our inequalities are used.

7 Conclusion

In this paper, we have studied the C-RSA problem. We have first proposed an integer programming formulation for the problem, and have investigated the polytope associated with this formulation. So, we have introduced several families of inequalities which define facets for the polytope. Then we have devised a B&C algorithm using the inequalities and conducted some computational experiments. Our study shows that the inequalities we have introduced in this paper are effective for solving real and realistic instances of the problem.

It is worth noting that the cut formulation presents some symmetry issues which have not been addressed in this work. Thus further work should be done for addressing these symmetry issues and improving the resolution of the real and realistic instances we have presented here. It could also be interesting to investigate combination of the B&C algorithm with other techniques like machine learning or Branch-and-Price (B&P) methods, in order to solve large-scale instances.

References

1. Balas, E., Zemel, E.: Facets of the knapsack polytope from minimal covers. SIAM J. Appl. Math. **34**, 119–148 (1978)
2. Bianchetti, M., Marenco, J.: Valid inequalities and a branch-and-cut algorithm for the routing and spectrum allocation problem. In: Proceedings of the XI Latin and American Algorithms, Graphs and Optimization Symposium, pp. 523–531 (2021)
3. Cheng, B., et al.: Routing and spectrum assignment algorithm based on spectrum fragment assessment of arriving services. In: 28th Wireless and Optical Communications Conference (WOCC), pp. 1–4 (2019)
4. The Network Cisco's Technology News Site: Cisco Predicts More IP Traffic in the Next Five Years Than in the History of the Internet. https://newsroom.cisco.com
5. Colares, R., Kerivin, H., Wagler, A.: An extended formulation for the Constraint Routing and Spectrum Assignment Problem in Elastic Optical Networks (2021). https://hal.uca.fr/hal-03156189
6. Diarrassouba, I., Hadhbi, Y., Mahjoub, A. R.: On the Facial Structure of the Constrained-Routing and Spectrum Assignment Polyhedron: Part II. https://hal.uca.fr/hal-03287205
7. Diarrassouba, I., Hadhbi, Y., Mahjoub, A. R.: Valid Inequalities and Branch-and-Cut Algorithm for the Constrained-Routing and Spectrum Assignment Problem. https://hal.uca.fr/hal-03287146
8. Goldberg, A.V., Tarjan, R.E.: A new approach to the maximum flow problem. In: Proceedings of the Eighteenth Annual Association for Computing Machinery Symposium on Theory of Computing, pp. 136–146 (1986)
9. Hadhbi, Y., Kerivin, H., Wagler, A.: A novel integer linear programming model for routing and spectrum assignment in optical networks. In: Federated Conference on Computer Science and Information Systems (FedCSIS), pp. 127–134 (2019)

10. Orlowski, S., Pióro, M., Tomaszewski, A., Wessäly, R.: SNDlib 1.0-survivable network design library. In: Proceedings of the 3rd International Network Optimization Conference (INOC 2007), Spa, Belgium (2007)
11. Ramaswami, R.: Optical Networks: A Practical Perspective, 3rd edn. Morgan Kaufmann Publishers Inc., Boston (2009)
12. The SCIP Optimization Suite 7.0 (2020). http://www.optimization-online.org/ DB_HTML/2020/03/7705.html
13. Talebi, S., Alam, F., Katib, I., Khamis, M., Salama, R., Rouskas, G. N.: Spectrum management techniques for elastic optical networks: a survey. In: Optical Switching and Networking, pp. 34–48 (2014)

Polyhedra and Combinatorics

Top-*k* List Aggregation: Mathematical Formulations and Polyhedral Comparisons

Sina Akbari$^{(\boxtimes)}$ and Adolfo R. Escobedo

School of Computing and Augmented Intelligence, Arizona State University,
Tempe, AZ 85281, USA
{Sina.Akbari,adres}@asu.edu

Abstract. Top-*k* lists are being increasingly utilized in various fields and applications including information retrieval, machine learning, and recommendation systems. Since multiple top-*k* lists may be generated by different algorithms to evaluate the same set of entities or system of interest, there is often a need to consolidate this collection of heterogeneous top-*k* lists to obtain a more robust and coherent list. This work introduces various exact mathematical formulations of the top-*k* list aggregation problem under the generalized Kendall tau distance. Furthermore, the strength of the proposed formulations is analyzed from a polyhedral point of view.

Keywords: Top-*k* list aggregation · Rank aggregation · Kendall tau distance · Mixed integer programming · Polyhedral analysis

1 Introduction

Top-*k* lists are a special form of item orderings (i.e., rankings) wherein out of n total items only a small number of them, k, are explicitly ordered. Top-*k* lists have many advantages that can overcome some of the practical drawbacks of the traditional full-list approach: a collection of items may be too large to rank or even present, processing the full list could present a massive computational/cognitive load, and it may be impossible or meaningless to compare and rank items beyond a certain point [7]. Examples of top-*k* lists are the top-250 movies on IMDB or the top-10 played songs on Spotify [22].

Due to the increased use of such lists, the top-*k* list aggregation problem (TOP-*k*-AGG) has attracted considerable attention. TOP-*k*-AGG seeks to find a top-*k* list or full list that best represents the input lists. This problem has been utilized in many different applications, including recommender systems [20], metasearch engines [12], and bioinformatics [17]. TOP-*k*-AGG is interrelated with many other problems such as top-*k* recommendation and top-*k* query.

TOP-*k*-AGG falls under the umbrella of the more general rank aggregation problem whose objective is to combine individual rankings over a set of items

© The Author(s), under exclusive license to Springer Nature Switzerland AG 2022
I. Ljubić et al. (Eds.): ISCO 2022, LNCS 13526, pp. 51–63, 2022.
https://doi.org/10.1007/978-3-031-18530-4_4

into one representative collective ranking [5]. Variants of this problem have been studied probabilistically [6,8] and deterministically [10,12]. In the probabilistic approach, it is assumed that the observed rankings are realizations of a probabilistic model on ranking data, such as Mallows model [16], and the goal is to recover the ground-truth ranking.

Deterministic approaches can be further categorized into score-based and distance-based methods. Approaches in the first category apply relatively simple and efficient functions to calculate the score of each item, and the aggregate ranking is obtained by sorting items based on their total scores. Score-based methods are relatively susceptible to errors and manipulation, and they may violate certain fundamental social choice properties [5]. Conversely, distance-based methods provide more robust aggregation mechanisms. The aim of these approaches is to find a *consensus list* that has the least cumulative disagreement with the input lists. They are typically founded on axiomatic frameworks, from which the aggregate solution is formally guaranteed to satisfy certain desirable properties [9]. However, their aggregation problems tend to be more computationally demanding and are often NP-hard [5].

Distance-based TOP-k-AGG techniques can be divided based on whether the output ranking is considered a full list or another top-k list. Dwork et al. [10], Ailon [1], and Nápoles et al. [19] fall into the first category; Fagin et al. [12] falls into the second category. The works referenced under the first category define TOP-k-AGG as finding a full list with the least cumulative distance to the input lists using the induced Kendall tau, Kendall tau, and Hausdorff distances, respectively. Fagin et al. [12]'s method provides higher flexibility, and it induces a far smaller solution space. Letting n denote the total number of items, there are $\binom{n}{k}k!$ possible top-k lists using the latter approach, which is $(n-k)!$ times smaller than $n!$ (the number of possible full strict lists over n).

There are various distance measures for comparing top-k lists including generalized Kendall tau, generalized Spearman's footrule, Hausdorff [12], and Goodman and Kruskal's gamma [14]. This paper focuses on the distance-based variant of TOP-k-AGG induced by the generalized Kendall tau distance [12]. This focus is motivated by its widespread use for comparing top-k lists, and more importantly, its flexibility at handling partial information from these lists. This distance measure has been used in this capacity for similarity search [21], search engines [18], and influence maximization [4]. Additionally, variants of this distance have been used for comparing and aggregating bucket orders [2,11] and top-k XML lists [23]. However, to the best of our knowledge, this distance measure has not been utilized for the purpose of aggregating top-k lists since its introduction in Fagin et al. [12], possibly due to a lack of existing exact methods. To facilitate this essential use of the distance measure, this paper studies various exact mathematical formulations.

Contributions. Section 3 introduces a binary nonlinear programming formulation and four mixed integer linear programming (MIP) formulations of TOP-k-AGG under the generalized Kendall tau distance. Two of these formulations result from the introduction of preference cycle-prevention constraints specific to

TOP-k-AGG. Section 4 compares the strengths of the MIP formulations using techniques from polyhedral theory. The mathematical formulations and polyhedral analyses presented herein can be extended to TOP-k-AGG using any other distance measure between top-k lists by modifying the objective functions accordingly.

2   Preliminaries

The rank aggregation problem was originally defined over strict rankings. Formally, a strict ranking π is a bijection of $[n] = \{1, 2, \ldots, n\}$ onto itself, which represents a strict order of the n items. The Kendall tau distance [15] is one of the most prominent measures of dissimilarity between rankings, which counts the number of distinct item-pairs whose relative order is different in two rankings. The Kendall tau distance between strict rankings π^1, π^2 is given by $K(\pi^1, \pi^2) = \sum_{i \in [n]} \sum_{j \in [n]} K_{i,j}(\pi^1, \pi^2)$, where $K_{i,j}(\pi^1, \pi^2)$ is set to 1 if the relative orderings of i and j are different in π^1 and π^2, and 0 otherwise. The rank aggregation problem under Kendall tau distance is known alternatively as Kemeny Aggregation (KEMENY-AGG).

A top-k list τ is a bijection from a domain $\mathcal{I}_\tau$ (the members of τ) to $[k] = \{1, \ldots, k\}$, where $k < n$. All items in τ are presumed to be ranked ahead of items not in τ; however, the exact ordering of items not in the list is unknown. Let $i \in \tau$ indicate that item i appears in the top-k list, and let $\tau(i)$ denote the rank or position of i therein. Additionally, let $i \succ_\tau j$ denote that item i is rank ahead of item j in τ, that is, if $(i \in \tau \wedge j \notin \tau)$ OR $(i, j \in \tau \wedge (\tau(i) < \tau(j)))$. Given top-$k$ lists τ^1 and τ^2, let $\Lambda(\tau^1, \tau^2)$ be the set of all unordered pairs of distinct items in $\mathcal{I}_{\tau^1} \bigcup \mathcal{I}_{\tau^2}$.

Definition 1 *(TOP-k-AGG). Let $\mathcal{L} = \{1, 2, \ldots, m\}$ be the set of indices of the input top-k lists, τ^l be the input top-k list $l \in \mathcal{L}$, $\mathcal{I} = \bigcup_{l \in \mathcal{L}} \mathcal{I}_{\tau^l}$ be the universe of items, $n := |\mathcal{I}|$ be the number of items in the universe $\mathcal{I}$, $\mathcal{T}$ be the set of all possible top-k lists over $\mathcal{I}$, and $d(.,.)$ be a distance measure between top-k lists. TOP-k-AGG seeks to find a top-k list $\tau^* \in \mathcal{T}$ with the lowest cumulative distance to the input lists; it can be written succinctly as*

$$\tau^* = \underset{\tau \in \mathcal{T}}{\operatorname{argmin}} \sum_{l \in \mathcal{L}} d(\tau, \tau^l). \tag{1}$$

The rest of this paper focuses on the generalized Kendall tau distance [12]. Accordingly, the distance is restated in the following. Let p be a fixed parameter, with $0 \le p \le 1$, and let $K_{i,j}^{(p)}(\tau^1, \tau^2)$ be the contribution to the distance function, for each item-pair $(i,j) \in \Lambda(\tau^1, \tau^2)$. The generalized Kendall tau distance with penalty parameter p, denoted by $K^{(p)}$, is defined as

$$K^{(p)}(\tau^1, \tau^2) = \sum_{(i,j) \in \Lambda(\tau^1, \tau^2)} K_{i,j}^{(p)}(\tau^1, \tau^2), \tag{2}$$

where

$$K_{i,j}^{(p)}(\tau^1, \tau^2) = \begin{cases} 1 & (i \succ_{\tau^1} j \ \wedge \ j \succ_{\tau^2} i) \ \vee \ (j \succ_{\tau^2} i \ \wedge \ i \succ_{\tau^1} j) \\ p & (i, j \in \tau^1 \ \wedge \ i, j \notin \tau^2) \ \vee \ (i, j \notin \tau^1 \ \wedge \ i, j \in \tau^2) \\ 0 & \text{otherwise.} \end{cases}$$

$K^{(p)}$ is a *near metric* since it satisfies a relaxed version of the triangle inequality [12]. TOP-k-AGG under $K^{(p)}$ is a combinatorial NP-hard problem [12], which includes KEMENY-AGG as a special case (when $k = n$).

3   Integer Programming Formulations

To the best of our knowledge, no efforts have been made to derive an explicit mathematical model of TOP-k-AGG. This section presents various formulations.

First, we define required parameters for defining the objective functions of the presented formulations. Let μ_{il} be an indicator parameter that is equal to 1 if $i \in \tau^l$, where $l \in \mathcal{L}$. Additionally, let s_{ij} denote the number of input lists where item i is ranked ahead of item j, which can be expressed as

$$\begin{aligned} s_{ij} &= \sum_{l \in \mathcal{L}} \mathbb{1}_{(i, j \in \tau^l \wedge (\tau^l(i) < \tau^l(j)) \vee (i \in \tau^l \wedge j \notin \tau^l)} \\ &= \sum_{l \in \mathcal{L}} \left[\mu_{il} \mu_{jl} \mathbb{1}_{\tau^l(i) < \tau^l(j)} + \mu_{il}(1 - \mu_{jl}) \right]. \end{aligned} \tag{3}$$

In words, s_{ij} tallies the number of input lists in which i is ranked ahead of j, that is, the number of input lists in which both items are present and i is ranked ahead of j, plus the number of inputs lists in which i is present but j is not.

Using these parameters, the cumulative $K^{(p)}$ distance between a given top-k list $\tau \in \mathcal{T}$ and all of the input top-k lists, i.e., $\sum_{\tau^l \in \mathcal{L}} \sum_{(i,j) \in \Lambda(\tau, \tau^l)} K_{ij}^{(p)}(\tau, \tau^l)$, can be expressed as $\sum_{(i,j) \in \Lambda} K_{ij}^{(p)}(\tau)$ where Λ is set of all unordered pairs of distinct items in $\mathcal{I}$, and

$$K_{ij}^{(p)}(\tau) = \begin{cases} s_{ji} + p \sum_{l \in \mathcal{L}} (1 - \mu_{il})(1 - \mu_{jl}) & \text{if } i, j \in \tau \ \wedge \ (\tau(i) < \tau(j)), \\ s_{ji} & \text{if } i \in \tau \ \wedge \ j \notin \tau, \\ p \sum_{l \in \mathcal{L}} \mu_{il} \mu_{jl} & \text{if } i, j \notin \tau. \end{cases} \tag{4}$$

Equation (4) states that, whenever item i and j are both present in τ (the solution top-k list) and i is ranked ahead of item j, the imposed $K^{(p)}$ distance between τ and all of the input lists for this pair of items equals the number of input lists where j is ranked ahead of i, plus p-times the number of input lists neither i nor j is present in the same list. Whenever i but not j is present in

$\boldsymbol{\tau}$, the imposed $K^{(p)}$ distance equals the number of input lists where j is ranked ahead of i. Finally, whenever neither i nor j is present in $\boldsymbol{\tau}$, the imposed $K^{(p)}$ distance equals p times the number of input lists where i and j are simultaneously present.

The first formulation is an MIP possessing an assignment problem-like structure, with which exactly k items are assigned to the k available positions of the solution top-k list. Its decisions variables are as follows:

$$u_{it} = \begin{cases} 1 & \text{if } i \text{ is assigned to position } t \in [k] \\ 0 & \text{otherwise;} \end{cases}$$

$$w_{ij} = \begin{cases} 1 & \text{if } i \text{ and } j \text{ are in the top-}k \text{ list, and } i \text{ is ranked ahead of } j \\ 0 & \text{otherwise;} \end{cases}$$

$$w'_{ij} = \begin{cases} 1 & \text{if } i \text{ is in the top-}k \text{ list, but not } j \\ 0 & \text{otherwise;} \end{cases}$$

$$w''_{ij} = \begin{cases} 1 & \text{if neither } i \text{ nor } j \text{ is present in the top-}k \text{ list, where } j > i \\ 0 & \text{otherwise.} \end{cases}$$

From the definitions, item i is present in the top-k list if $\sum_{t=1}^{k} u_{it} = 1$, and it is absent if $\sum_{t=1}^{k} u_{it} = 0$. The variables $\boldsymbol{w}, \boldsymbol{w}'$, and $\boldsymbol{w}''$ determine the relative ordering of the items; these are dependent variables, as their exact values are determined by the values of the $\boldsymbol{u}$-variables. The first formulation (MIP#1) is as follows.

$$\min_{u,w,w',w''} \quad \sum_{i \in \boldsymbol{I}} \sum_{j \in \boldsymbol{I}} \left[\left(s_{ji} + p \sum_{l \in \boldsymbol{L}} (1 - \mu_{il})(1 - \mu_{jl}) \right) w_{ij} + s_{ji} w'_{ij} \right] + \tag{5a}$$

$$p \sum_{i,j \in \boldsymbol{I}, j > i} \sum_{l \in \boldsymbol{L}} \mu_{il} \mu_{jl} w''_{ij}$$

$$\text{s.t.} \quad \sum_{i \in \boldsymbol{I}} u_{it} = 1 \qquad\qquad \forall t \in [k] \tag{5b}$$

$$\sum_{t \in [k]} u_{it} \leq 1 \qquad\qquad \forall i \in \boldsymbol{I} \tag{5c}$$

$$w_{ij} \geq \sum_{t'=1}^{t} u_{it'} + \sum_{t''=t+1}^{k} u_{jt''} - 1 \quad \forall i,j \in \boldsymbol{I}, \ i \neq j; \ \forall t \in [k-1] \tag{5d}$$

$$\sum_{i,j \in \boldsymbol{I}} w_{ij} \leq \frac{k(k-1)}{2} \tag{5e}$$

$$w'_{ij} \geq \sum_{t \in [k]} u_{it} - \sum_{t \in [k]} u_{jt} \qquad \forall i,j \in \boldsymbol{I}, \ i \neq j \tag{5f}$$

$$\sum_{i,j \in \boldsymbol{I}} w'_{ij} = k(n-k) \tag{5g}$$

$$w''_{ij} \geq 1 - \sum_{t\in[k]} u_{it} - \sum_{t\in[k]} u_{jt} \qquad \forall i,j \in \boldsymbol{I},\ i \neq j \qquad (5h)$$

$$\sum_{i,j\in\boldsymbol{I},j>i} w''_{ij} = \frac{(n-k)(n-k-1)}{2} \qquad (5i)$$

$$u_{it} \in \{0,1\} \qquad \forall i \in \boldsymbol{I};\quad \forall t \in [k] \qquad (5j)$$

$$w_{ij}, w'_{ij} \geq 0 \qquad \forall i,j \in \boldsymbol{I},\ i \neq j \qquad (5k)$$

$$w''_{ij} \geq 0 \qquad \forall i,j \in \boldsymbol{I},\ j > i. \qquad (5l)$$

Objective function (5a) minimizes the cumulative $K^{(p)}$ distance to the input lists according to Eq. (4). Constraint (5b) enforces that exactly one item must be assigned to each position of the top-k list. Constraint (5c) enforces that every item must be assigned to at most one position of the list. Constraint (5d) determines the respective values of the $\boldsymbol{w}$-variables. More specifically, $w_{ij} = 1$ if i occupies one of the first t positions ($\sum_{t'=t+1}^{t} u_{it'} = 1$) and j occupies position t'', where $t+1 \leq t'' \leq k$ ($\sum_{t''=t+1}^{k} u_{jt''} = 1$); otherwise, this constraint becomes redundant. Constraint (5d) and (5e) together impose preference transitivity (i.e., prevent preference cycles); this means that if h is ranked ahead of i, and i is ranked of j, then h must be ranked ahead of j as well (see Theorem 1). Constraint (5f) determines the respective values of $\boldsymbol{w}'$-variables; it enforces that $w'_{ij} = 1$ if i is present in the top-k list but not j; otherwise, this constraint becomes redundant. Constraint (5g) enforces that at most $k(n-k)$ of the $\boldsymbol{w}'$-variables can take a value of 1 as there are $k(n-k)$ distinct item-pairs where exactly one of the items appears in the list. Constraint (5h) enforces that $w''_{ij} = 1$ if neither i nor j is present in the top-k list; otherwise, this constraint becomes redundant. Constraint (5i) enforces that at most $(n-k)(n-k-1)/2$ of the $\boldsymbol{w}''$-variables can take a value of 1 as this is the number of distinct item-pairs where both items are absent from the list. Constraints (5j)–(5l) specify the domain of the variables.

Taking a closer look at the structure of the constraints, we can observe that even though variables $\boldsymbol{w}, \boldsymbol{w}'$ and $\boldsymbol{w}''$ are specified as binary, they can be treated as non-negative continuous variables since the constraints of the model alone enforce them to only take a value of 0 or 1. It is important also to remark that the reason for including constraints (5f) and (5g) is that the objective function coefficients are not necessarily positive. More specifically, if both i and j are present in the solution top-k list, constraint (5f) implies that $w'_{ij} \geq 0$; however, if the objective function coefficient s_{ij} is 0, then any value of w'_{ij} results in the same objective function value, which is not desirable.

Theorem 1. *Constraints* (5d)–(5e) *impose preference transitivity.*

Proof. Assume that items h,i,j are present in the solution top-k list with h placed in position $t \geq 1$, i in position $t' > t$, and j in position t'', where $k \geq t'' > t'$. Constraint (5d) enforces that $w_{hi} = w_{hj} = w_{ij} = 1$. However, this constraint only implies that $w_{jh} \geq -1$. In other words, the optimization model may have incentive to assign $w_{jh} = 1$, creating a preference cycle, in order to decrease the objective function value. Hence, Constraint (5d) on its own does not prevent preference cycles.

However, the total number of w-variables that *must* take a value of 1 is given by $(k-1) + (k-2) + \cdots + 1 + 0 = k(k-1)/2$—the first-ranked item is ahead of $k-1$ other items in the list, the second-ranked item is ahead of $k-2$ items, ..., and the item at the bottom of the list is not ranked ahead of any other items on the list. For this reason, constraint (5e) allows at most $k(k-1)/2$ of the w-variables to take a value of 1, forcing all other variables (including w_{jh}) to equal 0. Therefore, constraints (5d)–(5e) together impose preference transitivity on the solution top-k list returned by solving MIP#1. □

Since KEMENY-AGG is a special case of TOP-k-AGG, MIP#1 provides a novel formulation for that problem as well; however, it does not apply to the variant of the problem with ties (see Yoo and Escobedo [24]). It is important to mention that Cook [9] proposed a binary linear programming formulation of KEMENY-AGG using the structure of the assignment problem; however, their set of preference cycle prevention constraint is different from constraints (5d)–(5e).

Next, we present a binary non-linear programming formulation for TOP-k-AGG. The formulation uses the w-variables defined for MIP#1 as well as the following decision variables:

$$z_i = \begin{cases} 1 & \text{if } i \text{ is in the top-}k \text{ list} \\ 0 & \text{otherwise.} \end{cases}$$

The formulation is given by:

$$\min_{w,z} \sum_{i \in \mathcal{I}} \sum_{j \in \mathcal{I}} \left[\left(s_{ji} + p \sum_{l \in \mathcal{L}} (1 - \mu_{il})(1 - \mu_{jl})\right) w_{ij} + s_{ji} z_i (1 - z_j) \right] + \tag{6a}$$

$$p \sum_{i,j \in \mathcal{I}, j > i} \sum_{l \in \mathcal{L}} \mu_{il} \mu_{jl} (1 - z_i)(1 - z_j)$$

$$\text{s.t.} \quad \sum_{i \in \mathcal{I}} z_i = k \tag{6b}$$

$$w_{hi} + w_{ij} + w_{jh} \le 2 \qquad \forall h, i, j \in \mathcal{I}, \, i, j > h, \, i \ne j \tag{6c}$$

$$w_{ij} + w_{ji} = z_i z_j \qquad \forall i, j \in \mathcal{I}, \, j > i \tag{6d}$$

$$z_i, w_{ij} \in \{0, 1\} \qquad \forall i, j \in \mathcal{I}, \, i \ne j. \tag{6e}$$

Objective function (6a) minimizes the cumulative $K^{(p)}$ distance to the input lists. Constraint (6b) restricts k items to be present in the top-k list. Constraint (6c) imposes preference transitivity only whenever items h, i, j all appear in the list; otherwise it becomes redundant, with the help of constraint (6d). Constraint (6d) enforces that, when both i and j are present in the list, one must proceed the other. Constraint (6e) specifies the domains of the variables. Given a feasible solution, the output top-k items are defined by the set $\overline{\tau} := \{i \in \mathcal{I} \mid z_i = 1\}$, and the exact rank of item $i \in \overline{\tau}$ is obtained as $\overline{\tau}(i) := k - \sum_{j \in \overline{\tau}} w_{ij}$.

The above non-linear optimization model can be linearized using a technique from Glover and Woolsey [13]. Specifically, constraint (6d) can be replaced with three linear constraints for each distinct item pair (i, j): $w_{ij} + w_{ji} \leq z_i$, $w_{ij} + w_{ji} \leq z_j$, and $w_{ij} + w_{ji} \geq z_i + z_j - 1$. Similarly, the term $z_i(1 - z_j)$ in the objective function is replaced by auxiliary continuous variable x'_{ij} and constraints $x'_{ij} \geq z_i - z_j$ and $x'_{ij} \geq 0$; and the term $(1 - z_i)(1 - z_j)$ in the objective function is replaced by auxiliary continuous variable x''_{ij} and constraints $x''_{ij} \geq 1 - z_i - z_j$ and $x''_{ij} \geq 0$. The latter two cases use the fact the objective function coefficients of $z_i(1 - z_j)$ and $(1 - z_i)(1 - z_j)$ are non-negative, leading to a reduction in the number of constraints required by the linearization. The resulting formulation (MIP#2) is given by:

$$
\min_{w,x',x'',z} \sum_{i\in\mathcal{I}}\sum_{j\in\mathcal{I}}\left[\left(s_{ji} + p\sum_{l\in\mathcal{L}}(1 - \mu_{il})(1 - \mu_{jl}))w_{ij} + s_{ji}x'_{ij}\right] +
$$

$$
p\sum_{i,j\in\mathcal{I},j>i}\sum_{l\in\mathcal{L}}\mu_{il}\mu_{jl}x''_{ij} \tag{7a}
$$

$$
\text{s.t.}\quad (6b),(6c),(6e) \tag{7b}
$$

$$
w_{ij} + w_{ji} \geq z_i + z_j - 1 \qquad \forall i,j \in \mathcal{I},\, j > i \tag{7c}
$$

$$
w_{ij} + w_{ji} \leq z_i \qquad \forall i,j \in \mathcal{I},\, i \neq j \tag{7d}
$$

$$
x'_{ij} \geq z_i - z_j \qquad \forall i,j \in \mathcal{I},\, i \neq j \tag{7e}
$$

$$
\sum_{i,j\in\mathcal{I}} x'_{ij} = k(n - k) \tag{7f}
$$

$$
x''_{ij} \geq 1 - z_i - z_j \qquad \forall i,j \in \mathcal{I},\, j > i \tag{7g}
$$

$$
\sum_{i,j\in\mathcal{I},j>i} x''_{ij} = \frac{(n - k)(n - k - 1)}{2} \tag{7h}
$$

$$
x'_{ij} \geq 0 \qquad \forall i,j \in \mathcal{I},\, i \neq j, \tag{7i}
$$

$$
x''_{ij} \geq 0 \qquad \forall i,j \in \mathcal{I},\, j > i. \tag{7j}
$$

The rationale behind including constraints (7f) and (7h) is the same as constraints (5g) and (5i) in MIP#1.

Next, we define two variants of the preference transitivity constraints utilized in MIP#2.

Proposition 1. *Constraint* (6c) *can be replaced by non-linear constraints*

$$
w_{hi} + w_{ij} + w_{jh} \leq 3 - z_h z_i z_j \qquad \forall i,j > h,\, i \neq j, \quad \textbf{or} \tag{8}
$$

$$
w_{hi} + w_{ij} + w_{jh} \leq 1 + z_h z_i z_j \qquad \forall i,j > h,\, i \neq j. \tag{9}
$$

Furthermore, these constraints can be linearized respectively as

$$
w_{hi} + w_{ij} + w_{jh} \leq 3 - \frac{1}{3}(z_h + z_i + z_j) \qquad \forall h,i,j \in \mathcal{I},\, i,j > h,\, i \neq j, \tag{10}
$$

$$
w_{hi} + w_{ij} + w_{jh} \leq 1 + \frac{1}{3}(z_h + z_i + z_j) \qquad \forall h,i,j \in \mathcal{I},\, i,j > h,\, i \neq j. \tag{11}
$$

Proof. The right-hand side of constraints (8)–(11) becomes 2, as desired, when items h, i, j are all in the solution top-k list, i.e., when $z_h = z_i = z_j = 1$. For the remaining cases, these constraints become redundant, with the help of constraint (7d). In particular, assume i is not in the top-k list; constraint (7d) enforces that $w_{ij} + w_{ji} \leq 0$ and $w_{ih} + w_{hi} \leq 0$; hence, constraints (8)–(11) effectively reduce to $w_{jh} \leq 1$, which is redundant. □

Replacing constraint (6c) with constraints (10) and (11), respectively, induces two additional MIPs.

MIP#3:

$$\min_{w, x', x'', z} \quad \text{(7a)}$$

$$\text{s.t.} \quad (6b), (6e), (7c)–(7g)$$

$$w_{hi} + w_{ij} + w_{jh} \leq 3 - \frac{1}{3}(z_h + z_i + z_j) \quad \forall h, i, j \in \boldsymbol{I}, \, i, j > h, \, i \neq j.$$

MIP#4:

$$\min_{w, x', x'', z} \quad \text{(7a)}$$

$$\text{s.t.} \quad (6b), (6e), (7c)–(7g)$$

$$w_{hi} + w_{ij} + w_{jh} \leq 1 + \frac{1}{3}(z_h + z_i + z_j) \quad \forall h, i, j \in \boldsymbol{I}, \, i, j > h, \, i \neq j.$$

4   Polyhedral Comparison

Next, we compare the strength of the proposed MIPs based on their linear programming (LP) relaxation models. First, we compare the strength of MIPs #2, #3, and #4. To that end, notice that these three MIPs become equivalent when $k \leq 2$—when the preference transitivity relations are irrelevant—or when $n = k$—when all items appear in the solution top-k list. Afterwards, we show that each of these formulations is stronger than MIP#1. For the remainder of the paper, let $\mathcal{P}^1, \mathcal{P}^2, \mathcal{P}^3, \mathcal{P}^4$ be the polyhedral corresponding to the LP relaxations of MIPs #1, #2, #3, #4, respectively.

Theorem 2. *For any instance of TOP-k-AGG, $\mathcal{P}^4 \subseteq \mathcal{P}^2 \subseteq \mathcal{P}^3$, and these inclusions can be strict.*

Proof. Note that MIPs #2, #3, and #4 differ only in their preference transitivity constraints. First, we show that $\mathcal{P}^4 \subseteq \mathcal{P}^2 \subseteq \mathcal{P}^3$.

Since $0 \leq z_i \leq 1 \, \forall i \in \boldsymbol{I}$, for every feasible solution in $\mathcal{P}^2, \mathcal{P}^3, \mathcal{P}^4$, we have that $(z_h + z_i + z_j)/3 \leq 1 \, \forall h, i, j \in \boldsymbol{I}, \, i, j > h, \, i \neq j$. Letting $(\boldsymbol{w}, \boldsymbol{x}', \boldsymbol{x}'', \boldsymbol{z})^{(4)} \in \mathcal{P}^4$ be a feasible solution to MIP#4, we have that

$$w_{hi}^{(4)} + w_{ij}^{(4)} + w_{jh}^{(4)} \leq 1 + \frac{1}{3}(z_i^{(4)} + z_j^{(4)} + z_h^{(4)}) \leq 2 \leq 3 - \frac{1}{3}(z_i^{(4)} + z_j^{(4)} + z_h^{(4)}).$$

Therefore, all feasible solutions to MIP#4 are also feasible to MIPs #2 and #3. Using the same logic, all feasible solutions to MIP#2 are feasible to MIP#3. This gives that $\mathcal{P}^4 \subseteq \mathcal{P}^2 \subseteq \mathcal{P}^3$.

To show that the inclusion $\mathcal{P}^4 \subseteq \mathcal{P}^2$ can be strict, consider a small instance with $\boldsymbol{I} = \{1, 2, 3, 4\}$ and $k = 3$. Fix the solution $(\boldsymbol{w}, \boldsymbol{x}', \boldsymbol{x}'', \boldsymbol{z})^{(2)} \in \mathcal{P}^2$ as

$$x_{14}^{\prime(2)} = x_{24}^{\prime(2)} = x_{34}^{\prime(2)} = 0.24, \quad w_{12}^{(2)} = w_{23}^{(2)} = w_{31}^{(2)} = 0.62, \quad w_{14}^{(2)} = w_{24}^{(2)} = w_{34}^{(2)} = 0.38,$$

$$z_1^{(2)} = z_2^{(2)} = z_3^{(2)} = 0.81, \quad z_4^{(2)} = 0.57;$$

with all other variables equal to 0. By inspection, this solution satisfies all constraints of MIP#2. However, we have that

$$w_{12}^{(2)} + w_{23}^{(2)} + w_{31}^{(2)} = 1.86 \not\leq 1 + \frac{0.81 + 0.81 + 0.81}{3} = 1.81.$$

This indicates that this solution does not satisfy the preference transitivity constraints of MIP#4.

Next, we use a similar process to show that the inclusion $\mathcal{P}^2 \subseteq \mathcal{P}^3$ can be strict. Consider a small instance with $\boldsymbol{I} = \{1, 2, 3, 4\}$ and $k = 3$. Fix the solution $(\boldsymbol{w}, \boldsymbol{x}', \boldsymbol{x}'', \boldsymbol{z})^{(3)} \in \mathcal{P}^3$ as

$$x_{14}^{\prime(3)} = x_{24}^{\prime(3)} = x_{34}^{\prime(3)} = 0.4, \quad w_{12}^{(3)} = w_{23}^{(3)} = w_{31}^{(3)} = 0.7, \quad w_{14}^{(3)} = w_{24}^{(3)} = w_{34}^{(3)} = 0.3,$$

$$z_1^{(2)} = z_2^{(3)} = z_3^{(3)} = 0.85, \quad z_4^{(3)} = 0.45;$$

with all other variables equal to 0. By inspection, this solution satisfies all constraints of MIP#3. However, we have that

$$w_{12}^{(3)} + w_{23}^{(3)} + w_{31}^{(3)} = 2.1 \not\leq 2.$$

This indicates that this solution does not satisfy the preference transitivity constraints of MIP#2. $\qquad\qquad\qquad\qquad\qquad\qquad\qquad\qquad\qquad\qquad\qquad\qquad \square$

Theorem 3. *For any instance of TOP-k-AGG, $\mathrm{proj}_w \mathcal{P}^2, \mathrm{proj}_w \mathcal{P}^3, \mathrm{proj}_w \mathcal{P}^4 \subseteq \mathrm{proj}_w \mathcal{P}^1$, and these inclusions can be strict.*

Proof. First, we prove that $\mathrm{proj}_w \mathcal{P}^3 \subseteq \mathrm{proj}_w \mathcal{P}^1$. We show that, starting from an arbitrary solution $(\boldsymbol{w}, \boldsymbol{x}', \boldsymbol{x}'', \boldsymbol{z}) \in \mathcal{P}^3$, we can deduce a solution $(\boldsymbol{u}, \boldsymbol{w}, \boldsymbol{w}', \boldsymbol{w}'') \in \mathcal{P}^1$. To this end, we define the following affine mappings of variables from $\mathcal{P}^3$ to $\mathcal{P}^1$:

$$u_{it} = \frac{z_i}{k} \quad \forall i \in \boldsymbol{I}, \ \forall t \in \{1, \ldots, k\} \rightarrow \sum_{t=1}^{k} u_{it} = z_i \quad \forall i \in \boldsymbol{I}, \tag{12a}$$

$$w_{ij}' = x_{ij}' \quad \forall i, j \in \boldsymbol{I}, \ i \neq j, \tag{12b}$$

$$w_{ij}'' = x_{ij}'' \quad \forall i, j \in \boldsymbol{I}, \ j > i. \tag{12c}$$

Mapping (12b)–(12c) guarantees that the objective function values achieved by the respective feasible points are equal. To establish that $\mathrm{proj}_w \mathcal{P}^3 \subseteq \mathrm{proj}_w \mathcal{P}^1$, it is sufficient to show that, given a feasible solution in $\mathcal{P}^3$, the mapped variables are guaranteed to satisfy all constraints of MIP#1 (i.e., this point belongs to $\mathcal{P}^1$).

Consider constraint (5b). For any $t \in \{1, \dots, k\}$, we have

$$\sum_{i \in \mathcal{I}} u_{it} = \sum_{i \in \mathcal{I}} \frac{z_i}{k} = \frac{\sum_{i \in \mathcal{I}} z_i}{k} \xrightarrow{\sum_{i \in \mathcal{I}} z_i = k} \sum_{i \in \mathcal{I}} u_{it} = 1.$$

Therefore, mapping (12a) provides a solution that is guaranteed to satisfy constraint (5b).

Consider constraint (5c). For every $i \in \mathcal{I}$, we have

$$\sum_{t=1}^{k} u_{it} = \sum_{t=1}^{k} \frac{z_i}{k} = \frac{k z_i}{k} = z_i \leq 1.$$

The last inequality follows from the fact that the z-variables are binary. Therefore, mapping (12a) provides a solution that is guaranteed to satisfy constraint (5c).

Next, consider constraint (5d); we focus on the maximum value of the right-hand side of this constraint given mapping (12a). For any arbitrary item-pair (i, j) and any $t \in \{1, \dots, k-1\}$ we have

$$\sum_{t'=1}^{t} u_{it'} + \sum_{t''=t+1}^{k} u_{jt''} - 1 = \sum_{t'=1}^{t} \frac{z_i}{k} + \sum_{t''=t+1}^{k} \frac{z_j}{k} - 1$$
$$= \frac{t z_i}{k} + \frac{(k-t) z_j}{k} - 1$$
$$\leq \frac{t}{k} + \frac{k-t}{k} - 1 = \frac{k}{k} - 1 = 1 - 1 = 0.$$

The above equation states that using mapping (12a), the left-hand side values of constraint (5d) will be non-positive. Since $w_{ij} \geq 0$, mapping (12a) provides a solution that is guaranteed to satisfy constraint (5d).

Next, consider constraint (5e). By summing over constraint (7d), we have

$$2 \sum_{i,j \in \mathcal{I}} w_{ij} \leq (k-1) \sum_{i \in \mathcal{I}} z_i = k(k-1)$$
$$\rightarrow \sum_{i,j \in \mathcal{I}} w_{ij} \leq \frac{k(k-1)}{2},$$

which is exactly constraint (5e).

Finally, consider constraints (5f)–(5i). Mappings (12a)–(12c) imply that all feasible solutions to constraints (7e)–(7h) are feasible to constraints (5f)–(5i). Putting all pieces together, we have $\text{proj}_w \mathcal{P}^3 \subseteq \text{proj}_w \mathcal{P}^1$.

Note that the preference cycle-prevention constraints of MIP#3 have no counterpart in MIP#1. Therefore, we can show that the inclusion $\text{proj}_w \mathcal{P}^3 \subseteq \text{proj}_w \mathcal{P}^1$ can be strict by providing a solution that satisfies constraints (7c)–(7f) but violates preference cycle-prevention constraint (10), as this solution satisfies all constraints of MIP#1. There is an infinite number of such solutions; for example, consider a small instance with $\mathcal{I} = \{1,2,3,4\}$ and $k = 3$. Fix the solution $(\boldsymbol{w}, \boldsymbol{x}', \boldsymbol{x}'', \boldsymbol{z})^{(3)}$ as

$$x_{14}'^{(3)} = x_{24}'^{(3)} = x_{34}'^{(3)} = 0.44, \quad w_{12}^{(3)} = w_{23}^{(3)} = w_{31}^{(3)} = 0.72, \quad w_{14}^{(3)} = w_{24}^{(3)} = w_{34}^{(3)} = 0.28,$$
$$z_1^{(2)} = z_2^{(3)} = z_3^{(3)} = 0.86, \quad z_4^{(3)} = 0.42;$$

with all other variables equal to 0. By inspection, this solution satisfies constraints (7c)–(7f); however, it violates the preference transitivity constraints involved in MIP#3, as we have

$$w_{12} + w_{23} + w_{31} = 2.16 \nleq 3 - (0.86 + 0.86 + 0.86)/3 = 2.14.$$

Finally, from Theorem 2, we have that $\mathcal{P}^4 \subseteq \mathcal{P}^2 \subseteq \mathcal{P}^3$; therefore, we can conclude that $\text{proj}_{\boldsymbol{w}} \mathcal{P}^2, \text{proj}_{\boldsymbol{w}} \mathcal{P}^4 \subseteq \text{proj}_{\boldsymbol{w}} \mathcal{P}^1$, and these inclusions can be strict. $\square$

5   Concluding Remarks

This paper studies the top-k list aggregation problem, which includes Kemeny aggregation as a special case. It presents a binary non-linear and four mixed-integer linear programming formulations. Furthermore, it studies the strength of the four mixed-integer linear programming formulations using polyhedral analysis. Our findings shows that the presented formulations can be ordered based on the strength of their LP relaxations. The strongest formulation is induced by a novel set of preference cycle-prevention constraints tailored to the specific structure of the top-k list aggregation problem introduced herein.

Future research will explore heuristic and approximation algorithms for this problem. Additionally, investigating whether lower bounding techniques of Kemeny aggregation [3] can be modified for the top-k list aggregation problem can be another avenue of research.

References

1. Ailon, N.: Aggregation of partial rankings, p-ratings and top-m lists. Algorithmica **57**(2), 284–300 (2010). https://doi.org/10.1007/s00453-008-9211-1
2. Akbari, S., Escobedo, A.R.: Beyond Kemeny aggregation: theoretical and computational insights for robust ranking aggregation. Under review
3. Akbari, S., Escobedo, A.R.: Lower bounds on Kemeny rank aggregation with non-strict rankings. In: 2021 IEEE Symposium Series on Computational Intelligence (SSCI), pp. 1–8. IEEE (2021)
4. Aslay, C., Barbieri, N., Bonchi, F., Baeza-Yates, R.: Online topic-aware influence maximization queries. In: EDBT, pp. 295–306 (2014)
5. Brandt, F., Conitzer, V., Endriss, U., Lang, J., Procaccia, A.D.: Handbook of Computational Social Choice. Cambridge University Press, Cambridge (2016)

6. Chen, Y., Fan, J., Ma, C., Wang, K.: Spectral method and regularized MLE are both optimal for top-K ranking. Ann. Stat. **47**(4), 2204 (2019)

7. Chierichetti, F., Dasgupta, A., Haddadan, S., Kumar, R., Lattanzi, S.: Mallows models for top-k lists. In: Advances in Neural Information Processing Systems, pp. 4382–4392 (2018)

8. Collas, F., Irurozki, E.: Concentric mixtures of mallows models for top-k rankings: sampling and identifiability. In: International Conference on Machine Learning, pp. 2079–2088. PMLR (2021)

9. Cook, W.D.: Distance-based and ad hoc consensus models in ordinal preference ranking. Eur. J. Oper. Res. **172**(2), 369–385 (2006)

10. Dwork, C., Kumar, R., Naor, M., Sivakumar, D.: Rank aggregation methods for the web. In: Proceedings of the 10th International Conference on World Wide Web, pp. 613–622 (2001)

11. Fagin, R., Kumar, R., Mahdian, M., Sivakumar, D., Vee, E.: Comparing and aggregating rankings with ties. In: Proceedings of the Twenty-Third ACM SIGMOD-SIGACT-SIGART Symposium on Principles of Database Systems, pp. 47–58 (2004)

12. Fagin, R., Kumar, R., Sivakumar, D.: Comparing top k lists. SIAM J. Discrete Math. **17**(1), 134–160 (2003)

13. Glover, F., Woolsey, E.: Converting the 0-1 polynomial programming problem to a 0-1 linear program. Oper. Res. **22**(1), 180–182 (1974)

14. Goodman, L.A., Kruskal, W.H.: Measures of association for cross classifications. II: further discussion and references. J. Am. Stat. Assoc. **54**(285), 123–163 (1959)

15. Kendall, M.G.: A new measure of rank correlation. Biometrika **30**(1/2), 81–93 (1938)

16. Mallows, C.L.: Non-null ranking models. I. Biometrika **44**(1/2), 114–130 (1957)

17. Marbach, D., et al.: Wisdom of crowds for robust gene network inference. Nat. Methods **9**(8), 796–804 (2012)

18. McCown, F., Nelson, M.L.: Agreeing to disagree: search engines and their public interfaces. In: Proceedings of the 7th ACM/IEEE-CS Joint Conference on Digital Libraries, pp. 309–318 (2007)

19. Nápoles, G., Falcon, R., Dikopoulou, Z., Papageorgiou, E., Bello, R., Vanhoof, K.: Weighted aggregation of partial rankings using ant colony optimization. Neuro-computing **250**, 109–120 (2017)

20. Oliveira, S.E., Diniz, V., Lacerda, A., Merschmanm, L., Pappa, G.L.: Is rank aggregation effective in recommender systems? An experimental analysis. ACM Trans. Intell. Syst. Technol. (TIST) **11**(2), 1–26 (2020)

21. Pal, K., Michel, S.: Efficient similarity search across top-k lists under the Kendall's Tau distance. In: Proceedings of the 28th International Conference on Scientific and Statistical Database Management, pp. 1–12 (2016)

22. Pedroche, F., Conejero, J.A.: Corrected evolutive Kendall's τ coefficients for incomplete rankings with ties: application to case of spotify lists. Mathematics **8**(10), 1828 (2020)

23. Varadarajan, R., Farfán, F., Hristidis, V.: Comparing top-k XML lists. Inf. Syst. **38**(6), 820–834 (2013)

24. Yoo, Y., Escobedo, A.R.: A new binary programming formulation and social choice property for Kemeny rank aggregation. Decis. Anal. **18**(4), 296–320 (2021)

Bounded Variation in Binary Sequences

Christoph Buchheim[ID] and Maja Hügging[(✉)]

Technische Universität Dortmund, Vogelpothsweg 87, 44227 Dortmund, Germany
{christoph.buchheim,maja.huegging}@math.tu-dortmund.de

Abstract. In optimal control problems with binary switches varying over time, it often arises as a subproblem to optimize a linear function over the set of binary vectors of a given finite length satisfying certain practical constraints, such as a minimum dwell time or a bound on the number of changes over the entire time horizon. While the former constraint has been investigated polyhedrally, no results seem to exist for the latter, although it arises naturally when discretizing binary optimal control problems subject to a bounded total variation. We investigate two variants of the problem, depending on whether the number of changes in a switch is penalized in the objective function or whether it is bounded by a hard constraint. We show that, while the former variant is easy to deal with, the latter is more complex, but still tractable. We present a full polyhedral description of the set of feasible switchings for this case.

Keywords: Total variation · Binary switching · Unit commitment · Convex hull · Polyhedral combinatorics

1 Introduction

Many practical applications lead to optimization problems over binary sequences. One example is the unit commitment problem which deals with the optimal scheduling of on/off-decisions for a set of generating units over a discrete time horizon. This problem has been studied intensively; see [5] for a survey. In the literature, one can find both algorithmic and polyhedral results for different kinds of constraints on the scheduling of the units or, more precisely, on the binary incidence vectors corresponding to a unit. Lee, Leung, and Margot [4], and Rajan and Takriti [9] focus on a setting with a single unit and so-called *min-up/min-down constraints*, which bound from below the minimal time span that a unit has to stay on (off) after being switched on (off). The corresponding polytope is fully characterized by well-structured linear inequalities and a linear-time separation algorithm is presented [4]; Queyranne and Wolsey [8] extend these results to upper and lower bounded up/down times. Bendotti, Fouilhoux, and Rottner [1] analyze polyhedral aspects for the unit commitment problem with *min-up/min-down constraints* for multiple units, they provide valid inequalities and devise an efficient Branch-and-Cut algorithm. Damci-Kurt, Küçükyavuz, Rajan, and Atamturk [3] study the unit commitment problem from a polyhedral perspective

I. Ljubić et al. (Eds.): ISCO 2022, LNCS 13526, pp. 64–75, 2022.
https://doi.org/10.1007/978-3-031-18530-4_5

as well, but focus on *ramping constraints* which model the maximum change in production level from one time period to the next, and production limits. The authors completely describe the convex hull of feasible solutions for the two-period case, and they derive several classes of facet-defining inequalities for the general ramp-up/ramp-down polytope. Pan and Guan [6] extend the polyhedral investigation of the unit commitment problem by investigating the integrated polytope including both *min-up/min-down constraints* and *ramping constraints* simultaneously. They provide a convex hull description for the two-and three-period case, and for the multiple period case they derive strong valid inequalities which are facet-inducing and can be separated in polynomial time. The authors continue their polyhedral study of this problem in [7], in particular, they prove convex hull descriptions for two special cases for general time periods.

In this paper, we consider a different class of constraints: we assume that the overall number of changes in the binary sequence is bounded from above. To the best of our knowledge, no polyhedral investigation of the resulting problem exists. This is surprising, considering the extensive study of other constraints in the unit commitment context, but also considering that this type of constraint arises naturally in many applications where switchings are expensive. Such problems also arise when discretizing binary optimal control problems subject to a bounded total variation [2,10]. Here, the binary string encodes a dynamic control variable that may change its state only a limited number of times within the considered time horizon. In the context of optimal control, this is a particularly relevant constraint, since many approaches tend to produce a large number of switchings in order to approximate an optimal solution of the continuous relaxation. Such a chattering behaviour is not desirable from a practical point of view. Using finite-dimensional projections, the approach presented in [2] obtains an outer description of a tailored convex relaxation of some parabolic optimal control problem by exploiting the outer descriptions derived in the present paper.

More formally, the problems we investigate are defined as follows. We consider a binary string $x = (x_0, x_1 \ldots, x_n)^\top \in \{0,1\}^{n+1}$. Throughout this paper, we always assume that vectors in $\mathbb{R}^{n+1}$ are indexed by $0, \ldots, n$. We require that x starts with a zero and count the number of times two adjacent entries differ. We can interpret this string as the (binary) incidence vector of a discrete switch, which can either be turned on at time period i $(x_i = 1)$ or off $(x_i = 0)$, where at the beginning of time horizon the switch is turned off, i.e., $x_0 = 0$. In the first problem variant, we consider a variable S which is an upper bound on the overall number of switchings. We thus address the problem

$$
\left.
\begin{aligned}
\max \quad & c^\top x - \lambda S \\
\text{s.t.} \quad & \sum_{i=1}^n |x_{i-1} - x_i| \leq S \\
& x_0 = 0,\ x_i \in \{0,1\} \text{ for } i \geq 1 \\
& S \in \mathbb{R}
\end{aligned}
\right\} \qquad \text{(OPT-S)}
$$

where $\lambda \geq 0$ is a penalty parameter and $c \in \mathbb{R}^{n+1}$. If $\lambda > 0$, it follows that S is the number of switchings in any optimal solution of (OPT-S). In the second problem variant, we consider S a fixed natural number, i.e., we address

$$\left. \begin{array}{ll} \max & c^\top x \\ \text{s.t.} & \sum_{i=1}^n |x_{i-1} - x_i| \le S \\ & x_0 = 0, \ x_i \in \{0,1\} \ \text{for } i \ge 1 \end{array} \right\} \tag{OPT}$$

for fixed $S \in \mathbb{N}_0$. Although closely related, the two problems are not equivalent: while (OPT-S) can be reduced to (OPT) by enumerating all $S \in \{0, \ldots, n\}$, the reduction is not directly possible in the other direction. In fact, it might happen that the optimal solution of (OPT) cannot be obtained with (OPT-S), no matter how λ is chosen. Nevertheless, (OPT) (and hence also (OPT-S)) is tractable, as can be shown by a simple dynamic programming scheme: denote by $c^*(i, s, b)$ the optimal value of Problem (OPT) restricted to time points $\{0, \ldots, i\}$ with bound $s \in \{0, \ldots, S\}$ on the variation and $b \in \{0, 1\}$ such that $x_i = b$. Then $c^*(0, s, 0) = 0$ and $c^*(0, s, 1) = -\infty$ for all s and

$$c^*(i, s, b) = bc_i + \max \begin{cases} c^*(i - 1, s, b) \\ c^*(i - 1, s - 1, 1 - b) \ \text{if } s \ge 1. \end{cases}$$

The optimal value is $\max\{c^*(n, S, 0), c^*(n, S, 1)\}$ and an optimal solution can easily be constructed from the recursion. However, it is often required to solve the separation problem for (OPT) or (OPT-S) rather than the optimization problem; see, e.g., the approach devised in [2]. This motivates our polyhedral investigation of both problems presented in this paper. We will provide complete polyhedral descriptions for both problem variants.

This paper is organized as follows. In Sect. 2, we briefly discuss (OPT-S) and show that this problem is easy to deal with polyhedrally, either by a compact extended formulation or by a fast separation algorithm in the original space. The more challenging problem (OPT) is studied in Sect. 3 from a polyhedral perspective. We first present a class of valid inequalities called *alternating inequalities* and then prove that these together with the trivial inequalities provide a complete description of the convex hull of feasible solutions to Problem (OPT), thus we obtain a complete linear inequality description.

2   Penalized Variation

In order to investigate the Problem (OPT-S) polyhedrally, we first consider the following equivalent extended model (EXT-OPT-S) in which we introduce additional variables z which model the difference of consecutive entries in x:

$$\left. \begin{array}{ll} \max & c^\top x - \lambda S \\ \text{s.t.} & \sum_{i=1}^n z_i \le S \\ & x_i - x_{i-1} \le z_i \ \text{for } i \ge 1 \\ & x_{i-1} - x_i \le z_i \ \text{for } i \ge 1 \\ & x_i \in \{0,1\}, \ z_i \in \mathbb{R} \ \text{for } i \ge 1 \\ & x_0 = 0, \ S \in \mathbb{R} \end{array} \right\} \tag{EXT-OPT-S}$$

It can be shown by a straightforward proof that the polyhedron corresponding to the LP-relaxation of this formulation is integer, i.e., a complete linear inequality description of the convex hull of feasible points for (EXT-OPT-S) is obtained by relaxing the binary constraint. Moreover, with Fourier-Motzkin elimination we can derive a complete linear inequality description for (OPT-S).

Lemma 1. *The convex hull of feasible solutions of (OPT-S) is completely described by the linear constraints $x_0 = 0$, $x_i \in [0,1]$ for $i = 1, \ldots, n$, and*

$$x_1 + \sum_{i=2}^{n} (-1)^{\sigma(i)} (x_{i-1} - x_i) \leq S \tag{1}$$

for all $\sigma \colon \{2, \ldots, n\} \to \{0, 1\}$.

Observe that the number of inequalities in the convex hull description of (OPT-S) is exponential in n, while it is linear in n in the extended model (EXT-OPT-S). However, it can be easily seen that the inequalities (1) can be separated efficiently: for given $(\bar{x}, \bar{S})$, a most violated inequality is defined by setting $\sigma(i) := 1$ if and only if $\bar{x}_{i-1} \leq \bar{x}_i$.

In summary, Problem (OPT-S) is very easy to deal with from a polyhedral point of view, either by a compact extended formulation or by a fast separation algorithm in the original space. We will see in the following that Problem (OPT) is more challenging.

3 Bounded Variation

We now consider the problem variant (OPT) in which S is no longer a variable, but a fixed integer. We remark that although having found a linear convex hull description for (OPT-S), we cannot derive any polyhedral results for (OPT) from this. Intuitively, one might think that a convex hull description for the feasible set of (OPT) is obtained by simply fixing the variable S to the given value in the linear inequalities (1) describing the convex hull of (OPT-S). This does not work, however, since the resulting polytope is not integer, as illustrated by the following example.

Example 1. Consider Problem (OPT-S) for $n = 3$ and $c = (0, 1, -3, 1)^{\top}$, and any $\lambda \geq 0$. According to Lemma 1, this problem is equivalent to

$$
\left.
\begin{aligned}
\max \quad & x_1 - 3x_2 + x_3 - \lambda S \\
& 2x_1 - x_3 \leq S \\
& 2x_1 - 2x_2 + x_3 \leq S \\
& 2x_2 - x_3 \leq S \\
& x_3 \leq S \\
& x_0 = 0, \; x_1, x_2, x_3 \in [0,1], \; S \in \mathbb{R}
\end{aligned}
\right\} \tag{EX}
$$

Fixing $S = 2$ in (EX), we obtain $(0, 1/2, 0, 1)^\top$ as unique optimizer of the resulting linear program. In particular, the intersection of the feasible set of (EX) with the hyperplane given by $S = 2$ is not an integer polytope and, therefore, the convex hull of feasible solutions to (OPT) cannot be obtained in this way. □

Motivated by this, we focus on Problem (OPT) in the remainder of this section. In particular, we investigate the polyhedral structure of the convex hull of its feasible set and present a complete description of the latter by linear inequalities.

To begin, we remark that a polyhedral investigation of (OPT) for $S = 1$ is not particularly interesting. In this case, the feasible set X_1 consists of $n + 1$ affinely independent elements, so, $\text{conv}(X_1)$ is a simplex of dimension n and its facets correspond to n-element subsets of X_1. Therefore, the polytope $\text{conv}(X_1)$ is completely described by the linear constraints

$$0 = x_0 \le x_1 \le x_2 \le \ldots \le x_{n-1} \le x_n \le 1.$$

From now on let $S \ge 2$ and let X_S denote the set of feasible incidence vectors for (OPT). Apart from the trivial inequalities $0 \le x_i \le 1$, $i = 1, \ldots, n$, the following inequalities, called *alternating inequalities*, are valid for $\text{conv}(X_S)$.

Lemma 2. *For $2 \le S \le n$, consider any index set $I = \{i_1, \ldots, i_m\} \subseteq \{1, \ldots, n\}$ with $i_j < i_{j+1}$ for $j = 1, \ldots, m - 1$ such that m is odd if and only if S even. Then the following inequalities are valid for $\text{conv}(X_S)$:*

$$\sum_{j=1}^{m}(-1)^{j+1}x_{i_j} \le \left\lfloor \tfrac{S}{2} \right\rfloor \tag{ALT}$$

Proof. Consider any index set $\{i_1, \ldots, i_m\}$ having the required properties. It suffices to prove that the inequality $\sum_{j=1}^{m}(-1)^{j+1}x_{i_j} \le \left\lfloor \tfrac{S}{2} \right\rfloor$ is valid for each element $x \in X_S$. For $m < S$, this is obvious. So suppose that $m \ge S$ and that there is an element $\tilde{x} \in X_S$ such that

$$\sum_{j=1}^{m}(-1)^{j+1}\tilde{x}_{i_j} \ge \left\lfloor \tfrac{S}{2} \right\rfloor + 1.$$

In case S is odd, there must be at least $\left\lfloor \tfrac{S}{2} \right\rfloor + 1$ indices i_j, $j \in \{1, 3, 5, \ldots, m\}$, such that $\tilde{x}_{i_j} = 1$ and $\tilde{x}_{i_{j+1}} = 0$. Consequently, $\tilde{x}$ contains at least $S + 1$ switchings, which contradicts $\tilde{x} \in X_S$. In case S is even, the reasoning is similar: either there are at least $\tfrac{S}{2} + 1$ indices i_j, $j \in \{1, 3, 5, \ldots, m\}$, with $\tilde{x}_{i_j} = 1$ and $\tilde{x}_{i_{j+1}} = 0$, or there are only $\tfrac{S}{2}$ indices with this property, but $x_{i_m} = 1$. Either way $\tilde{x}$ contains more than S switchings. □

Now, we will prove that the polytope P defined by the trivial inequalities and by the alternating inequalities, i.e.,

$$P := \{x \in [0, 1]^{n+1} \mid x_0 = 0,\ x \text{ satisfies all alternating inequalities (ALT)}\}$$

is precisely the switching polytope $\text{conv}(X_S)$. In other words, the trivial constraints $x \in [0,1]^{n+1}$ and $x_0 = 0$ together with the alternating inequalities (ALT) yield a complete linear description of $\text{conv}(X_S)$. By Lemma 2, we already know that $\text{conv}(X_S) \subseteq P$, so it remains to show the reverse inclusion $P \subseteq \text{conv}(X_S)$. Clearly, every integer vector in P belongs to X_S, so we are left to prove that P is integer. We will show the latter by LP-duality. More precisely, we will prove the integrality of P by describing a primal-dual optimization algorithm. As a by-product, we will obtain an alternative algorithm for Problem (OPT).

First, we eliminate variable x_0 in (OPT), as it is fixed to zero and does not appear in any alternating inequality. Let p denote the total number of alternating inequalities and let $Ax \le \lfloor \frac{S}{2} \rfloor \mathbb{1}_p$ collect all of them, where $x := (x_1, \ldots, x_n)^\top$. We replace the feasible region of (OPT) with the requirement that the vector x lies in P and consider the resulting linear problem, which can be written in primal (left) and dual (right) form as follows:

$$\left. \begin{array}{c} \max \quad c^\top x \\ \\ Ax \le \lfloor \frac{S}{2} \rfloor \mathbb{1}_p \\ \\ x \le \mathbb{1}_n,\ x \in \mathbb{R}^n_{\ge 0} \end{array} \right\} \text{(OPT-P)} \qquad \begin{array}{c} \min \quad \lfloor \frac{S}{2} \rfloor \mathbb{1}_p^\top y + \mathbb{1}_n^\top z \\ \\ A^\top y + z \ge c \\ \\ y \in \mathbb{R}^p_{\ge 0}, z \in \mathbb{R}^n_{\ge 0} \end{array}$$

Observe that for $n \le S$ the integrality of P follows immediately, as there always exists an integer optimum solution for Problem (OPT-P). Consequently, in the following we restrict ourselves to instances of Problem (OPT-P) with $S < n$.

Our aim is to prove the integrality of P by constructing, for a given objective vector $c \in \mathbb{R}^n$, a primal and dual solution with the same objective value, where the primal solution is integer. The idea is to reduce the given problem instance step-by-step to a "smallest" non-trivial case, by merging or deleting variables. Then, we determine an optimum integer primal and dual solution for the resulting instance and use these to reverse the reduction process so as to obtain an optimal integer primal and optimum dual solution for the original instance. For a precise description of this reduction, we use the following method of indexing our variables so that we are able to (in a manner of speaking) save and reverse the operations performed in the reduction process.

Definition 1. *We refer to instances of (OPT-P) via (c, I), where c denotes the objective while the finite set $I = \{\{1\}, \ldots, \{n\}\}$ contains sets of indices of the binary sequence. For $1 \le i \le |I|$ let $I(i)$ denote the i-th element of I. The objective c as well as the vectors x and z are indexed with the corresponding elements $J \in I$. As each alternating inequality is uniquely determined by the elements contained in its support $\mathcal{J} \subseteq I$, we index the corresponding dual variable y with the subset $\mathcal{J}$.*

We remark that in the course of our construction, elements of I are merged recursively, so that the resulting set I will then correspond to a partition of $\{1, \ldots, n\}$. In order to illustrate our way of indexing the variables according to Definition 1, we provide the following example.

Example 2. Consider an instance of (OPT-P) with $n = 4$ and $S = 2$. According to our definition above, we have $I = \{\{1\}, \{2\}, \{3\}, \{4\}\}$, so the alternating inequality $x_{\{1\}} - x_{\{2\}} + x_{\{3\}} \leq 1$ is determined by $\mathcal{J} = \{\{1\}, \{2\}, \{3\}\} \subseteq I$, which also is the index of the corresponding dual variable. The (non-trivial) dual constraints are thus given by:

$$
\begin{aligned}
+y_{\{\{1\},\{2\},\{3\}\}} +y_{\{\{1\},\{2\},\{4\}\}} +y_{\{\{1\},\{3\},\{4\}\}} +z_{\{1\}} &\geq c_{\{1\}} \\
-y_{\{\{1\},\{2\},\{3\}\}} -y_{\{\{1\},\{2\},\{4\}\}} +y_{\{\{2\},\{3\},\{4\}\}} +z_{\{2\}} &\geq c_{\{2\}} \\
+y_{\{\{1\},\{2\},\{3\}\}} -y_{\{\{1\},\{3\},\{4\}\}} -y_{\{\{2\},\{3\},\{4\}\}} +z_{\{3\}} &\geq c_{\{3\}} \\
+y_{\{\{1\},\{2\},\{4\}\}} +y_{\{\{1\},\{3\},\{4\}\}} +y_{\{\{2\},\{3\},\{4\}\}} +z_{\{4\}} &\geq c_{\{4\}}
\end{aligned}
$$

We first prove that (OPT-P) always has an integer optimum solution if the objective c has the following Property 1. Later, we will discuss that the same also holds for an arbitrary objective.

Property 1. Given an instance (c, I) of (OPT-P), we assume that the objective c has the following properties:

(i) $c_J \neq 0$ for all $J \in I$,
(ii) $\mathrm{sgn}(c_{I(1)}) = 1$ and $\mathrm{sgn}(c_{I(i)}) \neq \mathrm{sgn}(c_{I(i+1)})$ for all $i = 1, \ldots, |I| - 1$, and
(iii) $\mathrm{sgn}(c_{I(|I|)}) = (-1)^S$.

As the main ingredient of our construction, we first describe the merging rule which we apply to reduce a given instance (c, I) of (OPT-P) with dimension $|I|$ to an instance $(\bar{c}, \bar{I})$ with dimension $|\bar{I}| = |I| - 2$.

Definition 2 (Merging Rule). *Given a vector c having Property 1 and an index set I, let $I(\hat{\imath})$ be an element of I such that $|c_{I(\hat{\imath})}|$ is minimal. We define a merged instance with objective $\bar{c}$ and index set $\bar{I}$, with $|\bar{I}| = |I| - 2$, as follows:*

(a) if $\hat{\imath} = 1$, set $\bar{I} := I \setminus \{I(1), I(2)\}$ and $\bar{c}_J := c_J$ for $J \in \bar{I}$
(b) if $\hat{\imath} = |I|$, set $\bar{I} := I \setminus \{I(|I| - 1), I(|I|)\}$ and $\bar{c}_J := c_J$ for $J \in \bar{I}$
(c) otherwise, set $\bar{I} := I \setminus \{I(\hat{\imath} - 1), I(\hat{\imath}), I(\hat{\imath} + 1)\} \cup \{I(\hat{\imath} - 1) \cup I(\hat{\imath}) \cup I(\hat{\imath} + 1)\}$ and define $\bar{c}_J := c_J$ for $J \in \bar{I} \cap I$ and

$$\bar{c}_{I(\hat{\imath}-1) \cup I(\hat{\imath}) \cup I(\hat{\imath}+1)} := c_{I(\hat{\imath}-1)} + c_{I(\hat{\imath})} + c_{I(\hat{\imath}+1)}.$$

Example 3. By applying the Merging Rule to an instance of (OPT-P) given by $S = 3$ and $I = \{\{1\}, \{2\}, \{3\}, \{4\}, \{5\}, \{6\}, \{7\}, \{8\}\}$, where the objective function $c = (4, -5, 1, -7, 10, -3, 8, -2)^\top$ has Property 1, we obtain $\hat{\imath} = 3$ and, as a result of applying (c),

$$\bar{I} = \{\{1\}, \{2, 3, 4\}, \{5\}, \{6\}, \{7\}, \{8\}\}, \quad \bar{c} = (4, -11, 10, -3, 8, -2)^\top.$$

Another application of the Merging Rule to the latter instance yields $\hat{\imath} = 6$ and thus, using (b),

$$\bar{\bar{I}} = \{\{1\}, \{2, 3, 4\}, \{5\}, \{6\}\}, \quad \bar{\bar{c}} = (4, -11, 10, -3)^\top.$$

It is easy to see that $(\bar{c}, \bar{I})$ constructed by the Merging Rule again has Property 1. In a primal-dual context, deleting two entries at the beginning of c as in (a) (or at the end as in (b), respectively) corresponds to deleting the first two corresponding rows from the dual problem (or last two, respectively), whereas merging three entries of c into a single one as in (c) can be interpreted as summing up the corresponding three dual constraints.

Theorem 1 (Reversing the merging process). *Let $\bar{c}$ and $\bar{I}$ be obtained from c and I by applying the Merging Rule. Let $\bar{A}$ denote the matrix corresponding to the merged problem and let $\bar{A}^{\top}_{J,\mathcal{J}}$ refer to the entry in row J and column $\mathcal{J}$ of $\bar{A}^{\top}$. Assume that, for the merged instance, an integer optimal solution $\bar{x}$ to the primal problem is given as well as an optimal solution $(\bar{y}, \bar{z})$ to the dual problem, where $\bar{y}$ and $\bar{z}$ satisfy the following conditions:*

(C1) If $\bar{y}_{\mathcal{J}} > 0$ and $\bar{A}^{\top}_{J,\mathcal{J}} \neq 0$, then $\operatorname{sgn}(\bar{A}^{\top}_{J,\mathcal{J}}) = \operatorname{sgn}(\bar{c}_{J})$ for all $J \in \bar{I}$.
(C2) The value of $\bar{y}_{\bar{I}}$ equals $\min_{J \in \bar{I}} |\bar{c}_{J}|$.

Then, for the original instance (c, I), there exist an integer optimal primal solution x and an optimal dual solution (y, z) satisfying (C1) and (C2) again.

Proof. The construction of x and (y, z) depends on how the merged instance $(\bar{c}, \bar{I})$ is obtained from the instance (c, I). Assume first that $\overset{*}{\imath} = 1$ as in case (a). Then we define a dual solution (y, z) for the original instance by

$$y_I := |c_{I(\overset{*}{\imath})}|, \ y_{\bar{I}} := \bar{y}_{\bar{I}} - |c_{I(\overset{*}{\imath})}|, \ y_{\mathcal{J}} := \bar{y}_{\mathcal{J}} \text{ for } \mathcal{J} \subseteq \bar{I} \cap I, \ y_{\mathcal{J}} := 0 \text{ else}$$
$$z_J := \bar{z}_J \text{ for } J \in \bar{I} \cap I, \quad z_{I(1)} := z_{I(2)} := 0 .$$

Since $(\bar{y}, \bar{z})$ satisfies (C2) and according to the Merging Rule we get $(y, z) \geq 0$. Dual constraints other than $I(1), I(2)$ are fulfilled by construction of (y, z). So, for feasibility of (y, z), we only need to verify that the defined dual solution fulfills those constraints—but this is true by definition of y_I and because of $\min_{J \in I} |c_J| = c_{I(1)}$. Thus, (y, z) is dual feasible and it is not difficult to see that (C1) and (C2) are satisfied. We define an (obviously feasible) integer primal solution x by extending $\bar{x}$ by two zeros in the front. By computing the corresponding objective function values for $\bar{x}, (\bar{y}, \bar{z}), x$, and (y, z) we see that all of them agree, thus, the constructed solutions x and (y, z) are optimal for the non-merged instance by weak duality.

Now assume that $(\bar{c}, \bar{I})$ is obtained by deleting the last two elements of c and I as in case (b), so $\overset{*}{\imath} = |I|$. We define the dual variables y exactly as in the first case above. The definition of x and z, however, depends on whether S is even or odd:

– If S is even, we define z and x via

$$z_J := \bar{z}_J \text{ for } J \in \bar{I} \cap I, \quad z_{I(|I|-1)} = z_{I(|I|)} := 0$$
$$x_J := \bar{x}_J \text{ for } J \in \bar{I}, \quad x_{I(|I|-1)} = x_{I(|I|)} := 0 .$$

The vector x as defined is clearly integer and feasible. By arguing as above, we obtain that the dual solution (y, z) is feasible for the original instance and that it satisfies (C1) and (C2). Because of $\sum_{J \subseteq \bar{I}} \bar{y}_J = \sum_{J \subseteq I} y_J$ and $\sum_{J \in \bar{I}} \bar{z}_J = \sum_{J \in I} z_J$, the objective values of all solutions agree same as in the previous case. Consequently, x and (y, z) are optimal by weak duality.

- If S is odd, we define z and x via

$$z_J := \bar{z}_J \text{ for } J \in \bar{I} \cap I, \quad z_{I(|I|-1)} := c_{I(|I|-1)} + c_{I(|I|)}, \quad z_{I(|I|)} := 0,$$
$$x_J := \bar{x}_J \text{ for } J \in \bar{I}, \quad x_{I(|I|-1)} = x_{I(|I|)} := 1.$$

The integrality and feasibility of x are obvious and all constraints of the dual problem for the non-merged instance other than $I(|I|-1)$ and $I(|I|)$ are satisfied by definition of y and z. Since c has Property 1 and S is odd, we have $c_{I(|I|-1)} > 0$, $c_{I(|I|)} < 0$, and because of the Merging Rule $|c_{I(|I|)}| = |c_{I(\hat{\imath})}| \leq |c_{I(|I|-1)}|$, which implies $z_{I(|I|-1)} \geq 0$. Therefore, (y, z) is dual feasible for the original instance. By definition of our dual solution we have $\sum_{J \subseteq \bar{I}} \bar{y}_J = \sum_{J \subseteq I} y_J$ and $\sum_{J \in \bar{I}} \bar{z}_J + z_{I(|I|-1)} + z_{I(|I|)} = \sum_{J \in I} z_J$. By exploiting this when computing the objective values of x and (y, z), we see that those values agree. Hence, weak duality again shows optimality of x and (y, z). The latter satisfies criteria (C1) and (C2) for the same reasons as in the first case.

It remains to consider the case in which $(\bar{c}, \bar{I})$ is obtained by applying the Merging Rule with $\hat{\imath} \in \{2, \ldots, |I| - 1\}$, as in (c). Consider the constraint $M := I(\hat{\imath} - 1) \cup I(\hat{\imath}) \cup I(\hat{\imath} + 1)$. Recall that we can understand this constraint as the sum of the corresponding three dual constraints $I(\hat{\imath}-1)$, $I(\hat{\imath})$, and $I(\hat{\imath}+1)$ in the non-merged dual problem. Now, we undo this summation such that we obtain a dual solution (y, z) for the original instance with the same objective function value as $(\bar{y}, \bar{z})$. First, define

$$y_J := \bar{y}_J \text{ for } J \subseteq I \cap \bar{I} \quad \text{and} \quad z_J := \bar{z}_J \text{ for } J \in I \cap \bar{I}.$$

For all $J \in \mathcal{K} := \{J \subseteq \bar{I} \mid \bar{y}_J > 0, M \in \mathcal{J}\}$, we distribute the value $\bar{y}_J$ among the three variables $y_{J \setminus \{M\} \cup L}$, $y_{J \setminus \{M\} \cup \{I(\hat{\imath}-1)\}}$, and $y_{J \setminus \{M\} \cup \{I(\hat{\imath}+1)\}}$, where $L := \{I(\hat{\imath} - 1), I(\hat{\imath}), I(\hat{\imath} + 1)\}$, and assign values to the variables $z_{I(\hat{\imath}-1)}, z_{I(\hat{\imath})}$, and $z_{I(\hat{\imath}+1)}$ as follows:

- Because of the Merging Rule and (C2) we can assign values to the variables $y_{J \setminus \{M\} \cup L}, J \in \mathcal{K}$ such that

$$\sum_{J \in \mathcal{K}} y_{J \setminus \{M\} \cup L} = |c_{I(\hat{\imath})}|, \quad y_{J \setminus \{M\} \cup L} \in [0, \bar{y}_J] \text{ for } J \in \mathcal{K},$$

and there is such an assignment with $y_I = |c_{I(\hat{\imath})}|$. We set $z_{I(\hat{\imath})} := 0$ and thus, the constraint $I(\hat{\imath})$ in the original instance is satisfied with equality.
- Define $\tilde{c}_{I(\hat{\imath}-1)} := |c_{I(\hat{\imath}-1)}| - |c_{I(\hat{\imath})}|$ and $\tilde{c}_{I(\hat{\imath}+1)} := |c_{I(\hat{\imath}+1)}| - |c_{I(\hat{\imath})}|$ for a better overview and assign values to the remaining variables identified above by

$$y_{J \setminus \{M\} \cup \{I(\hat{\imath}-1)\}} := (\bar{y}_J - y_{J \setminus \{M\} \cup L}) \frac{\tilde{c}_{I(\hat{\imath}-1)}}{\tilde{c}_{I(\hat{\imath}-1)} + \tilde{c}_{I(\hat{\imath}+1)}} \text{ for all } J \in \mathcal{K},$$
$$y_{J \setminus \{M\} \cup \{I(\hat{\imath}+1)\}} := (\bar{y}_J - y_{J \setminus \{M\} \cup L}) \frac{\tilde{c}_{I(\hat{\imath}+1)}}{\tilde{c}_{I(\hat{\imath}-1)} + \tilde{c}_{I(\hat{\imath}+1)}} \text{ for all } J \in \mathcal{K},$$
$$z_{I(\hat{\imath}-1)} := \bar{z}_M \frac{\tilde{c}_{I(\hat{\imath}-1)}}{\tilde{c}_{I(\hat{\imath}-1)} + \tilde{c}_{I(\hat{\imath}+1)}}, \quad z_{I(\hat{\imath}+1)} := \bar{z}_M \frac{\tilde{c}_{I(\hat{\imath}+1)}}{\tilde{c}_{I(\hat{\imath}-1)} + \tilde{c}_{I(\hat{\imath}+1)}}.$$

Then, the constraints $I(\check{i}-1)$ and $I(\check{i}+1)$ in the original instance are satisfied as well.

All other variables $y_{\mathcal{J}}$ are set to zero. By construction, the defined solution (y, z) satisfies criteria (C1) and (C2). Moreover,

$$y_{\mathcal{J}\backslash\{M\}\cup L} + y_{\mathcal{J}\backslash\{M\}\cup\{I(\check{i}-1)\}} + y_{\mathcal{J}\backslash\{M\}\cup\{I(\check{i}+1)\}} = \bar{y}_{\mathcal{J}} \text{ for } \mathcal{J} \in \mathcal{K} \,,$$

$$z_{I(\check{i}-1)} + z_{I(\check{i})} + z_{I(\check{i}+1)} = \bar{z}_M.$$

The first equation implies that all other constraints $I(i)$ with $i \notin \{\check{i}-1, \check{i}, \check{i}+1\}$ in the original instance are satisfied as well. Furthermore, by these two equations we have $\sum_{\mathcal{J} \subseteq \bar{I}} \bar{y}_{\mathcal{J}} = \sum_{\mathcal{J} \subseteq I} y_{\mathcal{J}}$ and $\sum_{J \in \bar{I}} \bar{z}_J = \sum_{J \in I} z_J$. Therefore, the objective values of (y, z) and $(\bar{y}, \bar{z})$ agree. Finally, define an integer primal solution x for the original instance by

$$x_J := \bar{x}_J \text{ for } J \in I\backslash L, \quad x_J := \bar{x}_M \text{ for } J \in L\,.$$

The feasibility of x is obvious and by definition of $\bar{c}_M$ the solution x has the same objective function value as $\bar{x}$. Optimality of x and (y, z) thus follows from weak duality again.   □

Recall that the idea for showing integrality of P is to start with an instance where the objective c has Property 1 and to then apply the Merging Rule until we obtain an instance (c, I) which is "small enough". This is achieved for $|I| = S+1$, which will serve as the root for our recursive application of Theorem 1. For this case, we directly construct an integer optimal primal solution as well as a dual optimum solution satisfying criteria (C1) and (C2) as follows:

Lemma 3 (Recursion root). *Consider an instance (c, I) with $|I| = S + 1$ such that c has Property 1. Let $\check{i} \in \arg\min_i |c_{I(i)}|$. Then, the solutions*

$$x_{I(l)} := \begin{cases} 1 & \text{if } c_{I(l)} > 0 \text{ and } l \neq \check{i} \\ 0 & \text{if } c_{I(l)} < 0 \text{ and } l \neq \check{i} \end{cases}, \quad x_{I(\check{i})} := \begin{cases} 1 & \text{if } c_{I(\check{i})} < 0 \\ 0 & \text{if } c_{I(\check{i})} > 0 \end{cases}, \quad \text{and}$$

$$y_I := |c_{I(\check{i})}|, \quad y_{\mathcal{J}} := 0 \text{ otherwise}, \quad z_{I(l)} := \begin{cases} c_{I(l)} - y_I & \text{if } c_{I(l)} > 0 \text{ and } l \neq \check{i} \\ 0 & \text{otherwise} \end{cases}$$

are primal and dual optimal.

Proof. Because of Property 1, the coefficient for y_I in every constraint of the dual problem is the same as the sign of the corresponding right side c, so (y, z) satisfies (C1) and (C2). The feasibility of the defined solution is obvious, their optimality follows from weak duality by computing the corresponding objective values.   □

If the objective of the given instance of (OPT-P) has Property 1, the integrality of P follows directly from Theorem 1 and Lemma 3. Otherwise, let (c, I) be an instance of (OPT-P), so that the objective function does not have Property 1. If we perform the following operations in the given order, the result is an instance $(c^{(4)}, I^{(4)})$ which has Property 1:

(1) Set $I^{(1)} := \{J \in I \mid c_J \neq 0\}$ and $c_J^{(1)} := c_J$ for all $J \in I^{(1)}$.
(2) For the first positive entry l in $c_{I^{(1)}}$, set $I^{(2)} := \{I^{(1)}(i) \mid i \geq l\}$ and $c_J^{(2)} := c_J^{(1)}$ for all $J \in I^{(2)}$.
(3) For all maximal sequences $c_{I^{(2)}(i)}, \ldots, c_{I^{(2)}(j)}$ of equal sign, let $M := \bigcup_{i \leq k \leq j} I^{(2)}(k)$ and set $I^{(3)} := I^{(2)} \setminus \bigcup_{i \leq k \leq j} \{I^{(2)}(k)\} \cup \{M\}$ as well as $c_M^{(3)} := \sum_{k=i}^{j} c_{I^{(2)}(k)}^{(2)}$, $c_J^{(3)} := c_J^{(2)}$ for $J \in I^{(3)}$.
(4) For the last l with $\operatorname{sgn}(c_{I^{(3)}(l)}) = (-1)^S$, set $I^{(4)} := \{I^{(3)}(i) \mid i \leq l\}$ and $c_J^{(4)} := c_J^{(3)}$ for all $J \in I^{(4)}$.

Since the objective $c^{(4)}$ has Property 1, according to Theorem 1 and Lemma 3, we may assume that we are given an optimum integer primal solution $x^{(4)}$ and an optimum dual solution $(y^{(4)}, z^{(4)})$. Starting with these solutions, we undo the modifications of steps (1) to (4) in reverse order similar to the proof of Theorem 1 and define step-by-step integer optimum solutions as well as dual optimum solutions for the instances $(c^{(3)}, I^{(3)})$, $(c^{(2)}, I^{(2)})$, $(c^{(1)}, I^{(1)})$, and, finally, for the original instance (c, I).

Altogether, we prove the existence of an integer optimum solution for (OPT-P) for an arbitrary objective as well and thus obtain our main result:

Theorem 2. *The polytope P is integer and hence agrees with $\operatorname{conv}(X_S)$.*

4 Conclusion and Future Work

In this paper, we studied two problem variants related to bounded variation in binary sequences. The penalized problem (OPT-S) we considered turned out to be very easy and we gave a complete linear inequality description of its feasible set as well as a fast separation algorithm. Our focus, however, was on the second problem variant (OPT) with a hard constraint on the variation of the binary sequence. This problem variant turned out to be more complex to handle, although it is still tractable as the presented dynamic programming algorithm showed. We investigated the problem from a polyhedral perspective and presented a description of the convex hull of feasible points of Problem (OPT) by linear inequalities. While proving completeness of our linear inequality description, we obtained a primal-dual optimization algorithm in addition to the presented dynamic programming algorithm.

For possible future research, one may study extensions of Problem (OPT). For example, we can interpret Problem (OPT) as a special case of the following optimization problem in graphs: given an undirected graph with weighted nodes, determine a cut containing at most S edges such that the total weight of all nodes on one side of the cut is maximal. Our Problem (OPT) studied here would correspond to a graph consisting of a single path. Investigating bounded variation for multiple independent switches or sequences would then correspond to a graph consisting of multiple disjoint paths. We remark that this problem remains tractable, as the dynamic programming algorithm can be adapted to this case. Even more so, it can be further generalized to work for trees or forests. Although this implies that the optimization problems with bounded variation stay tractable for these settings, they are still open for investigation from a polyhedral perspective, which is important when separation algorithms are needed. To the best of our knowledge, for more general classes of graphs, even the complexity of the problem is unknown.

References

1. Bendotti, P., Fouilhoux, P., Rottner, C.: The min-up/min-down unit commitment polytope. J. Comb. Optim. **36**(3), 1024–1058 (2018). https://doi.org/10.1007/s10878-018-0273-y
2. Buchheim, C., Grütering, A., Meyer, C.: Parabolic optimal control problems with combinatorial switching constraints - part I: convex relaxations. Technical report 2203.07121 [math.OC], arXiv (2022)
3. Damci-Kurt, P., Küçükyavuz, S., Rajan, D., Atamturk, A.: A polyhedral study of production ramping. Math. Program. **158**, 175–205 (2016)
4. Lee, J., Leung, J., Margot, F.: Min-up/min-down polytopes. Discret. Optim. **1**(1), 77–85 (2004)
5. Mallipeddi, R., Suganthan, P.: Unit commitment - a survey and comparison of conventional and nature inspired algorithms. Int. J. Bio-Inspir. Comput. **6**, 71–90 (2014)
6. Pan, K., Guan, Y.: A polyhedral study of the integrated minimum-up/-down time and ramping polytope. Technical report 1604.02184, arXiv Optimization and Control (2016)
7. Pan, K., Guan, Y.: Convex hulls for the unit commitment polytope. Technical report 1701.08943, arXiv Optimization and Control (2017)
8. Queyranne, M., Wolsey, L.A.: Tight MIP formulations for bounded up/down times and interval-dependent start-ups. Math. Program. **164**(4), 129–155 (2017)
9. Rajan, D., Takriti, S.: Minimum up/down polytopes of the unit commitment problem with start-up costs. Technical report RC23628, IBM Research Report (2005)
10. Sager, S., Zeile, C.: On mixed-integer optimal control with constrained total variation of the integer control. Comput. Optim. Appl. **78**(2), 575–623 (2021)

On Minimally Non-firm Binary Matrices

Réka Ágnes Kovács$^{(\boxtimes)}$ (iD)

Mathematical Institute, University of Oxford, Oxford, UK
reka.kovacs@maths.ox.ac.uk

Abstract. For a binary matrix $\mathbf{X}$, the Boolean rank $br(\mathbf{X})$ is the smallest integer k for which $\mathbf{X}$ equals the Boolean sum of k rank-1 binary matrices, and the isolation number $i(\mathbf{X})$ is the maximum number of 1s no two of which are in a same row, column and a 2×2 submatrix of all 1s. In this paper, we continue Lubiw's study of firm matrices. $\mathbf{X}$ is said to be firm if $i(\mathbf{X}) = br(\mathbf{X})$ and this equality holds for all its submatrices. We show that the stronger concept of superfirmness of $\mathbf{X}$ is equivalent to having no odd holes in the rectangle cover graph of $\mathbf{X}$, the graph in which $br(\mathbf{X})$ and $i(\mathbf{X})$ translate to the clique cover and the independence number, respectively. A binary matrix is *minimally non-firm* if it is not firm but all of its proper submatrices are. We introduce two matrix operations that lead to generalised binary matrices and use these operations to derive four infinite classes of minimally non-firm matrices. We hope that our work may pave the way towards a complete characterisation of firm matrices via forbidden submatrices.

Keywords: Boolean rank · Rectangle covering number · Firm matrices

1 Introduction

The *Boolean rank* of a binary matrix $\mathbf{X}$, $br(\mathbf{X})$, is the smallest integer k for which $\mathbf{X}$ equals the sum of k rank-1 binary matrices, using Boolean arithmetic in which $1 + 1 = 1$ holds [10]. A *rectangle* of $\mathbf{X}$ is a submatrix of all 1s. Note that the support of a rank-1 binary matrix is precisely a rectangle, hence $br(\mathbf{X})$ is the minimum number of rectangles needed to cover $\mathrm{supp}(\mathbf{X}) := \{(i,j) : x_{i,j} = 1\}$.

An *isolated set* of $\mathbf{X}$ is a set $S \subseteq \mathrm{supp}(\mathbf{X})$ such that for any distinct (i_1, j_1), (i_2, j_2) in S, it holds $i_1 \neq i_2$, $j_1 \neq j_2$ and $x_{i_1, j_2} = 0$ or $x_{i_2, j_1} = 0$. The *isolation number* of $\mathbf{X}$, $i(\mathbf{X})$, is the maximum cardinality of an isolated set [8]. In the field of communication complexity, quantities $br(\mathbf{X})$ and $i(\mathbf{X})$ are often referred to as the rectangle covering number and the fooling set bound [11].

In the bipartite graph whose biadjacency matrix is $\mathbf{X}$, $br(\mathbf{X})$ is the minimum number of bicliques (complete bipartite subgraphs) needed to cover the edge set,

Supported by The Alan Turing Institute, London, UK

while $i(\mathbf{X})$ is the maximum cardinality of a matching in which no two edges are in a 4-cycle. Both $br(\mathbf{X})$ and $i(\mathbf{X})$ are NP-hard to compute for general binary [16,17] and totally balanced matrices as well [14,15].

For any binary matrix $\mathbf{X}$, it can be readily checked that $i(\mathbf{X}) \leq br(\mathbf{X})$. This inequality may however be strict for many matrices. In fact, the complement of the identity matrix shows that the gap between $i(\mathbf{X})$ and $br(\mathbf{X})$ may be arbitrarily large [3]. We say $\mathbf{X}$ is *firm* if $i(\mathbf{X}) = br(\mathbf{X})$ and this equality also holds for all its submatrices. The concept of firmness, along with many results that form the basis of this paper were introduced by Lubiw in [13]. A key tool in Lubiw's work is to define the *rectangle cover graph* of $\mathbf{X}$ (the 1's graph in her words) in which $i(\mathbf{X})$ and $br(\mathbf{X})$ translate to the independence and clique cover number, respectively. Lubiw defines $\mathbf{X}$ to be *superfirm* if $\mathbf{X}$'s rectangle cover graph is perfect and demonstrates that superfirm matrices are a strict subset of firm matrices. In addition, she shows that covering rectilinear polygons by a minimum number of continuous rectangles is a special case of the rectangle cover problem on binary matrices [13]. In the bipartite setting, firmness is later redefined under the name 'edge-perfection' [15], while superfirmness is investigated under the name 'cross-perfection' from a polyhedral perspective [6]. The following important classes of matrices have been shown to be firm. *Interval* matrices, matrices whose columns can be permuted so the 1s appear consecutively in each row, are proved to be firm by a deep result of Győri [9]. *Linear* matrices, matrices that have no 2×2 submatrix of 1s, and matrices that can be decomposed into linear matrices via the matrix equivalent of split decomposition on bipartite graphs are shown to be superfirm by Lubiw [13]. The firmness of biadjacency matrices of domino-free bipartite graphs is implied by a result of Amilhastre et al. [1].

In this paper, we start the investigation of minimally non-firm matrices. A binary matrix $\mathbf{X}$ is *minimally non-firm* if $i(\mathbf{X}) < br(\mathbf{X})$ and $i(\mathbf{X}') = br(\mathbf{X}')$ for all proper submatrices $\mathbf{X}'$ of $\mathbf{X}$. Our main tool is looking at the problem through the rectangle cover graph. First, we extend a theorem of Lubiw and show that interestingly odd antiholes cannot appear without odd holes in rectangle cover graphs. Then we characterise the necessary and sufficient submatrices for 5-holes to appear. We define *simplicial* 1s and a procedure for their removal which leads to *generalised* binary matrices. We introduce the *stretching* matrix operation which then along with the simplicial 1 removal procedure are used to give a general recipe for the construction of minimally non-firm matrices. We then prove by using this general recipe that four infinite classes of matrices are minimally non-firm. To the best of our knowledge, minimally non-firm matrices have not been studied before. We believe that studying them is a natural approach to better understand firmness, akin to the study of perfect graphs via minimally imperfect graphs. We hope that our results may pave the way towards a complete characterisation of firm and superfirm matrices via forbidden submatrices.

This paper is organised as follows. Section 2 gives a brief recap on the work of Lubiw introducing the concept of rectangle cover graphs, superfirmness and generalised binary matrices. In Sect. 3, simplicial 1s and the stretching operation are introduced. In Sect. 4, we show that a matrix is superfirm if and only if it has

$$\mathbf{D_4} = \begin{bmatrix} 0 & 1 & 1 & 0 \\ 1 & 1 & 1 & 0 \\ 1 & 1 & 1 & 1 \\ 0 & 0 & 1 & 1 \end{bmatrix}$$

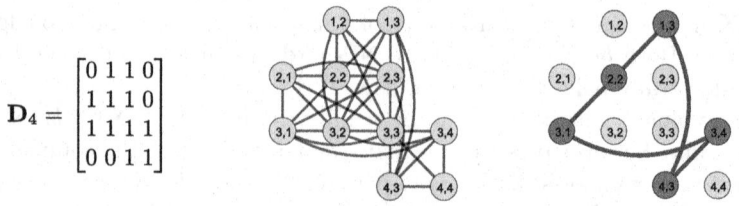

Fig. 1. $\mathbf{D_4}$, its rectangle cover graph $\mathcal{G}(\mathbf{D_4})$ and the 5-hole in $\mathcal{G}(\mathbf{D_4})$ highlighted

no odd holes in its rectangle cover graph. In Sect. 5 we prove our main theorem, which we then use to derive four infinite classes of minimally non-firm binary matrices. We conclude in Sect. 6 and mention two open problems.

2 Preliminaries

Let $\mathbf{X} \in \{0, 1\}^{m \times n}$. For $I \subseteq [n] := \{1, \ldots, n\}$ and $J \subseteq [m]$, a *submatrix* of $\mathbf{X}$ identified by $I \times J$ is obtained by deleting the rows not in I and the columns not in J. If $I \subsetneq [n]$ or $J \subsetneq [m]$ then $I \times J$ is a *proper submatrix* of $\mathbf{X}$. A submatrix is a *rectangle* if $I \times J \subseteq \text{supp}(\mathbf{X}) = \{(i, j) : x_{i,j} = 1\}$. As the 1s in a row or column form a rectangle, we have $br(\mathbf{X}) \leq \min\{m, n\}$. In addition, note that $br(\mathbf{X})$ is invariant under transposition and under duplicating rows and columns.

For an isolated set S and rectangle $I \times J$, we have $|S \cap (I \times J)| \leq 1$, hence $i(\mathbf{X}) \leq br(\mathbf{X})$. Recall that $\mathbf{X}$ is *firm* if $i(\mathbf{X'}) = br(\mathbf{X'})$ holds for all submatrices $\mathbf{X'}$ of $\mathbf{X}$, including $\mathbf{X}$. The *rectangle cover graph* $\mathcal{G}(\mathbf{X})$ of $\mathbf{X}$ is the graph on vertex set $\text{supp}(\mathbf{X})$, where two vertices are adjacent if they can be covered by a common rectangle of $\mathbf{X}$. We adopt the convention that vertices of $\mathcal{G}(\mathbf{X})$ are drawn in the positions of the corresponding 1s' of $\mathbf{X}$. See Fig. 1 for an example of $\mathcal{G}(\mathbf{X})$ for matrix $\mathbf{D_4}$. Clearly, the independent sets of $\mathcal{G}(\mathbf{X})$ are just the isolated sets of $\mathbf{X}$. Lubiw shows that maximal cliques of $\mathcal{G}(\mathbf{X})$ are in direct correspondence with maximal rectangles of $\mathbf{X}$ [13]. Therefore we have $i(\mathbf{X}) = \alpha(\mathcal{G}(\mathbf{X}))$ and $br(\mathbf{X}) = \theta(\mathcal{G}(\mathbf{X}))$, where $\alpha(G)$ and $\theta(G)$ denote the independence and clique cover number of a graph G, respectively. A graph G is *perfect* if $\alpha(H) = \theta(H)$ holds for every induced subgraph H of G. A *hole* is an induced chordless cycle of length at least four. An *odd hole* is a hole of odd length and an *odd antihole* is the complement of an odd hole. Perfect graphs are exactly those that have no odd holes and no odd antiholes by the Strong Perfect Graph Theorem [4]. $\mathbf{X}$ is said to be *superfirm* if $\mathcal{G}(\mathbf{X})$ is perfect [13]. Superfirm matrices are a strict subset of firm matrices [13], as for instance $\mathbf{D_4}$ is an interval matrix hence firm by Győri's Theorem [9] but not superfirm as $\mathcal{G}(\mathbf{D_4})$ contains a 5-hole as shown in Fig. 1. Note that this is because not every induced subgraph of $\mathcal{G}(\mathbf{X})$ corresponds to a submatrix of $\mathbf{X}$ and firmness requires $\alpha(H) = \theta(H)$ to hold for only those subgraphs H of $\mathcal{G}(\mathbf{X})$ where $H = \mathcal{G}(\mathbf{X'})$ for a submatrix $\mathbf{X'}$ of $\mathbf{X}$.

Replacing a 1 of $\mathbf{X}$ at (i, j) with a 0 does not necessarily correspond to the deletion of vertex (i, j) from $\mathcal{G}(\mathbf{X})$ as edges not incident to (i, j) may get deleted.

To represent all induced subgraphs of $\mathcal{G}(\mathbf{X})$ in matrix form, Lubiw introduces a new entry type ? which may be part of a rectangle but need not be covered in a feasible covering. A matrix over $\{0, 1, ?\}$ is called a *generalised binary matrix* [13]. A *rectangle* of a generalised binary matrix $\mathbf{Y}$ is a submatrix containing no 0s, while an *isolated set* of $\mathbf{Y}$ is a subset of $\mathrm{supp}(\mathbf{Y}) := \{(i, j) : y_{i,j} = 1\}$ in which no two elements are contained in a common rectangle of $\mathbf{Y}$. Then $i(\mathbf{Y})$, $br(\mathbf{Y})$ and firmness are analogously defined as for standard binary matrices. For $\mathbf{X} \in \{0,1\}^{m \times n}$ and $P \subseteq \mathrm{supp}(\mathbf{X})$, let $\mathbf{X}^P$ be the generalised binary matrix obtained from $\mathbf{X}$ by replacing all 1s in P by ?s, i.e. $x_{i,j}^P = ?$ for $(i, j) \in P$ and $x_{i,j}^P = x_{i,j}$ otherwise. For $\mathbf{X}^P$ define its rectangle cover graph $\mathcal{G}(\mathbf{X}^P)$ to be the subgraph of $\mathcal{G}(\mathbf{X})$ induced by $\mathrm{supp}(\mathbf{X}) \backslash P$. Superfirmness of $\mathbf{X}$ is then equivalent to the requirement that $i(\mathbf{X}^P) = br(\mathbf{X}^P)$ for all $P \subseteq \mathrm{supp}(\mathbf{X})$ [13].

3 Simplicial 1s and Stretching

Let $\mathbf{Y}$ be a generalised binary matrix. We say $(\ell, k) \in \mathrm{supp}(\mathbf{Y})$ is a *simplicial 1* of $\mathbf{Y}$ if $I \times J$ with $I = \{i : y_{i,k} \in \{1, ?\}\}$ and $J = \{j : y_{\ell, j} \in \{1, ?\}\}$ satisfies $I \times J \subseteq \{(i, j) : y_{i,j} \in \{1, ?\}\}$, that is $I \times J$ is a rectangle of $\mathbf{Y}$. Note that $I \times J$ is a maximal rectangle and the only maximal rectangle of $\mathbf{Y}$ that covers the simplicial 1 at (ℓ, k). To *remove the simplicial* 1 at (ℓ, k) of $\mathbf{Y}$ we delete row ℓ and column k and set all remaining entries that are in $I \times J$ to ?s.

Lemma 1. *If* $\mathbf{Y}'$ *is obtained by removing a simplicial 1 of a generalised binary matrix* $\mathbf{Y}$, *then* $i(\mathbf{Y}) = i(\mathbf{Y}') + 1$ *and* $br(\mathbf{Y}) = br(\mathbf{Y}') + 1$.

Proof. Let (ℓ, k) be the simplicial 1 and $I \times J$ its unique maximal rectangle. For a maximum isolated set S' and a minimum rectangle cover $\mathcal{R}'$ of $\mathbf{Y}'$, $S' \cup \{(\ell, k)\}$ and $\mathcal{R}' \cup (I \times J)$ are clearly feasible for $\mathbf{Y}$. Conversely, if S is a maximum isolated set of $\mathbf{Y}$, then $S \cap (I \times J) = \{(i, j)\}$ for some $(i, j) \in I \times J$, as otherwise $S \cup \{(\ell, k)\}$ would be a larger isolated set of $\mathbf{Y}$. So $S \backslash \{(i, j)\}$ is a feasible isolated set of $\mathbf{Y}'$. As (ℓ, k) is a simplicial 1, we may assume that $I \times J$ is used in a minimum cover $\mathcal{R}$ of $\mathbf{Y}$. Then $\mathcal{R} \backslash \{I \times J\}$ is a feasible cover of $\mathbf{Y}'$. $\square$

Our definition of simplicial 1s for a standard binary matrix $\mathbf{X}$ is identical to the definition of *bisimplicial edges* [7] in the bipartite graph whose biadjacency matrix is $\mathbf{X}$. The key difference is how we remove a simplicial 1 and transition into generalised binary matrices.

We have seen that not every induced subgraph of $\mathcal{G}(\mathbf{X})$ corresponds to a submatrix of $\mathbf{X}$, but by turning 1s to ?s we can consider arbitrary induced subgraphs of $\mathcal{G}(\mathbf{X})$ in matrix form. The idea behind the next matrix operation is to expose induced subgraphs of rectangle cover graphs without explicitly setting matrix entries to ?s. Let $\mathbf{X} \in \{0,1\}^{m \times n}$. By *stretching* a 1 at $(\ell, k) \in \mathrm{supp}(\mathbf{X})$ we get the $(m + 1) \times (n + 1)$ binary matrix $\mathcal{S}^{(\ell, k)}(\mathbf{X})$ which satisfies

$$\mathcal{S}^{(\ell, k)}(\mathbf{X})_{i,j} = x_{i,j} \qquad i \in [m], j \in [n], \tag{1}$$

$$\mathcal{S}^{(\ell, k)}(\mathbf{X})_{i,j} = 1 \qquad (i, j) \in \{(\ell, n + 1), (m + 1, k), (m + 1, n + 1)\}, \tag{2}$$

and $\mathcal{S}^{(\ell,k)}(\mathbf{X})_{i,j} = 0$ otherwise. For instance, if $(m,n) \in \mathrm{supp}(\mathbf{X})$ then by stretching (m,n) we obtain

$$\mathcal{S}^{(m,n)}(\mathbf{X}) = \begin{bmatrix} x_{1,1} & \cdots & & x_{1,n} & 0 \\ \vdots & \ddots & & \vdots & \vdots \\ & & & x_{m-1,n} & 0 \\ x_{m,1} & \cdots & x_{m,n-1} & 1 & 1 \\ 0 & \cdots & 0 & 1 & 1 \end{bmatrix}. \tag{3}$$

Stretching (ℓ,k) adds in a simplicial 1 at position $(m+1, n+1)$ whose unique maximal rectangle covers only (ℓ,k) from $\mathrm{supp}(\mathbf{X})$. By Lemma 1, removing the simplicial 1 at $(m+1, n+1)$, we get

$$i(\mathcal{S}^{(\ell,k)}(\mathbf{X})) = i(\mathbf{X}^{(\ell,k)}) + 1, \quad br(\mathcal{S}^{(\ell,k)}(\mathbf{X})) = br(\mathbf{X}^{(\ell,k)}) + 1, \tag{4}$$

where $\mathbf{X}^{(\ell,k)}$ is a shorter notation for $\mathbf{X}^P$ with $P = \{(\ell,k)\}$.

For a non-empty set $Q \subseteq \mathrm{supp}(\mathbf{X})$, the matrix obtained by stretching each 1 in Q is denoted by $\mathcal{S}^Q(\mathbf{X})$. We adopt the convention to stretch 1s in Q in non-decreasing order of row and then column index, so $\mathcal{S}^Q(\mathbf{X})$ may be written in block form as

$$\mathcal{S}^Q(\mathbf{X}) = \begin{bmatrix} \mathbf{X} & \mathbf{U} \\ \mathbf{L} & \mathbf{I}_{|Q|} \end{bmatrix} \tag{5}$$

where $\mathbf{U}$ is an $m \times |Q|$ matrix with $|Q|$ 1s exactly one in each column that have non-decreasing row index from left to right, $\mathbf{L}$ is an $|Q| \times n$ matrix with $|Q|$ 1s exactly one in each row and $\mathbf{I}_t$ is the $t \times t$ identity matrix.

If (ℓ,k) is a simplicial 1 of $\mathbf{X}$ then we say that $\mathcal{S}^{(\ell,k)}(\mathbf{X})$ is obtained by *simplicial stretching*. Looking at $\mathcal{G}(\mathbf{X})$ and using Lemma 1 and the Clique Cutset Lemma [2], the following can be proved.

Lemma 2. *Let $\mathbf{X}$ be superfirm. Then $\mathcal{S}^{(\ell,k)}(\mathbf{X})$ is firm. Furthermore, if (ℓ,k) is a simplicial 1 of $\mathbf{X}$, then $\mathcal{S}^{(\ell,k)}(\mathbf{X})$ is superfirm.*

This lemma is tight in two ways. First, non-simplicial stretching may destroy superfirmness. Second, both simplicial and non-simplicial stretching do not preserve firmness. In Sect. 5, we will exploit the superfirmness and firmness destroying properties of stretching to create minimally non-firm matrices.

For $n \geq 3$, let $\mathbf{C}_n \in \{0,1\}^{n \times n}$ be the n-th *cycle matrix* with exactly two 1s in each row and column such that no proper submatrix has this property. A binary matrix is *totally balanced* if it has no $\mathbf{C}_n$ submatrices for any $n \geq 3$. Totally balanced matrices are exactly those that have a Γ-free ordering [12], where $\Gamma = \left[\begin{smallmatrix} 1 & 1 \\ 1 & 0 \end{smallmatrix}\right]$. The following result can be verified by a Γ-free ordering.

Lemma 3. *If $\mathbf{X}$ is totally balanced then so is $\mathcal{S}^Q(\mathbf{X})$ for any $Q \subseteq \mathrm{supp}(\mathbf{X})$.*

4   Superfirm Matrices and Odd Holes

The Strong Perfect Graph Theorem [4] tells us that a binary matrix $\mathbf{X}$ is superfirm if and only if $\mathcal{G}(\mathbf{X})$ has no odd holes and no odd antiholes. But which are

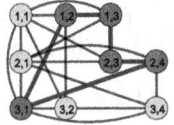 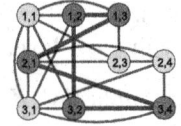 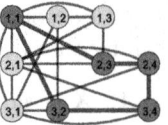

Fig. 2. The three 5-holes in $\mathcal{G}(\mathbf{H}_3)$

the necessary submatrices so that odd holes or odd antiholes appear in $\mathcal{G}(\mathbf{X})$? In this section, we show that forbidding odd antiholes in $\mathcal{G}(\mathbf{X})$ is unnecessary. Then we study when a 5-hole in $\mathcal{G}(\mathbf{X})$ exists.

A theorem of Lubiw in [13] states that for $\mathcal{G}(\mathbf{X})$ to have an odd antihole of size 7 or more, $\mathbf{X}$ needs to have the 3×3 cycle matrix $\mathbf{C}_3$ as a submatrix. Note that $\mathbf{C}_3$ is superfirm. Let $\mathbf{1}$ be the all 1s column vector of appropriate size and define $\mathbf{W} := \begin{bmatrix} \mathbf{C}_3 & 1 \\ \mathbf{1}^\top & 1 \end{bmatrix}$ and $\bar{\mathbf{I}}_4 := \begin{bmatrix} \mathbf{C}_3 & 1 \\ \mathbf{1}^\top & 0 \end{bmatrix}$ in $\{0,1\}^{4\times 4}$. Considering a slight extension of Lubiw's proof, we show that these two larger matrices are necessary for the appearance of odd antiholes.

Lemma 4. *If $\mathcal{G}(\mathbf{X})$ contains an odd antihole of size 7 or more then $\mathbf{X}$ has $\mathbf{W}$ or $\bar{\mathbf{I}}_4$ as a submatrix.*

Proof. Following the proof structure of [13, Theorem 6.3], suppose that $\mathbf{X}$ has no such submatrices but $\mathcal{G}(\mathbf{X})$ contains an antihole $A \subseteq \mathrm{supp}(\mathbf{X})$ of odd size $k = |A| \geq 7$. By duplicating rows and columns of $\mathbf{X}$, we may assume that no two 1s in A are in the same row or column. Note that row and column duplication cannot introduce $\mathbf{W}$ or $\bar{\mathbf{I}}_4$ submatrices into $\mathbf{X}$. Then the submatrix $\mathbf{X}'$ of $\mathbf{X}$ that consists of the rows and columns of the 1s in A is of dimension $k \times k$ and may be permuted so that the vertices of A appear on the main diagonal and are non-adjacent to the two vertices that are directly above and below them. Then $\mathbf{X}'$ has the form as below where each undecided entry pair $(i,j), (j,i)$ denoted by $*$s satisfies $|\mathrm{supp}(\mathbf{X}') \cap \{(i,j),(j,i)\}| \leq 1$ so that A is indeed an antihole in $\mathcal{G}(\mathbf{X})$.

$$\mathbf{X}' = \begin{bmatrix} 1 & * & 1 & 1 & \cdots & 1 & 1 & * \\ * & 1 & * & 1 & & 1 & 1 & 1 \\ 1 & * & 1 & * & & 1 & 1 & 1 \\ 1 & 1 & * & 1 & & 1 & 1 & 1 \\ \vdots & & & & \ddots & & & \vdots \\ 1 & 1 & 1 & 1 & & 1 & * & 1 \\ 1 & 1 & 1 & 1 & & * & 1 & * \\ * & 1 & 1 & 1 & \cdots & 1 & * & 1 \end{bmatrix} \Rightarrow \begin{bmatrix} 1 & 0 & 1 & 1 & \cdots & 1 & 1 & * \\ 1 & 1 & 1 & 1 & & 1 & 1 & 1 \\ 1 & 0 & 1 & 0 & & 1 & 1 & 1 \\ 1 & 1 & 1 & 1 & & 1 & 1 & 1 \\ \vdots & & & & \ddots & & & \vdots \\ 1 & 1 & 1 & 1 & & 1 & * & 1 \\ 1 & 1 & 1 & 1 & & * & 1 & * \\ * & 1 & 1 & 1 & \cdots & 1 & * & 1 \end{bmatrix} \quad (6)$$

Assume without loss of generality that $x'_{1,2} = 0$. Suppose that $x'_{2,3} = 0$. If $x'_{5,6} = x'_{6,5} = 0$ then the submatrix $I \times J$ of $\mathbf{X}'$ with $I = \{1,2,5,6\}$, $J = \{2,3,5,6\}$ is $\bar{\mathbf{I}}_4$. Moreover, if $x'_{5,6} + x'_{6,5} = 1$ then $I \times J$ is $\mathbf{W}$. Hence, $x'_{2,3} \neq 0$. In general, exactly one of (i,j) and $(i+1, j+1)$ can be a 0 for all $*$s, so the zeros of $\mathbf{X}'$ must zigzag as shown in the right of Equation (6). But as k is odd, this is impossible. $\square$

The importance of Lemma 4 over Lubiw's theorem, is that both $\mathbf{W}$ and $\bar{\mathbf{I}}_4$ contain the submatrix $\mathbf{H}_3 := [1, \mathbf{C}_3]$ and $\mathcal{G}(\mathbf{H}_3)$ contains three 5-holes as shown in Fig. 2, whereas $\mathbf{C}_3$ is superfirm. This shows that a rectangle cover graph cannot contain an odd antihole of size 7 or larger if it does not contain an odd hole. Recalling that a 5-antihole is just a 5-hole, we obtain the following result.

Theorem 1. X *is superfirm if and only if* $\mathcal{G}(X)$ *has no odd holes.*

Theorem 1 motivates us to study when $\mathcal{G}(\mathbf{X})$ has odd holes. We initialise this by characterising when a 5-hole exists in $\mathcal{G}(\mathbf{X})$. The proof is skipped but it is of similar nature to that of Lemma 4. Let $\mathbf{K}_5 \in \{0,1\}^{5\times5}$ be the circulant matrix with exactly three 1s per row and column and recall $\mathbf{D}_4$ from Fig. 1.

Theorem 2. $\mathcal{G}(X)$ *contains a 5-hole if and only if* X *has at least one of* $\mathbf{D}_4$, $\mathbf{H}_3$, $\mathbf{H}_3^\top$ *or* $\mathbf{K}_5$ *as a submatrix.*

5   Four Infinite Classes of Minimally Non-firm Matrices

In this section we prove a theorem which shows how minimally non-firm matrices may arise by using the stretching operation. Then using this theorem we show that four infinite classes of matrices are minimally non-firm.

Recall that a standard binary matrix $\mathbf{X}$ is *minimally non-firm (mnf)* if it is not firm but all proper submatrices of it are. This definition naturally extends to generalised binary matrices $\mathbf{Y}$, $\mathbf{Y}$ is mnf if $i(\mathbf{Y}) < br(\mathbf{Y})$ and $i(\mathbf{Y}') = br(\mathbf{Y}')$ for all proper submatrices $\mathbf{Y}'$ of $\mathbf{Y}$. Note that as $br(\mathbf{Y})$ and $i(\mathbf{Y})$ are invariant under transposition, the transpose of any mnf matrix is mnf as well. The following two simple results apply to both standard and generalised mnf matrices.

Lemma 5. *Each row and column of an mnf matrix has at least two non-zeros.*

Proof. Suppose $\mathbf{Y}$ is mnf and its i-th row only has a single nonzero at entry (i,j). If $y_{i,j} =?$ then row i can clearly be dropped without changing $i(\mathbf{Y})$ or $br(\mathbf{Y})$. If $y_{i,j} = 1$ then (i,j) is a simplicial 1. By Lemma 1, removing it we obtain a firm submatrix $\mathbf{Y}'$ with $i(\mathbf{Y}) - 1 = i(\mathbf{Y}') = br(\mathbf{Y}') = br(\mathbf{Y}) - 1$, a contradiction.   □

Lemma 6. *If* Y *is mnf then* $i(Y) = br(Y) - 1$.

Proof. Let $\mathbf{Y}$ be mnf. By deleting a single row or column of $\mathbf{Y}$ we get a submatrix $\mathbf{Y}'$ which by definition is firm and satisfies $i(\mathbf{Y}) - 1 \le i(\mathbf{Y}') \le i(\mathbf{Y})$ and $br(\mathbf{Y}) - 1 \le br(\mathbf{Y}') \le br(\mathbf{Y})$ as a row or column forms a rectangle and may contain at most one element of an isolated set. So, we must have $br(\mathbf{Y}') = br(\mathbf{Y}) - 1$ as otherwise $\mathbf{Y}$ is firm. But then $i(\mathbf{Y}') = br(\mathbf{Y}') = br(\mathbf{Y}) - 1 \le i(\mathbf{Y})$ which together with $i(\mathbf{Y}) < br(\mathbf{Y})$ implies $br(\mathbf{Y}) - 1 = i(\mathbf{Y})$.   □

By Theorem 1, $\mathbf{X}$ is superfirm if $\mathcal{G}(\mathbf{X})$ has no odd holes, so for $\mathbf{X}$ to be mnf $\mathcal{G}(\mathbf{X})$ must contain odd holes. Using Theorem 2 one can show that the smallest mnf standard binary matrices are of dimension 4×4 and there are exactly two of them: $\bar{\mathbf{I}}_4$ and $\bar{\mathbf{I}}_4'$, where $\bar{\mathbf{I}}_4'$ is obtained from $\bar{\mathbf{I}}_4$ by turning a single 1 to a 0 (for instance at $(1,4)$, but due to symmetry any other 1 would work).

Let $\mathbf{X}$ be a standard binary matrix with an odd hole C in $\mathcal{G}(\mathbf{X})$ of size $|C| = 2k + 1$. Stretching all 1s at $Q = \mathrm{supp}(\mathbf{X})\setminus C$ of $\mathbf{X}$, by Lemma 1 we get

$$i(\mathcal{S}^Q(\mathbf{X})) - |Q| = i(\mathbf{X}^Q) = k < k + 1 = br(\mathbf{X}^Q) = br(\mathcal{S}^Q(\mathbf{X})) - |Q|, \quad (7)$$

so $\mathcal{S}^Q(\mathbf{X})$ is non-firm. This recipe however, does not guarantee that $\mathcal{S}^Q(\mathbf{X})$ is *minimally* non-firm. By adding extra conditions on Q, minimality can be enforced.

Theorem 3. *Let* $\mathbf{X} \in \{0,1\}^{m \times n}$. *If* $\mathbf{X}^Q$ *is a minimally non-firm generalised binary matrix for some non-empty* $Q \subset \mathrm{supp}(\mathbf{X})$ *and* $\mathbf{X}^P$ *is firm for all* $P \subsetneq Q$, *then* $\mathcal{S}^Q(\mathbf{X}) \in \{0,1\}^{(m+|Q|) \times (n+|Q|)}$ *is minimally non-firm.*

Proof. $\mathcal{S}^Q(\mathbf{X})$ may be written as a block matrix with four blocks $\mathbf{X}, \mathbf{L}, \mathbf{U}$ and $\mathbf{I}_{|Q|}$ as in Eq. (5). By construction all 1s in block $\mathbf{I}_{|Q|}$ are simplicial, hence removing them we obtain the mnf generalised binary matrix $\mathbf{X}^Q$. By Lemma 1 then $i(\mathcal{S}^Q(\mathbf{X})) = i(\mathbf{X}^Q) + |Q| < br(\mathbf{X}^Q) + |Q| = br(\mathcal{S}^Q(\mathbf{X}))$.

Suppose that not all proper submatrices of $\mathcal{S}^Q(\mathbf{X})$ are firm and let $\mathbf{Y}$ be the smallest non-firm proper submatrix indexed by $I \times J$. Then $\mathbf{Y}$ is mnf. Note that the four block matrices of $\mathcal{S}^Q(\mathbf{X})$ are all firm: (1) $\mathbf{X}$ is firm as it is just $\mathbf{X}^\emptyset$. (2) $\mathbf{I}_{|Q|}$ is clearly firm. (3) $\mathbf{U}$ has exactly one 1 per column, so it can be obtained from an identity matrix by duplicating columns and adding zero rows, and thus firm. (4) Similarly, as $\mathbf{L}$ has exactly one 1 per row, it is firm. Hence $\mathbf{Y}$ cannot be fully contained in any of the four blocks. As $\mathbf{Y}$ is a mnf standard binary matrix it has at least two 1s in each row and column by Lemma 5. Since block $[\,\mathbf{L}\ \mathbf{I}_{|Q|}\,]$ has exactly two 1s in each row, if $\mathbf{Y}$ has a row from this block, then $\mathbf{Y}$ must also contain the columns of both 1s in this row. Similarly, if $\mathbf{Y}$ contains a column from block $\begin{bmatrix} \mathbf{U} \\ \mathbf{I}_{|Q|} \end{bmatrix}$, it must contain the rows of both 1s in this column. Therefore, the rows in I from block $[\,\mathbf{L}\ \mathbf{I}_{|Q|}\,]$ and the columns in J from block $\begin{bmatrix} \mathbf{U} \\ \mathbf{I}_{|Q|} \end{bmatrix}$ come in pairs and may be identified with their 1 in block $\mathbf{I}_{|Q|}$. Let P be the subset of Q whose stretching created the 1s in block $\mathbf{I}_{|Q|}$ which are in $\mathbf{Y}$. Removing all $|P|$ simplicial 1s present in $\mathbf{Y}$ from block $\mathbf{I}_{|Q|}$ we obtain a generalised binary matrix which is fully contained in block $\mathbf{X}$ and is just a submatrix $\mathbf{Z}$ of $\mathbf{X}^P$. By Lemma 1, $\mathbf{Z}$ satisfies $i(\mathbf{Z}) + |P| = i(\mathbf{Y})$ and $br(\mathbf{Z}) + |P| = br(\mathbf{Y})$. If $P = Q$, then I contains all the rows and columns from block $\mathbf{I}_{|Q|}$ so $\mathbf{Z}$ must be a proper submatrix of $\mathbf{X}^Q$, hence firm. If $P \neq Q$, then $\mathbf{Z}$ is a submatrix of the firm matrix $\mathbf{X}^P$. In both cases $i(\mathbf{Z}) = br(\mathbf{Z})$ which implies $i(\mathbf{Y}) = br(\mathbf{Y})$, a contradiction. $\square$

One can see that a partial converse of the above theorem also holds, i.e. if a standard binary mnf matrix has some simplicial 1s then by removing those we obtain a generalised binary mnf matrix for which the theorem's conditions hold. Note however, that not all mnf matrices have simplicial 1s, e.g. $\bar{\mathbf{I}}_4$, hence certainly not all mnf matrices arise via Theorem 3.

Recall $\mathbf{C}_n$ is the $n \times n$ cycle matrix. For $n \geq 3$, let $\mathbf{M}_{n+1} := \mathcal{S}^{(n,n)}(\mathbf{C}_n)$ be the $(n+1) \times (n+1)$ matrix and $\mathbf{H}_n := [\mathbf{1}, \mathbf{C}_n]$ be the $n \times (n+1)$ matrix,

$$
\mathbf{M}_n = \begin{bmatrix} 1 & 1 & & & \\ & 1 & 1 & & \\ & & \ddots & \ddots & \\ & & & 1 & 1 \\ 1 & & & & 1 \\ & & & & 1 & 1 \end{bmatrix}, \qquad
\mathbf{H}_n = \begin{bmatrix} 1 & 1 & 1 & & & \\ 1 & & 1 & 1 & & \\ \vdots & & & \ddots & \ddots & \\ \vdots & & & & 1 & 1 \\ 1 & 1 & & & & 1 \end{bmatrix}. \tag{8}
$$

84   R. Á. Kovács

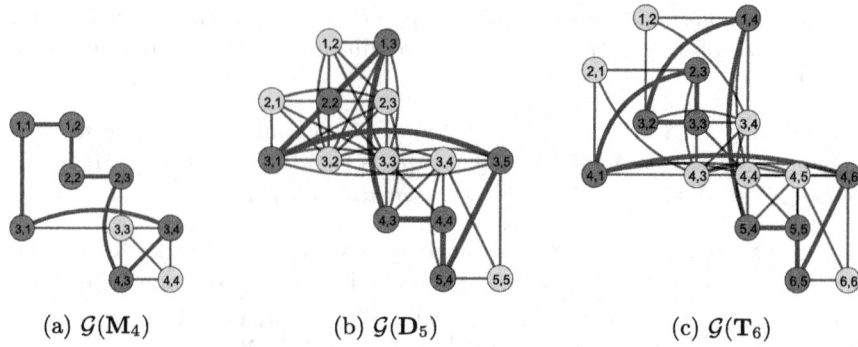

(a) $\mathcal{G}(\mathbf{M}_4)$ (b) $\mathcal{G}(\mathbf{D}_5)$ (c) $\mathcal{G}(\mathbf{T}_6)$

Fig. 3. Odd holes highlighted in the rectangle cover graphs of $\mathbf{M}_4, \mathbf{D}_5$ and $\mathbf{T}_6$

Matrices $\mathbf{M}_n$ appear in the work of Lubiw [13] as forbidden submatrices for a subset of superfirm matrices that can be decomposed into linear matrices by applying the matrix equivalent of split decomposition [5] on bipartite graphs.

Recall matrix $\mathbf{D}_4$ from Fig. 1 and for $n \geq 5$, let $\mathbf{D}_n := \mathcal{S}^{(3,n-1)}(\mathbf{D}_{n-1})$. In addition, let $\mathbf{T}_5 \in \{0,1\}^{5 \times 5}$ as below and for $n \geq 6$ define $\mathbf{T}_n := \mathcal{S}^{(4,n-1)}(\mathbf{T}_{n-1})$,

$$
\mathbf{D}_n = \begin{bmatrix} 1\,1 \\ 1\,1\,1 \\ 1\,1\,1 & 1 & \cdots & 1 \\ & 1\,1 \\ & & \ddots & \ddots \\ & & & 1\,1 \end{bmatrix}, \quad
\mathbf{T}_5 = \begin{bmatrix} 1 \\ 1\,1 \\ 1\,1\,1 \\ 1\,1\,1\,1 \\ 1\,1 \end{bmatrix}, \quad
\mathbf{T}_n = \begin{bmatrix} 1\,1 \\ 1\,1\,1 \\ 1 & 1\,1\,1 & 1 & \cdots & 1 \\ & 1\,1 \\ & & \ddots & \ddots \\ & & & 1\,1 \end{bmatrix}. \quad (9)
$$

All these matrices contain odd holes in their rectangle cover graph as shown in Fig. 2 for $\mathbf{H}_3$ and Fig. 3 for $\mathbf{M}_4, \mathbf{D}_5$ and $\mathbf{T}_6$. In the remaining parts of this section, we will prove that by choosing the set Q in Theorem 3 to be $\{(n,n)\}$ for $\mathbf{M}_n$, $G_n = \{(n,2),(n,n+1)\}$ for $\mathbf{H}_n$, $Q_n = \{(1,2),(2,1),(n,n)\}$ for $\mathbf{D}_n$ and $\mathbf{T}_n$, the conditions of Theorem 3 are satisfied and thus we get our main theorem.

Theorem 4. *For* $n \geq 4$, $\mathcal{S}^{(n,n)}(\mathbf{M}_n)$, $\mathcal{S}^{G_{n-1}}(\mathbf{H}_{n-1})$, $\mathcal{S}^{Q_n}(\mathbf{D}_n)$ *and* $\mathcal{S}^{Q_{n+1}}(\mathbf{T}_{n+1})$ *are mnf standard binary matrices. In addition,* $\mathcal{S}^{Q_n}(\mathbf{D}_n)$ *and* $\mathcal{S}^{Q_{n+1}}(\mathbf{T}_{n+1})$ *are totally balanced.*

The claim of total balancedness is immediate by Lemma 3 as $\mathbf{D}_4$ is an interval matrix and $\mathbf{T}_5$ is Γ-free [12]. Lubiw observed that $\mathcal{S}^{Q_4}(\mathbf{D}_4)$ and $\mathcal{S}^{Q_5}(\mathbf{D}_5)$ are non-firm [13]. Her observation served as a motivation to us to define the stretching operation and matrices $\mathbf{D}_n$.

For the first two classes, $\mathbf{M}_n$ and $\mathbf{H}_n$, the proofs that Theorem 3's conditions hold are similar because both are minimally non-superfirm. A standard binary matrix is *minimally non-superfirm (mnsf)* if it is not superfirm but all proper submatrices of it are. Next, we show that the conditions hold for the class $\mathbf{H}_n$.

Lemma 7. *For* $n \geq 3$, $\mathbf{H}_n^P$ *is firm for all* $P \subsetneq G_n = \{(n,2),(n,n+1)\}$ *and* $\mathbf{H}_n^{G_n}$ *is a mnf generalised binary matrix. In addition,* $\mathbf{H}_n$ *is mnsf.*

Proof. For $P \subseteq G_n$, at least n rectangles are needed to cover $\mathbf{H}_n^P$ as

$$\mathcal{C}_n := \operatorname{supp}(\mathbf{H}_n) \backslash (\{(i,1) : i \in [n-1]\} \cup G_n), \tag{10}$$

is a $2n-1$-hole in $\mathcal{G}(\mathbf{H}_n^P)$. As $\mathbf{H}_n^P$ only has n rows, $br(\mathbf{H}_n^P) = n$.

Note that submatrix $[1, n-1] \times [2, n+1]$ (where $[\ell, k] := \{\ell, \ell+1, \ldots, k\}$) has two isolated sets of size $n-1$. For $P \subsetneq G_n$, $(i,j) \in G_n \backslash P$ may be added to one of these two isolated sets to get an isolated set of size n for $\mathbf{H}_n^P$. For $\mathbf{H}_n^{G_n}$ however, none of the 1s can be added to these two isolated sets, so we only have $i(\mathbf{H}_n^{G_n}) \geq n-1$. Suppose $\mathbf{H}_n^{G_n}$ has an isolated set T_n of size n. Then T_n needs to contain a 1 from each row, so $(n,1) \in T_n$. But then T_n cannot contain $(1,2)$ and $(n-1, n+1)$, the only 1s in columns 2 and $n+1$, as they are in a rectangle with $(n,1)$. Hence T_n has n elements from $n-1$ distinct columns, which is impossible.

$\mathcal{G}(\mathbf{H}_n)$ has no odd antiholes of size 7 or more by Lemma 4 but it contains n $2n-1$-holes, one of which is $\mathcal{C}_n$. Any other hole in $\mathcal{G}(\mathbf{H}_n)$ is either contained in the submatrix $\mathbf{C}_n$ and hence it is the $2n$-hole, or contains at most two vertices from column 1. Note that if $(\ell, 1)$ is a vertex of a hole then the hole cannot have another vertex from row ℓ. If a hole contains a single vertex from column 1 then it is easy to see that it must be one of the n $2n-1$-holes. If the hole has two vertices from column 1, then it must contain an even number of vertices from submatrix $\mathbf{C}_n$, so it is an even hole. Therefore, the n $2n-1$-holes are the only odd holes in $\mathcal{G}(\mathbf{H}_n)$ which all have a vertex from every row and column. For $P \subseteq G_n$, $\mathcal{G}(\mathbf{Y})$ for any proper submatrix $\mathbf{Y}$ of $\mathbf{H}_n^P$ then has no odd holes and no odd antiholes, so $\mathcal{G}(\mathbf{Y})$ is perfect by the Strong Perfect Graph Theorem [4]. □

Observe that $\mathcal{G}(\mathbf{M}_n)$ contains a single odd-hole of size $2n-1$ as shown in Fig. 3a for $m=4$. To prove that conditions of Theorem 3 are satisfied by class $\mathbf{M}_n$, the same structure of proof as for $\mathbf{H}_n$ may be applied to get the following.

Lemma 8. *For $n \geq 4$, $\mathbf{M}_n$ is firm and mnsf, and $\mathbf{M}_n^{(n,n)}$ is a mnf generalised binary matrix.*

Although $\mathbf{D}_4$ and $\mathbf{T}_5$ are mnsf, for larger n as both $\mathbf{D}_n$ and $\mathbf{T}_n$ are defined recursively, they have proper submatrices which are not superfirm. Hence the argument used in the proof of the previous two classes does not work for $\mathbf{D}_n$ and $\mathbf{T}_n$. Next we prove that class $\mathbf{D}_n$ satisfies the conditions of Theorem 3.

Lemma 9. *For $n \geq 4$, $\mathbf{D}_n^P$ is firm for all $P \subsetneq Q_n = \{(1,2), (2,1), (n,n)\}$ and $\mathbf{D}_n^{Q_n}$ is a mnf generalised binary matrix. In addition, $\mathbf{D}_4$ is mnsf.*

Proof. **I.** For all $P \subseteq Q_n$, $\mathcal{G}(\mathbf{D}_n^P)$ contains the $2n-3$-hole

$$\mathcal{C}_n = \{(3,1), (2,2), (1,3), (4,3), \ldots, (3,n)\}, \tag{11}$$

and thus $br(\mathbf{D}_n^P) \geq n-1$. On the other hand, $\mathbf{D}_n$ has a feasible cover using $n-1$ rectangles in which each row $i \neq 3$ is covered by a distinct rectangle.

II. In $\mathcal{G}(\mathbf{D}_n)$, each $(i,j) \in Q_n$ is adjacent to two consecutive vertices of $\mathcal{C}_n$, and not adjacent to the others. For $P \subsetneq Q_n$, let $(\ell, k) \in Q_n \backslash P$ and S_n be an independent set of $\mathcal{C}_n$ of size $n-2$ which does not use the two vertices of $\mathcal{C}_n$ that are adjacent to (ℓ, k). Then $S_n \cup \{(\ell, k)\}$ is a feasible isolated set of $\mathbf{D}_n^P$.

For $\mathbf{D}_n^{Q_n}$, S_n is a feasible isolated set. Suppose that $\mathbf{D}_n^{Q_n}$ has an isolated set T_n of size $n-1$. Then as $\mathbf{D}_n^{Q_n}$ is of size $n \times n$, there is exactly one row and one column that does not have a 1 in T_n. Since columns 1 and n each have a single 1 which are both in row 3, exactly one of these 1s must be in T_n. (a) Suppose that $(3,1) \in T_n$. Then $(3,j) \notin T_n$ for any $j \neq 1$. Observe that $(2,2)$ can also not be in T_n as it is adjacent to $(3,1)$. But column 2 only has the 1s at $(2,2)$ and $(3,2)$, so T_n contains no 1s from column 2 and n and it has $n-1$ isolated 1s from $n-2$ columns, which is a contradiction. (b) Suppose that $(3,n) \in T_n$. Then $(3,j) \notin T_n$ for any $j \neq n$. As $(3,2) \notin T_n$, we must have the only available 1 at $(2,2)$ from column 2 in T_n. But then as $(2,2)$ is in a rectangle with $(1,3)$, we cannot have $(1,3)$ in T_n. As $(1,3)$ is the only 1 in row 1, T_n has no 1s from row 1. But T_n can also not have any 1s from row n, as row n only has a 1 at $(n, n-1)$ which is adjacent to $(3,n) \in T_n$. Hence T_n has $n-1$ isolated 1s from $n-2$ rows, which is impossible. Therefore, $i(\mathbf{D}_n^{Q_n}) = n-2$.

III. We use induction on n. For the base case take $n=4$ and observe that $\mathcal{G}(\mathbf{D}_4)$ has the 5-hole $\mathcal{C}_4$ as an only odd hole and $\mathcal{C}_4$ contains a vertex from each row and column of $\mathbf{D}_4$. Therefore, any proper submatrix of $\mathbf{D}_4^P$ is superfirm for any $P \subseteq Q_4$. Assume that for $k < n$, all proper submatrices of $\mathbf{D}_k^{P'}$ are firm for any $P' \subseteq Q_k$. Let $P \subseteq Q_n$, and suppose that not every proper submatrix of $\mathbf{D}_n^P$ is firm and let $\mathbf{Y}$ be a smallest non-firm proper submatrix indexed by $I \times J$. Note that we have $n \in I$ or $n \in J$, as otherwise $\mathbf{Y}$ is a submatrix of $\mathbf{D}_k^{P'}$ for some $k < n$ and $P' \subseteq \{(1,2),(2,1)\}$ and firm by either the induction hypothesis or by parts **I.** and **II.** of this proof as $P' \subsetneq Q_k$. By the minimality of $\mathbf{Y}$ it must be mnf. So $\mathbf{Y}$ has at least two non-zero entries in each row and column by Lemma 5. Hence $n \in I$ implies $n-1, n \in J$ and $n \in J$ implies $3, n \in I$. Thus we must have $3, n \in I$ and $n, n-1 \in J$. Similarly, if $i \in I$ for some $i > 3$ then $i-1, i \in J$; if $1 \in I$ then $2, 3 \in J$ and if $1 \in J$ then $2, 3 \in I$.

If $I = [n]$, then by the above we must have $J = [n]\backslash\{1\}$. Then $(3,2) \cup S_n$ and $\{\{1,2,3\} \times \{2,3\}\} \cup \mathcal{R}_n$ with $S_n := \{(i, i-1) : i \in [4,n] := \{4, \ldots, n\}\}$ and $\mathcal{R}_n := \{\{3, i\} \times \{i-1, i\} : i \in [4,n]\}$ give a feasible isolated set and rectangle cover of size $n-2$ of $\mathbf{Y}$, hence we cannot have $I = [n]$.

So let ℓ be the largest row index of $\mathbf{D}_n^P$ for which $\ell \notin I$. (a) If $\ell = 1$, then $I = [n]\backslash\{1\}$. Then $[4,n] \subset I$ implies $[3,n] \subseteq J$, and $2 \in I$ implies that column 1 or 2 are in J, so let $k \in J \cap \{1,2\}$. Then $(3,k) \cup S_n$ and $\{\{2,3\} \times (J \cap \{1,2,3\})\} \cup \mathcal{R}_n$ give a feasible isolated set and rectangle cover of size $n-2$ of $\mathbf{Y}$, so $\ell \neq 1$.

(b) If $\ell = 2$, then we have $1 \notin J$. If $1 \in I$, then $2, 3 \in J$ must hold, so we have $I = [n]\backslash\{2\}$ and $J = [2,n]$. Then $(3,2) \cup S_n$ and $\{\{1,3\} \times \{2,3\}\} \cup \mathcal{R}_n$ give a feasible isolated set and rectangle cover of size $n-2$ of $\mathbf{Y}$. If $1 \notin I$, then $2 \notin J$, so we have $I = [3,n]$ and $J = [3,n]$. Then S_n and $\mathcal{R}$ give a feasible isolated set and rectangle cover of size $n-3$ of $\mathbf{Y}$.

(c) If $\ell > 3$, then $(\ell+1, \ell)$ is a simplicial 1 of $\mathbf{Y}$ and its unique maximal rectangle is $\{3, \ell+1\} \times \{\ell, \ell+1\}$. Remove this simplicial 1 at $(\ell+1, \ell)$. But then $(\ell+2, \ell+1)$ becomes a simplicial 1, so it can also be removed. We may repeat this process until at last $(n, n-1)$ becomes a simplicial 1 and can be removed. Once $(n, n-1)$ is removed, column n only consist of 0s and a single ?, hence can be dropped. Let the resulting matrix be $\mathbf{Y}'$. As dropping a column which does not have any 1s does not impact the isolation number and Boolean rank, by Lemma 1 $\mathbf{Y}'$ satisfies $i(\mathbf{Y}') + n - \ell = i(\mathbf{Y})$ and $br(\mathbf{Y}') + n - \ell = br(\mathbf{Y})$. But then $\mathbf{Y}'$ is just a proper submatrix of $\mathbf{Y}$ formed by rows $(I \cap [\ell-1]) \times (J \cap [\ell-1])$, so firm. Hence $i(\mathbf{Y}) = br(\mathbf{Y})$ which contradicts $\mathbf{Y}$ being mnf. □

A proof which is very similar to the above may be applied to class $\mathbf{T}_n$ to get our final lemma below and by this completing the proof of Theorem 4.

Lemma 10. *For $n \geq 5$, $\mathbf{T}_n^P$ is firm for all $P \subsetneq Q_n = \{(1,2), (2,1), (n,n)\}$ and $\mathbf{T}_n^{Q_n}$ is a mnf generalised binary matrix. In addition, $\mathbf{T}_5$ is mnsf.*

6 Conclusion

In this paper, we studied firm and superfirm binary matrices. We showed that superfirmness is equivalent to having no odd holes in the rectangle cover graph. Then we presented four infinite classes of minimally non-firm binary matrices.

We close with two future research directions. We suspect that every minimally non-superfirm matrix is firm and any minimally non-firm matrix $\mathbf{X} \in \{0,1\}^{m \times n}$ satisfies $|m - n| \leq 1$.

Acknowledgements. I am very grateful to Ahmad Abdi for helping me begin studying firm matrices and for all the invaluable comments during our discussions.

References

1. Amilhastre, J., Vilarem, M., Janssen, P.: Complexity of minimum biclique cover and minimum biclique decomposition for bipartite domino-free graphs. Discrete Appl. Math. **86**(2), 125–144 (1998)
2. Berge, C.: Hypergraphs - Combinatorics of Finite Sets, vol. 45. North-Holland Mathematical Library, North-Holland (1989)
3. de Caen, D., Gregory, D., Pullman, N.J.: The Boolean rank of zero-one matrices. In: Proceedings of 3rd Caribbean Conference on Combinatorics and Computing, pp. 169–173 (1981)
4. Chudnovsky, M., Robertson, N., Seymour, P., Thomas, R.: The strong perfect graph theorem. Ann. Math. **164**, 51–229 (2006)
5. Cunningham, W.H., Edmonds, J.: A combinatorial decomposition theory. Can. J. Math. **32**(3), 734–765 (1980)
6. Dawande, M.: A notion of cross-perfect bipartite graphs. Inf. Process. Lett. **88**(4), 143–147 (2003)
7. Golumbic, M.C.: Algorithmic Graph Theory and Perfect Graphs. Annals of Discrete Mathematics, 2nd edn, vol. 57. Elsevier (2004)
8. Gregory, D.A., Pullman, N.J.: Semiring rank: Boolean rank and nonnegative rank factorisations. J. Comb. Inf. Syst. Sci. **8**(3), 223–233 (1983)

9. Győri, E.: A minimax theorem on intervals. J. Comb. Theory. Ser. B **37**(1), 1–9 (1984)
10. Kim, K.: Boolean matrix theory and applications. In: Monographs and Textbooks in Pure and Applied Mathematics. Dekker (1982)
11. Kushilevitz, E., Nisan, N.: Communication Complexity. Cambridge University Press, New York (1997)
12. Lubiw, A.: Doubly lexical orderings of matrices. SIAM J. Comput. **16**(5), 854–879 (1987)
13. Lubiw, A.: The Boolean basis problem and how to cover some polygons by rectangles. SIAM J. Discrete Math. **3**(1), 98–115 (1990)
14. Müller, H.: Alternating cycle-free matchings. Order **7**(1), 11–21 (1990). https://doi.org/10.1007/BF00383169
15. Müller, H.: On edge perfectness and classes of bipartite graphs. Discrete Math. **149**(1), 159–187 (1996)
16. Orlin, J.: Contentment in graph theory: covering graphs with cliques. Indag. Math. (Proc.) **80**(5), 406–424 (1977)
17. Pulleyblank, W.: Alternating cycle free matchings. Technical report, CORR 82-18, Department of Combinatorics and Optimization, University of Waterloo (1982)

Few Induced Disjoint Paths for H-Free Graphs

Barnaby Martin[1]([✉]), Daniël Paulusma[1], Siani Smith[1],
and Erik Jan van Leeuwen[2]

[1] Department of Computer Science, Durham University, Durham, UK
{barnaby.d.martin,daniel.paulusma,siani.smith}@durham.ac.uk
[2] Department of Information and Computing Sciences, Utrecht University, Utrecht,
The Netherlands
e.j.vanleeuwen@uu.nl

Abstract. Paths $P^1, \ldots, P^k$ in a graph $G = (V, E)$ are mutually
induced if any two distinct P^i and P^j have neither common vertices
nor adjacent vertices. For a fixed integer k, the k-INDUCED DISJOINT
PATHS problem is to decide if a graph G with k pairs of specified vertices (s_i, t_i) contains k mutually induced paths P^i such that each P^i
starts from s_i and ends at t_i. Whereas the non-induced version is well-
known to be polynomial-time solvable for every fixed integer k, a classical
result from the literature states that even 2-INDUCED DISJOINT PATHS is
NP-complete. We prove new complexity results for k-INDUCED DISJOINT
PATHS if the input is restricted to H-free graphs, that is, graphs without
a fixed graph H as an induced subgraph. We compare our results with a
complexity dichotomy for INDUCED DISJOINT PATHS, the variant where
k is part of the input.

Keywords: Induced disjoint paths · H-free graph · Complexity
dichotomy

1 Introduction

We consider problems related to finding paths connecting pre-specified pairs of
vertices. A path between vertices s and t in an undirected graph G is an s-t path
with *terminals* s and t. Terminal pairs $(s_1, t_1), \ldots, (s_k, t_k)$ are *pairwise disjoint* if
$\{s_i, t_i\} \cap \{s_j, t_j\} = \emptyset$ for $i \neq j$. The well-known problem k-DISJOINT PATHS is to
decide for a graph G and pairwise disjoint terminal pairs $(s_1, t_1) \ldots, (s_k, t_k)$, if
there are pairwise vertex-disjoint paths $P^1, \ldots, P^k$ such that P^i is an (s_i, t_i)-path
for $i \in \{1, \ldots, k\}$; here k is *fixed*, that is, k is not part of the input.

Shiloach [18] proved that 2-DISJOINT PATHS is polynomial-time solvable.
Robertson and Seymour [17] even gave a polynomial-time algorithm for k-
DISJOINT PATHS for every integer $k \geq 2$. In contrast, DISJOINT PATHS, the
variant where k is part of the input, appeared on Karp's list of NP-complete
problems.

Our Focus. We consider the *induced* variant of k-DISJOINT PATHS. We say that
paths $P^1, \ldots, P^k$ in a graph $G = (V, E)$ are *mutually induced* if any two distinct

I. Ljubić et al. (Eds.): ISCO 2022, LNCS 13526, pp. 89–101, 2022.
https://doi.org/10.1007/978-3-031-18530-4_7

P^i and P^j have neither common vertices nor adjacent vertices, that is, if $i \neq j$ then $V(P^i) \cap V(P^j) = \emptyset$ and $uv \notin E$ for every $u \in V(P^i)$ and $v \in V(P^j)$. This leads to the following problem, where k is a fixed constant.

---

k-INDUCED DISJOINT PATHS
> *Instance:* a graph G and pairwise disjoint terminal pairs $(s_1, t_1) \ldots$, (s_k, t_k).
> *Question:* Does G have mutually induced paths $P^1, \ldots, P^k$ such that P^i is an s_i-t_i path for $i \in \{1, \ldots, k\}$?

In contrast to the previous setting, even 2-INDUCED DISJOINT PATHS is NP-complete, as shown both by Bienstock [3] and Fellows [4]. Restricting the input to some special graph class might help improve our understanding of the hardness of the problem. To do this systematically we focus on hereditary graph classes.

A class of graphs is *hereditary* if it is closed under vertex deletion. This is a natural property and non-surprisingly hereditary graph classes provide a framework that captures many well-known graph classes. In particular, it is not difficult to see that a graph class $\mathcal{G}$ is hereditary if and only if it can be characterized by a (unique) set $\mathcal{F}_\mathcal{G}$ of forbidden induced subgraphs. For example, if $\mathcal{G}$ is the class of bipartite graphs, then $\mathcal{F}_\mathcal{G}$ is the set of all odd cycles.

The characterization by $\mathcal{F}_\mathcal{G}$ allows for a systematic study, which usually starts with the case where $\mathcal{F}_\mathcal{G}$ has size 1, say $\mathcal{F}_\mathcal{G} = \{H\}$ for some graph H. A graph is *H-free* if it cannot be modified to H by a sequence of vertex deletions, and if $\mathcal{F}_\mathcal{G} = \{H\}$ we obtain the class of *H-free graphs*, which we consider in our paper.

1.1 Related Work

We first discuss existing results for INDUCED DISJOINT PATHS (where k is part of the input). All the positive results hold for a slightly more general problem definition (see Sect. 4). Golovach et al. [7,8] proved that INDUCED DISJOINT PATHS is linear-time solvable for circular-arc graphs and polynomial-time solvable for AT-free graphs, respectively. Belmonte et al. [2] showed the latter for chordal graphs, and Jaffke et al. [9] did so for any graph class of bounded mim-width. In contrast, INDUCED DISJOINT PATHS stays NP-complete even for claw-free graphs [5], line graphs of triangle-free chordless graphs [16] and thus for (theta, wheel)-free graphs, and for planar graphs; to prove the latter, use a result of Lynch [14] (see [8]).

The following recent dichotomy is immediately relevant for our paper. Let $G_1 + G_2$ be the disjoint union of two vertex-disjoint graphs G_1 and G_2, and let sG denote the disjoint union of s copies of a graph G. We write $F \subseteq_i G$ if F is an *induced* subgraph of a graph G, that is, F can be obtained from G by a sequence of vertex deletions. We let P_r denote the path on r vertices. A *linear forest* is the disjoint union of one or more paths.

Theorem 1 ([15]). *For a graph H, INDUCED DISJOINT PATHS on H-free graphs is polynomial-time solvable if $H \subseteq_i sP_3 + P_6$ for some $s \geq 0$; NP-complete if H is not a linear forest; and quasipolynomial-time solvable otherwise.*

We return to Theorem 1 later, and we now fix k. Radovanović et al. [16] proved that k-INDUCED DISJOINT PATHS is polynomial-time solvable for (theta, wheel)-free graphs. Fiala et al. [5] proved the same result for claw-free graphs. Note that both results complement the aforementioned hardness results when k is part of the input. Golovach et al. [6] showed that INDUCED DISJOINT PATHS is even FPT with parameter k for claw-free graphs. The same holds for planar graphs [10], and even for graph classes of bounded genus, as shown by Kobayashi and Kawarabayashi [12]. Let C_r denote the r-vertex cycle. It follows (using Lemma 3) from a result of Leveque et al. [13] that 2-INDUCED DISJOINT PATHS is NP-complete for H-free graphs if $H = C_r$ for every $r \geq 3$ with $r \neq 6$.

The generalization from paths to connected subgraphs joining sets of terminals instead of pairs has also been considered, but these results do not impact upon our work in this paper; we refer to [15] for further details. Moreover, the restriction to H-free graphs has also been studied for DISJOINT PATHS (recall that if k is fixed this problem is polynomial in general [17]); see [11] for a complexity classification of DISJOINT PATHS for H-free graphs, subject to a set of three unknown cases.

1.2 Our Results

To explain our results we first introduce some extra terminology. For $r \geq 1$, the graph $K_{1,r}$ is the $(r+1)$-vertex *star*, i.e., the graph with vertices $x, y_1, \ldots, y_r$ and edges xy_i for $i = 1, \ldots, r$. The graph $K_{1,3}$ is known as the *claw*. The *subdivision* of an edge uw removes uw and replaces it with a new vertex v and edges uv, vw. A *subdivided claw* is a tree with one vertex x of degree 3 and exactly three leaves. For $1 \leq h \leq i \leq j$, let $S_{h,i,j}$ be the subdivided claw whose three leaves are of distance h, i and j from the vertex of degree 3. Note that $S_{1,1,1} = K_{1,3}$. The graph $S_{1,1,2}$ is called the *chair* (or *fork*). Let $\mathcal{S}$ be the set of graphs, each connected component of which is a path or a subdivided claw.

Using the above terminology we can now present our main theorem, which we prove in Sect. 3.

Theorem 2. *Let $k \geq 2$. For a graph H, k-INDUCED DISJOINT PATHS is polynomial-time solvable if H is a subgraph of the disjoint union of a linear forest and a chair, and it is NP-complete if H is not in $\mathcal{S}$.*

Comparing Theorems 1 and 2 shows that the problem becomes tractable for an infinite family of graphs H after fixing k. As the class of claw-free graphs is contained in the class of chair-free graphs, Theorem 2 extends the aforementioned polynomial-time result of Fiala et al. [5] for claw-free graphs. Moreover, the case $H = C_6$ (the 6-vertex cycle) fills a gap in the aforementioned result of Leveque et al. [13]. As we shall explain in Sect. 3, the NP-hardness construction relies on their gadget but also requires significant additional work. Before doing this we first prove the polynomial-time part of Theorem 2 in Sect. 2. In Sect. 4 we summarize our findings and give a number of relevant open problems.

2   Polynomial-Time Algorithms

In this section we prove the polynomial-time part of Theorem 2. We need the following lemma (proof omitted).

Lemma 1. *For every linear forest F, if the k-INDUCED DISJOINT PATHS problem is polynomial-time solvable for H-free graphs for some graph H, then it is so for $(F + H)$-free graphs.*

We need two known results for proving the polynomial part of Theorem 2 in Lemma 2.

Theorem 3 ([5]). *For every $k \geq 2$, k-INDUCED DISJOINT PATHS is polynomial-time solvable for claw-free graphs.*

Theorem 4 ([1]). *If a connected chair-free graph G contains an induced claw and an induced path P on at least eight vertices, then G has a vertex adjacent to all vertices of P.*

Lemma 2. *Let $k \geq 2$. For every linear forest F, k-INDUCED DISJOINT PATHS is polynomial-time solvable for $(F + chair)$-free graphs.*

Proof. By Lemma 1, it remains to consider chair-free graphs. Let (G, T) be an instance of INDUCED DISJOINT PATHS, where G is a chair-free graph on n vertices and $T = \{(s_1, t_1), \ldots, (s_k, t_k)\}$ is a set of terminal pairs. Let $(P^1, \ldots, P^k)$ be a solution for (G, T) (if it exists). We call a path P^i *long* if it has at least eight vertices; else we call it *short*. We first guess which of the paths of a solution for (G, T) will be short. There are 2^k options for doing this, which is a constant number as k is a constant. We will consider each of these options one by one.

Suppose we consider the option where $T' \subseteq T$ is the subset of terminal pairs that will be in short solution paths. Let $|T'| = k' \leq k$. We guess all $O(n^{5k'}) = O(n^{5k})$ options of choosing the inner vertices of the solution paths for the terminal pairs in T'. We discard an option if two of the guessed solution paths contain an edge between them or if a guessed solution path contains a vertex with a neighbour in some $(s_i, t_i) \notin T'$. Otherwise, we continue as follows.

We first delete all vertices of the guessed solution paths and also their neighbours from G. We denote the new instance by (G, T) again and also write $T = \{(s_1, t_1), \ldots, (s_k, t_k)\}$. Assuming our guess was correct, (G, T) only has solutions $(P^1, \ldots, P^k)$ in which each P^i is long. Hence, from G, we can safely remove for every $i \in \{1, \ldots, k\}$, every vertex that is adjacent to both s_i and t_i.

We now check in polynomial time if there are two terminal s_i and t_i that belong to different connected components of the resulting graph G'. If so, then we can discard this branch. Else, we let $(G'_1, T'_1), \ldots, (G'_r, T'_r)$ be the connected components of G', together with the terminal pairs subsets of T they contain. We consider each (G'_j, T'_j) as a separate instance. If G'_j has an induced claw, consider a path P^i in a solution. As P^i must be long, Theorem 4 tells us that G'_j must contain a vertex adjacent to all vertices of P^i. However, by construction,

G'_j contains no vertices adjacent to both s_i and t_i which both belong to P^i, a contradiction. We now check in polynomial time if G'_j is claw-free. If it is not, then we may discard the branch, as just argued. Otherwise, we apply Theorem 3 to check in polynomial time if (G'_j, T'_j) has a solution. If for some (G'_j, T'_j) no solution exists, then we move to the next branch; otherwise, we return a yes-answer.

As the number of branches is polynomial and processing each branch takes polynomial time, the total running time of our algorithm is polynomial.   □

3   Completing the Proof of Theorem 2

We first prove the NP-completeness part of Theorem 2. We base our proof on a hardness result of Leveque et al. [13] for 2-INDUCED CYCLE, which is to decide if a graph has an induced cycle (which may be assumed, without loss of generality, to be a *hole*, that is, on at least four vertices) containing two pre-specified vertices x and y. Namely, we derive the following relation (proof omitted).

Lemma 3. *An instance (G, x, y) of* 2-INDUCED CYCLE*, where x and y have degree 2, can be transformed in polynomial time into an instance of* 2-INDUCED DISJOINT PATHS *on a graph G'. Any vertex that is introduced has degree at most 3 and its incident edges can be subdivided an arbitrary number of times.*

Our first two results (proof details omitted) require a single change to the construction of [13]. The first rectifies a potential issue with the same claim made in [6] by using Lemma 3.

Lemma 4. 2-INDUCED DISJOINT PATHS *is* NP-*complete for $K_{1,4}$-free graphs.*

Lemma 5. *For every $s \geq 3$ with $s \neq 6$,* 2-INDUCED DISJOINT PATHS *is* NP-*complete for C_s-free graphs.*

Our third result requires a significant overhaul of the construction in [13].

3.1   Omitting "H"-Graphs and Six-Vertex Cycles

Let H_1 be the "H"-graph on six vertices formed by an edge joining the middle vertices of two paths on three vertices. For $\ell \geq 2$, let H_ℓ be the graph obtained from H_1 by subdividing the crossing edge (i.e., the edge whose endpoints both have degree 3) $\ell - 1$ times.

We prove that for every $\ell \geq 1$, 2-INDUCED DISJOINT PATHS is NP-complete for (C_6, H_ℓ)-free graphs. To this end, we consider the hardness reduction by Leveque et al. [13] for 2-INDUCED CYCLE in more detail. We very closely follow their notation and the proof of our main Lemma 6 mimics the proof of their Lemma 2.6. We show how their construction can be modified so that it becomes H_ℓ-free for any fixed $\ell \geq 1$ and C_6-free.

Let ϕ be an instance of 3-SATISFIABILITY consisting of m clauses $C_1, \ldots, C_m$ on n variables $z_1, \ldots, z_n$. For each clause C_j of the form $y_{3j-2} \vee y_{3j-1} \vee y_{3j}$ then

y_i, $i \in [3m]$, is a literal from $\{z_1, \ldots, z_n, \overline{z}_1, \ldots, \overline{z}_n\}$. Let $\ell \geq 1$ be given. We will construct a graph G_ϕ^ℓ with two specified vertices x and y of degree 2 so that G_ϕ^ℓ has a hole containing x and y if and only if there is a truth assignment satisfying ϕ.

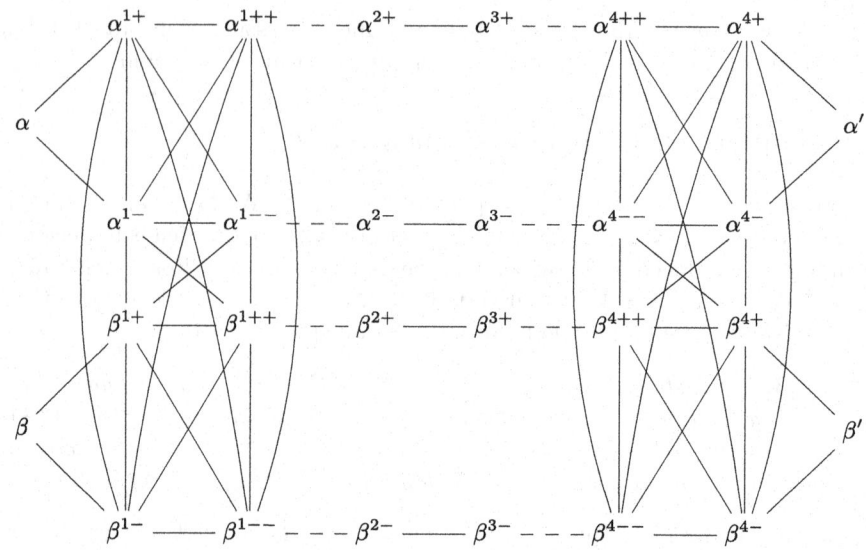

Fig. 1. The literal gadget (dashed lines indicate paths of length ℓ).

For each literal y_j, prepare a graph $G^\ell(y_j)$ as drawn in Fig. 1 where the corresponding labelled vertices inherit a subscript j. Numerous vertices on paths will remain unlabelled. Our literal gadget is more elaborate than that in [13] as we need to forbid, as induced subgraphs, the C_6 and for any fixed ℓ, every H_ℓ.

For each clause C_j, prepare a graph $G^\ell(C_j)$ as drawn in the bottom of Fig. 2 where the corresponding labelled vertices inherit a subscript j. Numerous vertices on paths will remain unlabelled. Our clause gadget is exactly the same as in [13] except we replaced the edges by paths.

For each variable z_i, prepare a graph $G^\ell(z_i)$ as in Fig. 3 consisting of two internally disjoint paths P_i^+ (top) and P_i^- (bottom). The idea in Fig. 3 is that full edges and dashed edges alternate on this diagram and the length is enough for m full edges. The end points of the full edges are labelled $(p_{i,1}^+, p_{i,1}^{++})$, $\ldots$, $(p_{i,2m}^+, p_{i,2m}^{++})$ on the top; and $(p_{i,1}^-, p_{i,1}^{--})$, $\ldots$, $(p_{i,2m}^-, p_{i,2m}^{--})$ on the bottom. Our variable gadget is exactly the same as in [13] except we lengthened some paths.

The final graph G_ϕ^ℓ is constructed in a manner similar to Leveque et al. [13] from the disjoint union of all the graphs $G^\ell(y_j)$ (literals), $G^\ell(C_j)$ (clauses) and $G^\ell(x_i)$ (variables) with the modifications as below. We indicate specifically where the modifications go beyond the construction of Lemma 2.6 in [13]. The top of

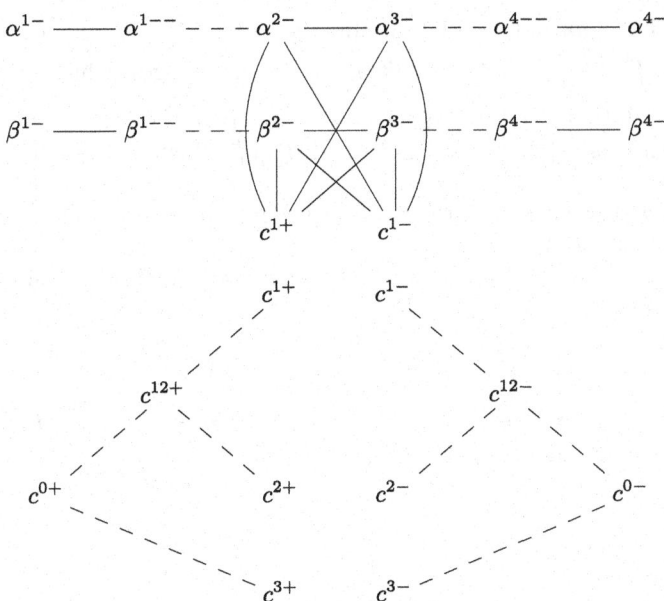

Fig. 2. The clause gadget together with its interface with the literal gadget (drawn above). Dashed lines indicate paths of length ℓ.

Fig. 2 shows how a clause gadget interacts with a literal gadget. Note that a variable gadget interacts with a clause gadget in a similar way.

1. In [13], for $j = 1, \ldots, 3m - 1$, they added the edges $\alpha'_j \alpha_{j+1}$ and $\beta'_j \beta_{j+1}$. We will instead add paths of length ℓ in place of these edges.
2. In [13], for $j = 1, \ldots, m - 1$, they added the edges $c_j^{0-} c_{j+1}^{0+}$. We will instead add paths of length ℓ in place of these edges.
3. In [13], for $i = 1, \ldots, m - 1$, they add the edges $d_i^- d_{i+1}^+$. We will instead add paths of of length ℓ in place of these edges.
4. For $i = 1, \ldots, n$, let $y_{n_1}, \ldots y_{n_{z_i^-}}$ be the occurrences of $\overline{z}_i$ over all literals. We have slightly different vertex names from [13]. For $j = 1, \ldots, z_i^-$, delete the edge $p_{i,j}^+ p_{i,j}^{++}$ and add the four edges $p_{i,j}^+ \alpha_{n_j}^{2+}$, $p_{i,j}^+ \beta_{n_j}^{2+}$, $p_{i,j}^{++} \alpha_{n_j}^{3+}$, $p_{i,j}^{++} \beta_{n_j}^{3+}$. Additionally to these edges, which were in [13], we also add: $p_{i,j}^+ \alpha_{n_j}^{3+}$, $p_{i,j}^+ \beta_{n_j}^{3+}$, $p_{i,j}^{++} \alpha_{n_j}^{2+}$, $p_{i,j}^{++} \beta_{n_j}^{2+}$.
5. For $i = 1, \ldots, n$, let $y_{n_1}, \ldots y_{n_{z_i^+}}$ be the occurrences of z_i over all literals. We have slightly different vertex names from [13]. For $j = 1, \ldots, z_i^+$, delete the edge $p_{i,j}^- p_{i,j}^{--}$ and add the four edges $p_{i,j}^- \alpha_{n_j}^{2+}$, $p_{i,j}^- \beta_{n_j}^{2+}$, $p_{i,j}^{--} \alpha_{n_j}^{3+}$, $p_{i,j}^{--} \beta_{n_j}^{3+}$. Additionally to these edges, which were in [13], we also add: $p_{i,j}^- \alpha_{n_j}^{3+}$, $p_{i,j}^- \beta_{n_j}^{3+}$, $p_{i,j}^{--} \alpha_{n_j}^{2+}$, $p_{i,j}^{--} \beta_{n_j}^{2+}$.

6. For $i = 1, \ldots, m$ and $j = 1, 2, 3$, add the edges $\alpha^{2-}_{3(i-1)+j}c^{j+}_i$, $\alpha^{3-}_{3(i-1)+j}c^{j-}_i$, $\beta^{2-}_{3(i-1)+j}c^{j+}_i$, $\beta^{3-}_{3(i-1)+j}c^{j-}_i$. Additionally to these edges, which were in [13], we also add: $\alpha^{3-}_{3(i-1)+j}c^{j+}_i$, $\alpha^{2-}_{3(i-1)+j}c^{j-}_i$, $\beta^{3-}_{3(i-1)+j}c^{j+}_i$, $\beta^{2-}_{3(i-1)+j}c^{j-}_i$.

7. In [13], they add the edges $\alpha'_{3m}d^+_1$ and $\beta'_{3m}c^{0+}_1$. Instead we will add a path of length ℓ.

8. Add the vertex x. In [13], they add the edges $x\alpha_1$ and $x\beta_1$. Instead we will add paths of length ℓ.

9. Add the vertex y. In [13], they add the edges yc^{0-}_m and yd^-_n. Instead we will add paths of length ℓ.

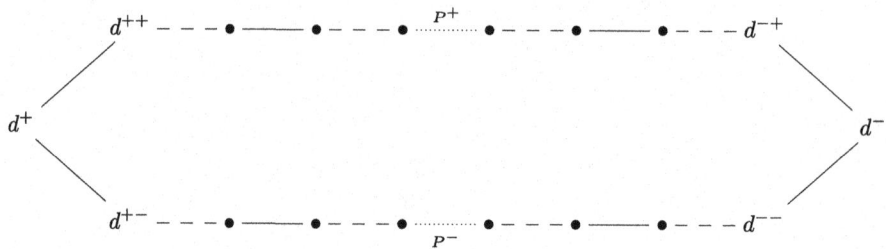

Fig. 3. The variable gadget. Dashed lines indicate paths of length ℓ. Dotted lines indicate a continuation of the gadget.

Claim 1. ϕ *is satisfied by a truth assignment if and only if* G^ℓ_ϕ *contains a hole passing through* x *and* y.

Proof. The idea is that any hole emanating from x and moving rightwards towards y (see Fig. 1) must traverse the literal gadgets in precisely one of two ways (upper path on top and bottom; or bottom path on the top and bottom). Now, subsequently the paths building the hole may return to these literal gadgets but they can never leave them as each $\alpha, \beta, \alpha', \beta'$ are already traversed. Hence, the paths do indeed not return to the literal gadgets to ensure their consistent evaluation with the variables and that one in each clause is true.

More formally, first assume that ϕ is satisfied by a truth assignment $\xi \in \{0,1\}^n$. We pick a set of vertices that induce a hole containing x and y.

1. Pick vertices x and y.
2. For $i = 1, \ldots, 3m$, pick $\alpha_i, \alpha'_i, \beta_i, \beta'_i$.
3. For $i = 1, \ldots, 3m$, if y_i is satisfied by ξ, then pick α^{1+}_i, α^{1++}_i, α^{2+}_i, α^{3+}_i, α^{4++}_i, α^{4+}_i and any vertices on a direct path between these. Else, pick α^{1-}_i, α^{1--}_i, α^{2-}_i, α^{3-}_i, α^{4--}_i, α^{4-}_i and any vertices on a direct path between these.
4. For $i = 1, \ldots, n$, if $\xi(i) = 1$, then pick all vertices of P^+_i and all the neighbours of the vertices in P^+_i of the form α^{2+}_k (or one could choose α^{3+}_k, but only one among the two) for any k. Additionally pick any vertices on a direct path between these.

5. For $i = 1, \ldots, n$, if $\xi(i) = 0$, then pick all the vertices of P_i^- and all the neighbours of the vertices in P_i^- of the form α_k^{2+} (or one could choose α_k^{3+}, but only one among the two) for any k. Additionally pick any vertices on a direct path between these.

6. For $i = 1, \ldots, m$, pick the vertices c_i^{0+} and c_i^{0-}. Choose any $j \in \{3i-2, 3i-1, 3i\}$ such that ξ satisfies y_j. Pick vertices α_j^{2-} and α_j^{3-}.
    - If $j = 3i-2$, then pick $c_i^{12+}, c_i^{1+}, c_i^{1-}, c_i^{12-}$ as well as all vertices on a path between: c^{0+} and c_i^{12+}; c_i^{12+} and c_i^{1+}; c^{0-} and c_i^{12-}; c_i^{12-} and c_i^{1-}.
    - If $j = 3i-1$, then pick $c_i^{12+}, c_i^{2+}, c_i^{2-}, c_i^{12-}$ as well as all vertices on a path between: c^{0+} and c_i^{12+}; c_i^{12+} and c_i^{2+}; c^{0-} and c_i^{12-}; c_i^{12-} and c_i^{2-}.
    - If $j = 3i$, then pick c_i^{3+}, c_i^{3-} as well as all vertices on a path between: c^{0+} and c_i^{3+}; c_i^{0-} and c_i^{3-}.

It suffices to show that the chosen vertices induce a hole in G_ϕ^ℓ containing x and y. The only potential problem is that for some k, one of the vertices $\alpha_k^{2+}, \alpha_k^{3+}, \alpha_k^{2-}, \alpha_k^{3-}$ was chosen more than once. If α_k^{2+} and α_k^{3+} were picked in Step 3, then y_k is satisfied by ξ. Therefore, α_k^{2+} and α_k^{3+} were not chosen in Step 4 or Step 5. Similarly, if α_k^{2-} and α_k^{3-} were picked in Step 6, then y_k is satisfied by ξ. Therefore, α_k^{2-} and α_k^{3-} were not chosen in Step 3. Thus, the chosen vertices induce a hole in G_ϕ^ℓ containing x and y.

Now assume that G_ϕ^ℓ has a hole including x and y. The hole must contain α_1 and β_1 since they are the only neighbours of x. Next, either both α_1^{1+} and β_1^{1+} are in the hole or both α_1^{1-} and β_1^{1-}. W.l.o.g., let α_1^{1+} and β_1^{1+} be in the hole (the same reasoning will apply in the other case). Since α_1^{1-}, β_1^{1-}, α_1^{1--}, β_1^{1--} are all neighbours of two vertices in the hole, they cannot themselves be in the hole. Thus, α_1^{2+}, β_1^{2+}, and the paths that lead to them, must be in the hole. Since α_1^{2+}, β_1^{2+} have the same neighbourhood outside of $G(y_1)$ it follows that α_1^{3+}, β_1^{3+} must be in the hole. Indeed, so must also α_1^{4++}, β_1^{4++}, α_1^{4+}, β_1^{4+} and the path in between. Note that α_1^{4-}, β_1^{4-} are not in the hole, as they are adjacent to both α_1^{4++} and β_1^{4++}. So it must contain instead α_1', β_1', α_2, β_2. By induction, we see that for $i \in [3m]$ that the hole must contain α_i, β_i, α_i', β_i'. Also, for each i, the hole must contain $\alpha_i^{1+}, \alpha_i^{1++}, \ldots, \alpha_i^{2+}, \alpha_i^{3+}, \ldots, \alpha_i^{4++}, \alpha_i^{4+}$ or $\alpha_i^{1-}, \alpha_i^{1--}, \ldots, \alpha_i^{2-}, \alpha_i^{3-}, \ldots, \alpha_i^{4--}, \alpha_i^{4-}$. Hence, the hole contains d_1^+ and c_1^{0+}.

By symmetry we may assume the hole contains d_1^{++}, and the path to $p_{1,1}^+$, and α_k^{2+} for some k. As α_k^{1++} is adjacent to two vertices in the hole, the hole must contain one of α_k^{2+} and α_k^{3+}. Similarly, the hole cannot proceed on a path to α_k^{4++}, so it must contain $p_{1,2}^+$ and $p_{1,2}^{++}$. By induction, we see that the hole contains $p_{1,i}^+, p_{1,i}^{++}$, for $i \in [n]$, and d_1^-. If the hole contains d_1^{--}, then the hole must contain $p_{1,i}^-, p_{1,i}^{--}$, for $i \in [n]$, and eventually d_1^{+-}, a contradiction. Thus, the hole must contain d_2^+. By induction, for $i \in [n]$, the hole contains all the vertices of the path P_i^+ or P_i^- and, by symmetry, we assume that the hole contains neighbours of the vertices in P_i^+ or P_i^-, one among α_k^{2+} and α_k^{3+}, for each k.

Similarly, for $i \in [m]$, it follows that the hole must contain c_i^{0+} and c_i^{0-}. The hole also contains one of the following:

- c_i^{12+}, c_i^{1+}, c_i^{1-}, c_i^{12-}, and the paths between, and either one of α_j^{2-}, α_j^{3-}; or one of β_j^{2-}, β_j^{3-}.
- c_i^{12+}, c_i^{2+}, c_i^{2-}, c_i^{12-}, and the paths between, and either one of α_j^{2-}, α_j^{3-}; or one of β_j^{2-}, β_j^{3-}.
- c_i^{3+}, and the path between, and either one of α_j^{2-}, α_j^{3-}; or one of β_j^{2-}, β_j^{3-}.

For $i \in [n]$, set $\xi(i) = 1$ if the vertices of P_i^+ are in the hole; otherwise set $\xi(i) = 0$. By construction, at least one literal in every clause is satisfied by ξ. □

Claim 2. *The graph G_ϕ^ℓ is C_6-free and H_i-free for every $i \in [\ell]$.*

Proof. Owing to the length of the ℓ paths that populate our construction and are drawn as dashed edges in our figures, we need only verify the omission of the relevant graphs on the connected components of the graph G_ϕ after the removal of these ℓ paths that are dashed edges. That would suffice for C_6, but H_i has a pendant edge, so for these we must leave a pendant edge from the corresponding connected component at the extremities of an instance of these ℓ paths that are drawn as dashed edges. In this fashion, we only need to check for omission of the given graphs in the non-trivial cases drawn in Fig. 4. It can be readily observed that these graphs are P_7 free (but they are not P_6-free). Hence, we need not test beyond H_3. This task was accomplished by a program testing subgraph isomorphism whose code we provide a link to.[1]

We will give an explicit argument for the case of C_6-freeness, which is simpler as C_6 has numerous symmetries (a transitive automorphism group). Let us begin with the graph depicted on the left-hand side of Fig. 4. This graph has an automorphism that swaps α and β at the same time as $+$ and $-$. It also has an automorphism that only swaps $+$ and $-$. Any subgraph that induces a C_6 cannot contain any of the unlabelled vertices, nor α nor β. This leaves eight vertices that may be involved. We will consider the case where the C_6 contains α^{1+}. Owing to the two automorphisms we have described, this argument would equally apply to α^{1-} and β^{1-}. But any C_6 must involve one of these vertices as there were only eight to choose from. Thus, when we have considered this case, our work is done:

Subcase A. The C_6 contains α^{1+} and α^{1++}. All other neighbours of α^{1+} (except α) are adjacent to α^{1++}. No C_6 can be formed here.
Subcase B. The C_6 contains α^{1+} and α^{1--}. Any C_6 involving a path α^{1--} to α^{1+} must next go to β^{1-}. We cannot continue this cycle.
Subcase C. The C_6 contains α^{1+} and β^{1--}. Any C_6 involving a path β^{1--} to α^{1+} must next go to α^{1-} or α^{1--}. We cannot continue this cycle.
Subcase D. The C_6 contains α^{1+} and β^{1-}. Now, β^{1-} can have as the next in the cycle either of β^{1+} or β^{1++}. We cannot continue this cycle.
Subcase E. The C_6 contains α^{1+} and α^{1-}. Now, α^{1-} can have as the next in the cycle either of β^{1+} or β^{1++}. We cannot continue this cycle.

---

[1] See https://github.com/barnabymartin/InducedSubgraph.

Now we consider the graph depicted on the right-hand side of Fig. 4. Any C_6 cannot contain any of the unlabelled vertices. It follows that it must use all six remaining vertices. But this induced graph has a triangle, so we are finished. □

We note that the construction in [13] omits all cycles other than C_6, and they note specifically this lacuna, which we have remedied.

We now prove our result and afterwards we prove Theorem 2.

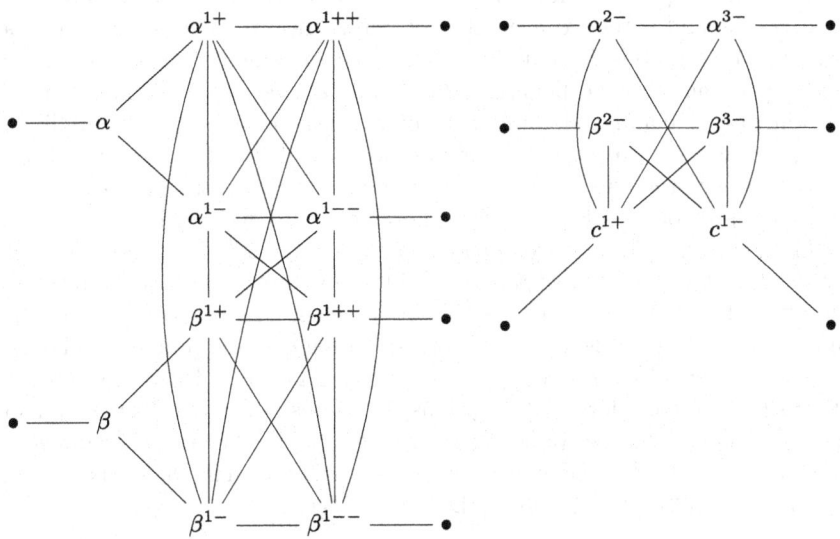

Fig. 4. Cases that need to be checked for omission of the graphs C_6 and H_i ($1 \leq i \leq \ell$).

Lemma 6. *For every integer $\ell \geq 1$, 2-INDUCED DISJOINT PATHS is NP-complete for (C_6, H_ℓ)-free graphs.*

Proof. We give a reduction from an instance ϕ of 3-SATISFIABILITY. First, we construct G^ℓ_ϕ. By Claim 1, G^ℓ_ϕ has a hole through x and y if and only if ϕ is satisfiable. Moreover, G^ℓ_ϕ is (C_6, H_ℓ)-free by Claim 2. We now apply the reduction of Lemma 3. As we can subdivide any number of times the edges incident on the newly created vertices, the resulting graph is still (C_6, H_ℓ)-free.   □

Theorem 2 (restated). *Let $k \geq 2$. For a graph H, k-INDUCED DISJOINT PATHS is polynomial-time solvable if H is a subgraph of the disjoint union of a linear forest and a chair, and it is NP-complete if H is not in $\mathcal{S}$.*

Proof. If H has a cycle C_s, apply Lemma 5 for $s \neq 6$ or Lemma 6 for $s = 6$. Then we may assume H is a forest. If H has a vertex of degree at least 4, then every $K_{1,4}$-free graph is H-free, so apply Lemma 4. Suppose H has maximum degree at most 3. If H has a connected component with at least two vertices of degree 3, then H has an induced H_ℓ, so apply Lemma 6 again. Else, H is in $\mathcal{S}$. If H is a subgraph of the disjoint union of a linear forest and a chair, apply Lemma 2.   □

4   Conclusions

We showed new tractable and hard results for k-INDUCED DISJOINT PATHS for H-free graphs and extended a number of known results in this way. The open cases all involve graphs H that are not the disjoint union of some linear forest and the chair but that do belong to the family $\mathcal{S}$; we recall that $\mathcal{S}$ consists of all graphs, every connected component of which is a path P_r or a subdivided claw $S_{h,i,j}$. In order to make further progress on k-INDUCED DISJOINT PATHS and these other problems we must better understand $S_{h,i,j}$-free graphs.

In some previous works, a slightly more general definition is used (see also Sect. 1). Given a graph G, *vertex-disjoint* paths $P^1, \ldots, P^k$, for some integer $k \geq 1$ are *flexibly mutually induced paths* of G if there is no edge between two vertices from different P^i and P^j except possibly between the endpoints of the paths. If k is in the input, the complexity of the corresponding decision problem and ours is most likely different for P_r-free graphs. Namely, FLEXIBLY INDUCED DISJOINT PATHS is NP-complete for P_{14}-free graphs [15], whilst INDUCED DISJOINT PATH is quasipolynomial-time solvable for P_{14}-free graphs by Theorem 1. However, it is readily seen that all polynomial-time results in Theorem 2 (so, for fixed $k \geq 2$) also hold for FLEXIBLE k-INDUCED DISJOINT PATHS.

References

1. Alekseev, V.E.: Polynomial algorithm for finding the largest independent sets in graphs without forks. Discrete Appl. Math. **135**, 3–16 (2004)
2. Belmonte, R., Golovach, P.A., Heggernes, P., van't Hof, P., Kaminski, M., Paulusma, D.: Detecting fixed patterns in chordal graphs in polynomial time. Algorithmica **69**, 501–521 (2014). https://doi.org/10.1007/s00453-013-9748-5
3. Bienstock, D.: On the complexity of testing for odd holes and induced odd paths. Discrete Math. **90**, 85–92 (1991)
4. Fellows, M.R.: The Robertson-Seymour theorems: a survey of applications. In: Proceedings of AMS-IMS-SIAM Joint Summer Research Conference (1989). Contemp. Math. **89**, 1–18
5. Fiala, J., Kamiński, M., Lidický, B., Paulusma, D.: The k-in-a-path problem for claw-free graphs. Algorithmica **62**, 499–519 (2012). https://doi.org/10.1007/s00453-010-9468-z
6. Golovach, P.A., Paulusma, D., van Leeuwen, E.J.: Induced disjoint paths in claw-free graphs. SIAM J. Discrete Math. **29**, 348–375 (2015)
7. Golovach, P.A., Paulusma, D., van Leeuwen, E.J.: Induced disjoint paths in circular-arc graphs in linear time. Theor. Comput. Sci. **640**, 70–83 (2016)
8. Golovach, P.A., Paulusma, D., van Leeuwen, E.J.: Induced disjoint paths in AT-free graphs. J. Comput. Syst. Sci. **124**, 170–191 (2022)

9. Jaffke, L., Kwon, O., Telle, J.A.: Mim-width I. induced path problems. Discrete Appl. Math. **278**, 153–168 (2020)
10. Kawarabayashi, K., Kobayashi, Y.: A linear time algorithm for the induced disjoint paths problem in planar graphs. J. Comput. Syst. Sci. **78**, 670–680 (2012)
11. Kern, W., Martin, B., Paulusma, D., Smith, S., van Leeuwen, E.J.: Disjoint paths and connected subgraphs for H-free graphs. Theor. Comput. Sci. **898**, 59–68 (2022)
12. Kobayashi, Y., Kawarabayashi, K.: Algorithms for finding an induced cycle in planar graphs and bounded genus graphs. In: Proceedings of the SODA 2009, pp. 1146–1155 (2009)
13. Lévêque, B., Lin, D.Y., Maffray, F., Trotignon, N.: Detecting induced subgraphs. Discrete Appl. Math. **157**, 3540–3551 (2009)
14. Lynch, J.: The equivalence of theorem proving and the interconnection problem. SIGDA Newslett. **5**, 31–36 (1975)
15. Martin, B., Paulusma, D., Smith, S., van Leeuwen, E.J.: Induced disjoint paths and connected subgraphs for H-free graphs. In: Bekos, M.A., Kaufmann, M. (eds.) WG 2022. LNCS, vol. 13453, pp. 398–411. Springer, Cham (2022). https://doi.org/10.1007/978-3-031-15914-5_29
16. Radovanović, M., Trotignon, N., Vušković, K.: The (theta, wheel)-free graphs Part IV: induced paths and cycles. J. Comb. Theory Ser. B **146**, 495–531 (2021)
17. Robertson, N., Seymour, P.D.: Graph minors. XIII. The disjoint paths problem. J. Comb. Theory Ser. B **63**, 65–110 (1995)
18. Shibi, N.: Algorithme de recherche d'un stable de cardinalité maximum dans un graphe sans étoile. Discrete Math. **29**, 53–76 (1980)

On Permuting Some Coordinates of Polytopes

Hans Raj Tiwary[✉]

Faculty of Mathematics and Physics, Charles University, Prague, Czech Republic
hansraj@kam.mff.cuni.cz

Abstract. Let $P \subseteq \mathbb{R}^d$ be a polytope with coordinates labeled $x_1, \ldots,$ x_d. Define $\operatorname{perm}_I(P)$ to be the polytope obtained by taking every permutation σ whose set of fixed-points is $[d] \setminus I$, permuting the coordinates of every point in P according to σ and taking the convex hull of all such points. Also, define $\operatorname{sort}(P)$ to be the polytope obtained by taking each vertex of P in "sorted order".

In this article we study the extension complexity of $\operatorname{perm}_I(P)$ and $\operatorname{sort}(P)$ in terms of the extension complexity of P. A result by Kaibel and Pashkovich states that if $\operatorname{sort}(P) \subseteq P$ and $I = [d]$ then the extension complexity of $\operatorname{perm}_I(P)$ is bounded above by a polynomial of the extension complexity of P. We show that the extension complexity of $\operatorname{perm}_I(P)$ can increase exponentially if $I \neq [d]$ even if the vertices of P contain only three values, say $0, 1$, or 2 at each of the coordinates x_i for $i \in I$. Furthermore, the extension complexity of $\operatorname{sort}(P)$ can be exponentially larger than that of P. We also discuss the implications for the $0/1$ case.

1 Introduction and Motivation

A polytope $P \subseteq \mathbb{R}^d$ is a bounded intersection of finitely many halfspaces. Equivalently, a polytope can also be described as the convex hull of finitely many points. The minimal description as a convex hull is unique for any polytope and the elements of such representations are exactly the vertices of the polytope. We will denote the set of vertices of a polytope P by $\operatorname{vert}(P)$. In what follows we will assume familiarity with polytopes and direct the reader to the excellent textbooks by Ziegler [17] or Grünbaum [9].

Perhaps the most common place, a computer scientist may encounter polytopes is the area of Linear Programming (LP) where the feasible region is often a polytope The number of constraints in a Linear Program has a clear impact on the efficiency of LP solvers and so the number of inequalities needed to describe a polytope can serve as a natural measure of its "size". For a polytope P the size of P – denoted by $\operatorname{size}(P)$ – is defined to be the minimum number of inequalities needed to describe P.

A well known and well studied phenomenon in Linear Programming is that the size of an LP can sometimes be dramatically reduced if new variables are

© The Author(s), under exclusive license to Springer Nature Switzerland AG 2022
I. Ljubić et al. (Eds.): ISCO 2022, LNCS 13526, pp. 102–114, 2022.
https://doi.org/10.1007/978-3-031-18530-4_8

added and connected to the older variables by means of new inequalities (see [3,11] for some surveys). For this method to be effective it suffices for the old feasible region to be a projection of the new feasible region.

An *extension* or an *extended formulation* of a polytope $P \subseteq \mathbb{R}^d$ is another polytope $Q \subseteq \mathbb{R}^{d+r}, r \geqslant 0$ such that P is the projection of Q under a suitably chosen affine map $\pi : \mathbb{R}^{d+r} \to \mathbb{R}^d$. That is, $P = \pi(Q)$. We will use $P \uparrow Q$ or $Q \downarrow P$ as a shorthand for "Q is an extension of P". The *extension complexity* of P – denoted by $\mathrm{xc}(P)$ – is the minimum size of any extension of P. That is, $\mathrm{xc}(P) = \min_{Q \downarrow P} \mathrm{size}(Q)$.

If one is only interested in combinatorial properties like the size or the extension complexity of a polytope, one may always assume that the projection map is actually an orthogonal projection that simply drops some of the coordinates. That is, $P = \mathrm{conv}(\{x \in \mathbb{R}^d \mid \exists y \in \mathbb{R}^r : (x, y) \in Q\})$. For general introduction to extension complexity and related notions we recommend the reader to start with survey by Kaibel [11] and Conforti et al. [3].

Finally, we would like to remark that we will need to speak about both graphs and polytopes at the same time in this paper and the use of terms like "vertex" can be confusing without a lot of extra text. Therefore, we will reserve the word "vertices" for vertices of polytopes and use "nodes" for vertices of graphs.

In this paper we are interested in studying the behavior of extension complexity when a subset of the coordinates of the vertices of a polytope are permuted in all possible ways leaving the rest of the coordinates unchanged, or when every vertex of a polytope is "sorted".

Let $P \subseteq \mathbb{R}^d$ be a polytope with coordinates marked $x_1, \ldots, x_d$. Let $I \subseteq [d]$ and let S_d denote the set of all permutations of the set $[d]$. Define

$$\mathrm{perm}_I(P) = \mathrm{conv}\left(\left\{ \left(x_{\sigma(1)}, \ldots, x_{\sigma(d)}\right) \;\middle|\; \begin{array}{l} (x_1, \ldots, x_d) \in \mathrm{vert}(P), \quad \wedge \\ \sigma \in S_d : \sigma(i) = i \; \forall i \notin I \end{array} \right\}\right),$$

$$\mathrm{sort}(P) = \mathrm{conv}\left(\left\{ \left(x_{\sigma(1)}, \ldots, x_{\sigma(d)}\right) \;\middle|\; \begin{array}{l} (x_1, \ldots, x_d) \in \mathrm{vert}(P), \quad \wedge \\ \sigma \in S_d : x_{\sigma(1)} \leqslant x_{\sigma(2)} \cdots \leqslant x_{\sigma(d)} \end{array} \right\}\right),$$

where $\mathrm{vert}(P)$ denotes the set of vertices of a polytope P.

The case when all the coordinates are permuted, $\mathrm{perm}_{[d]}(P)$, was considered by Kaibel and Pashkovich [12] who showed that permuting all coordinates does not blow up the extension complexity assuming that $\mathrm{sort}(P) \subseteq P$. This leads to a few natural questions:

Question 1. Can one drop the requirement that $\mathrm{sort}(P) \subseteq P$ without causing a large blowup in the extension complexity of $\mathrm{perm}_{[d]}(P)$?

Question 2. How does the extension complexity change if we "sort the coordinate values" of every vertex? (Perhaps to satisfy the requirements of Kaibel and Pashkovich in case it is not already satisfied for a polytope.)

Question 3. How does the extension complexity change if one only permutes a subset of the coordinates?

Apart from being natural questions arising from the result of Kaibel and Pashkovich, there is another motivation for studying the last two questions. Partial permutation of a polytope, that is, $\text{perm}_I(P)$ provides a way of applying parity constraints efficiently over a subset of binary-valued coordinates I of P. We discuss this more precisely in Sect. 3.

2   (More) Background and Related Work

Polytopes and polyehdra are widely studied objects in mathematics as well as computer science, particularly in combinatorial optimization. We refer the reader to any of the books by Grünbaum et al. [9], Ziegler [17], or Grötschel, Lovász and Schrijver [8] for background on polytopes.

Extended formulations have also been used for a long time in combinatorial optimization to construct small linear programs by means of extra variables. We recommend the survey articles by Kaibel [11] and by Conforti, Cornuéjols and Zambelli [3]. Lower bounds in the area are relatively new but have been studied quite extensively in recent years. The article that is most relevant to the present one is by Kaibel and Pashkovich [12] where they study the extension complexity when building new polytopes from a given polytope by means of a large class of operations that they call reflection relations. One particular result in their work is the proof of the following:

Theorem 1 ([12]). *Let $P \subseteq \mathbb{R}^d$ be a polytope such that $\text{sort}(P) \subseteq P$. Then, $\text{xc}(\text{perm}_{[d]}(P))$ is bounded by $\text{xc}(P)$ plus an additive polynomial term.*

Even though Theorem 1 will not be used in this paper, most of the work here was motivated by the question of whether it can be improved by allowing partial permutations, that is, permutations over arbitrary subsets of coordinates, or by removing the condition that $\text{sort}(P) \subseteq P$. Unfortunately, the answer to such questions appears to be negative. We prove some of the concrete negative results in Sect. 3 and in Sect. 4 we mention an open case that would still allow proving other useful results.

Two results related to extended formulations will be used crucially in this paper. The first is the following lemma by Balas.

Lemma 1 ([2]). *Let $P_1, \ldots, P_n$ be polytopes, then $\text{xc}\left(\bigcup_{i=1}^n P_i\right) \leqslant \sum_{i=1}^n \text{xc}(P_i) + \mathcal{O}(n)$.*

The second result that we will use concerns the *glued product* of two polytopes. Let $P \subseteq \mathbb{R}^{d_1}$ and $Q \subseteq \mathbb{R}^{d_2}$ be two polytopes, and let I, J be two ordered sets with $|I| = |J| = k$. That is, $I = \{a_1, \ldots, a_k\}, J = \{b_1, \ldots, b_k\}$. The polytope obtained by *gluing* coordinates I of P with the coordinates J of Q – denoted by $P_I \times_J Q$ – is defined as $\text{conv}(\{(\mathbf{u}, \mathbf{v}) \mid \mathbf{u} \in \text{vert}(P), \mathbf{v} \in \text{vert}(Q), \mathbf{u}_{a_i} = \mathbf{v}_{b_i} \; \forall i \in [k]\})$.

Glued products were studied by Margot [15] who gave a complete characterization of when the glued product can be obtained by simply taking the product of the two polytopes and adding equations enforcing $\mathbf{x}_{a_i} = \mathbf{y}_{b_i}$ for $\mathbf{x} \in P$, $\mathbf{y} \in Q$.

Later Conforti and Pashkovich [4] studied the same problem and gave an equivalent characterization. Here we state a much simpler special case that we will use and which appeared in [13].

Lemma 2. *Let $P \subseteq \mathbb{R}^{d_1}$ and $Q \subseteq \mathbb{R}^{d_2}$ be two polytopes, and let I, J be two ordered sets with $|I| = |J|$. If the projection of P onto coordinates indexed by I, and that of Q indexed by J are simplistic, then $\mathrm{xc}(P_I \times_J Q) \leqslant \mathrm{xc}(P) + \mathrm{xc}(Q)$.*

In particular, Lemma 2 ensures that two polytopes can be glued freely along two coordinates where the entries in any vertex only take binary values.

We will also use the following facts.

Fact 1. Let Q be an extension of P. Then, $\mathrm{xc}(P) \leqslant \mathrm{xc}(Q)$.

Fact 2. Let P be a face of Q. Then, $\mathrm{xc}(P) \leqslant \mathrm{xc}(Q)$.

Fact 3. Let P and Q be polytopes. Then, $\mathrm{xc}(P \times Q) \leqslant \mathrm{xc}(P) + \mathrm{xc}(Q)$, where $P \times Q$ is the cartesian product of P and Q.

2.1 Relevant Polytopes

Matchings. Let $G = (V, E)$ be a graph with $|V| = n, |E| = m$, and let k be a natural number. Three natural polytopes related to matching in G can be defined to be the convex hull of all perfect matchings, all matchings, and all matchings with k edges – denoted by $\mathrm{PMP}(G), \mathrm{MP}(G)$, and $\mathrm{CMP}(G, k)$ respectively.

While linear optimization over each of these polytopes can be done efficiently by Edmond's algorithm, the extension complexities are exponential in n for general graphs [16]. However when G is bipartite, these polytopes have polynomial extension complexity. The linear inequalities describing $\mathrm{PMP}(K_{n,n})$ and $\mathrm{MP}(K_{n,n})$ can be found in any textbook about combinatorial optimization. However, the author could not find a published source that $\mathrm{CMP}(K_{n,n}, k)$ has polynomial extension complexity. We therefore state it here with a simple proof.

Theorem 2. *Let k, n be positive integers. Then, $\mathrm{xc}(\mathrm{CMP}(K_{n,n}, k)) \leqslant \mathcal{O}(n^2)$.*

Proof. Consider the bipartite graph $G = (V, E)$ obtained by adding $n - k$ new nodes to each of the parts of $K_{n,n}$ and adding all possible edges except between the newly introduced nodes. We can observe that $\mathrm{CMP}(K_{n,n}, k)$ is obtained by projecting out the edges $\{i, j\}$ with either $i > n$ or $j > n$ – or equivalently projecting onto edge $\{i, j\}$ with $i, j \leqslant n$ – from $\mathrm{PMP}(G)$. $\square$

Independent Sets. Let $G = (V, E)$ be a graph. The Stable Set Polytope of G – denoted by $\mathrm{STAB}(G)$ – is the convex hull of the characteristic vectors of independent sets of G.

Stable set polytopes have super-polynomial extension complexity in general. This was first shown by Fiorini et al. [5] who proved the existence of n-node graphs whose stable set polytope had extension complexity $2^{\Omega(\sqrt{n})}$. Avis and the present author later showed that there exist n-node 3-regular graphs whose stable set polytope has extension complexity $2^{\Omega(\sqrt[4]{n})}$ [1]. Göös et al. later proved the existence of an n-node graph with linear number of edges whose stable set polytope has extension complexity $2^{\Omega(\frac{n}{\log n})}$ [7].

Here we will use the result of Avis and the present author due to small degree of every node in the resulting graph.

Theorem 3 ([1]). *For every natural number n there exists a 3-regular n-node graph G_n such that* $\mathrm{xc}(\mathrm{STAB}(G_n)) \geqslant 2^{\Omega(\sqrt[4]{n})}$.

Cliques. Let $G = (V, E)$ be a graph with $|V| = n$, and let k be a natural number. Define $\mathrm{CLIQUE}(G)$ to be the convex hull of (the characteristic vector of) all subsets of edges of G that define a clique in G, and similarly $\mathrm{CLIQUE}(G, k)$ to be the convex hull of all cliques with $\binom{k}{2}$ edges.

We will use the following lower bound for $\mathrm{CLIQUE}(K_n)$ by Fiorini et al. [5].

Theorem 4. $\mathrm{xc}(\mathrm{CLIQUE}(K_n)) \geqslant 2^{\Omega(n)}$.

The author could not find any published lower bound for $\mathrm{CLIQUE}(G, k)$ and so we state it here with a proof.

Theorem 5. *For every natural number $2n$,* $\mathrm{xc}(\mathrm{CLIQUE}(K_{2n}, n)) \geqslant 2^{\Omega(n)}$.

Proof. Recall that $\mathrm{CLIQUE}(G)$ is the convex hull of all edges of all induced cliques of G and $\mathrm{CLIQUE}(G, k)$ is the convex hull of induced cliques in G that have k nodes. Now notice that projecting out the coordinates corresponding to edges i, j with $i, j \leqslant n + 1$ in $\mathrm{CLIQUE}(K_{2n}, n)$ produces exactly $\mathrm{CLIQUE}(K_n)$. The theorem follows.   □

Balanced Complete Bipartite Subgraphs. Let $G = (V, E)$ be a graph and let k be a natural number. Define $\mathrm{BCB}(G, k)$ to be the convex hull of all balanced complete bipartite subgraphs of G with k nodes on each side.

Linear optimization over this polytope is NP-hard (stated in Garey and Johnson [6] and proved later in Johnson [10]). In Sect. 3, we will use this NP-hardness to prove an exponential lower bound on the extension complexity of these polytopes and at the same time show how to construct an extension using permutations of some of the coordinates of a polytope with polynomial size.

3   Results

We begin by describing how partial permutations of a polytope P may be used to obtain small extension for the convex hull of those vertices of P that satisfy parity requirements over a chosen subset of coordinates. As stated in the introductory sections, this was another motivation why we considered the extension complexity of $\mathrm{perm}_I(P)$ for arbitrary subsets of coordinates to be interesting.

3.1   Parity Constraints via Partial Permutations

Let $P \subseteq \mathbb{R}^d$ be a polytope, and let $I \subseteq [d]$ be a set of indices such that for every vertex $\mathbf{v} \in \text{vert}(P)$, $\mathbf{v}_i \in \{0,1\}$ for all $i \in I$. That is, P attains only binary values of either zero or one in every vertex at the selected coordinates. Consider the polytope obtained from the convex hull of exactly those vertices of P whose coordinate values at indices I have a given parity, say, odd. More formally, $\text{parity}_I(P) = \text{conv}\left(\{v \in \text{vert}(P) \mid \bigoplus_{i \in I} v_i = 1\}\right)$, where $\oplus$ indicates the xor. One can easily show the following:

Theorem 6. *Let $P \subseteq \mathbb{R}^d$ be a polytope and let $I \subseteq [d]$ be such that vertices of P attain only values in $\{0,1\}$ at coordinates I. Then, $\text{xc}(\text{parity}_I(P)) \leqslant \mathcal{O}(|I| \cdot \text{xc}(\text{perm}_I(P)))$.*

Proof. Let $k = |I|$. Without loss of generality we may assume that the first k coordinates have been picked by I, that is, $I = [k]$. Consider the polytope Q obtained by embedding P in $\mathbb{R}^{d+k}$ with $x_{d+i} = x_i$. That is, $Q = \{(x_1, \ldots, x_d, x_1, \ldots, x_k) \mid (x_1, \ldots, x_d) \in P\}$.

Let $J = \{d+1, \ldots, d+k\}$, and consider the face R_i of $\text{perm}_J(Q)$ containing points x with $x_j = 1$ for $d+1 \leqslant j \leqslant d+i$ and $x_j = 0$ for $d+i+1 \leqslant j \leqslant d+k$. It is easy to see that $v = (v_1, \ldots, v_{d+k})$ is a vertex of R_i if and only if $(v_1, \ldots, v_k)$ is a vertex of P with exactly i ones among the coordinate values $v_1, \ldots, v_k$.

Therefore, $\text{parity}_I(P)$ is affinely isomorphic to the convex hull of the union $\bigcup_{i \text{ odd}} R_i$. and so using Lemma 1 we get that $\text{xc}(\text{parity}_I(P)) \leqslant \sum_{i \in I} \text{xc}(R_i) + \mathcal{O}(|I|)$. Since R_i forms a face of $\text{perm}_J(Q)$, and $\text{perm}_J(Q)$ is affinely isomorphic to $\text{perm}_I(Q)$, we get that $\text{xc}(\text{parity}_I(P)) \leqslant \sum_{i \in I} \text{xc}(\text{perm}_I(P)) + \mathcal{O}(|I|)$ and the claim follows. □

It should be easy to see that the required parity need not be odd here, and even parity can be subjected to the same analysis. Furthermore, if instead of selecting vertices satisfying a given parity condition, we were required to append an extra coordinate to P specifying whether a vertex has odd of even parity at the chosen locations, one can just separate the odd and even parity vertices, append the right value, and take the union again.

Parity constraints are an important class of constrains in combinatorial optimization and so in light of previous theorem, a polynomial bound on $\text{xc}(\text{perm}_I(P))$ in terms of $\text{xc}(P)$ would allow us to handle such constrains in an efficient way.

Another usefulness of an efficient bound on $\text{xc}(\text{perm}_I(P))$ would be in obtaining better extended formulations for cut polytopes of bounded genus graphs. The cut polytope of a graph G with genus g can be obtained by applying a number of parity constraints to the perfect matching polytope of another graph G' with the same genus, and the number of parity constraints depends only on the genus g [14]. Thus if it were possible to permute only some coordinates of arbitrary 0/1-polytopes, this would yield polynomial extensions for the cut polytope of bounded genus graphs. Currently polynomial formulations are unknown even for toroidal graphs.

Naturally, for the purposes of adding parity constraints we would only be interested in polytopes where the interesting coordinates do not attain any values other than zero or one. But if we forget this particular utility of our quest and ask whether partial permutations of arbitrary polytopes preserve extension complexity (up to polynomial blow up), then we find out that the answer turns out to be negative. We will see this in the next subsections.

3.2   Partial Permutation over Quad-Valued Coordinates

We first show that for a polytope $P \subseteq \mathbb{R}^d$ even if the vertices of P contain only values from $\{0, 1, 2, 3\}$ at coordinates x_i for $i \in I, I \subseteq [d]$ then $\mathrm{xc}(\mathrm{perm}_I(P))$ can have super-polynomial extension complexity in terms of $\mathrm{xc}(P)$. We do this using the fact that extension complexity of the Stable Set Polytope of an arbitrary 3-regular graph on n vertices can be $2^{\Omega(\sqrt[4]{n})}$ (Theorem 3). Later on we will strengthen this result for values $0, 1, 2$ by choosing a different problem. We do this because the independent set problem is fairly well-known and so we believe that the reader can focus on the main part of the argument. The strengthening to three values requires essentially the same argument but with a different problem and more involved reduction steps.

Theorem 7. *For each natural number n, there exists a polytope $P \subset \mathbb{R}^{10n}$ and $I \subset [7n]$ with $|I| = n$ such that $\mathrm{xc}(P) \leqslant 9n$ but $\mathrm{xc}(\mathrm{perm}_I(P)) \geqslant \frac{2^{\Omega(\sqrt[4]{n})}}{n+1} - 1$.*

Proof. Let $G = (V, E)$ be a 3-regular graph with $|V| = n$. For each edge $e = v_i, v_j \in E$ take the polytope $P_e = \mathrm{conv}(\{(0, 0), (0, 1), (1, 0)\})$. We will name these coordinates x_i^e and x_j^e respectively and the values indicate whether nodes v_i and v_j are picked to be in the subset or not. Since $e \in E$, we do not include both the nodes at the same time as that would make the choice contradictory to the requirement of being an independent set.

Now take the product $Q = \prod_{e \in E} P_e$. Due to Fact 3 we have that $\mathrm{xc}(Q) \leqslant 9n$. For every node $v_i \in V$ that is incident to, say, $e_1, e_2, e_3 \in E$, consider the values at coordinates $x_i^{e_1}, x_i^{e_2}, x_i^{e_3}$. Interpreting the 0/1 values at these coordinates as whether or not the node v_i is picked in the corresponding subset, we can observe that if $x_i^{e_1} = x_i^{e_2} = x_i^{e_3}$ then these choices are consistent. We would like to remove all vertices of Q that represent an inconsistent choice for any node. To this end, we introduce new coordinates $y_i = \sum_{v_i \in e} x_i^e$ getting a new polytope that is just an affine embedding of Q in $\mathbb{R}^{10n}$ thus the extension complexity remains unchanged. We will name this polytope Q'.

Finally, let I be the subset of coordinate indices of Q' that correspond to coordinates y_i's. Without loss of generality, for example by reordering the coordinates, we may assume that $I = [n]$. The possible values at any of these coordinates in any vertex of Q' are $0, 1, 2$, or 3. Consider the polytope $\mathrm{perm}_I(Q')$ and its face F where $y \in \{0, 3\}^n$ and furthermore $y_1 \leqslant \cdots \leqslant y_n$. We will now show that $\mathrm{xc}(F) \leqslant (n + 1)\,\mathrm{xc}(\mathrm{perm}_I(Q')) + 1$ and furthermore F is an extended formulation for the stable set polytope of G. This will conclude the proof.

To bound the extension complexity of F, let F_j be the face of Q' where $y_1 = \cdots = y_j = 0$ and $y_{j+1} = \cdots = y_n = 3$. Then $F = \bigcup_{i=0}^{n} F_i$. Since $\operatorname{xc}(F_j) \leqslant \operatorname{xc}(Q')$, applying Lemma 1 we get that $\operatorname{xc}(F) \leqslant (n+1) \operatorname{xc}(\operatorname{perm}_I(Q')) + 1$.

To see why F forms an extended formulation of the stable set polytope of G, we note the following. In any vertex w of F, for any node $v_i \in V$ of G, $y_i \in \{0,3\}$. This means that x_i^e is the same for all edges e incident to v_i. Thus, w represents a consistent choice of node v_i whether or not it is picked in the corresponding subset and w restricted to x_i^e coordinates for an arbitrary $e \ni v_i$ for each v_i, gives the characteristic vector of a stable set of G. Conversely, if $S \subseteq V$ forms a stable set of G then there exists a vertex w of F such that $y_i(w) = 0$ if $v_i \notin S$, $y_i(w) = 3$ if $v_i \in S$, and $x_i^e = y_i/3 = 1$ for each e incident to v_i. □

3.3   Partial Permutation over Three-Valued Coordinates

The basic argument in the previous subsection was that permuting an arbitrary subset of coordinates allows one to isolate vertices where each of the coordinates contains only the minimum or the maximum possible value. Forcing these minimum and maximum values to represent some binary choice - such as whether or not to pick a graph node to satisfy some global requirement - allows one to construct extended formulations for complicated polytopes. We will use the same argument now but construct an extended formulation for the convex hull of all balanced bi-cliques in a graph.

The Polytope of All Balanced Bi-Cliques. Let $G = (V_1 \dot\cup V_2, E)$ be a connected bipartite graph and let k be a natural number. For any $E' \subseteq E$, let $V(E')$ denote the set of nodes incident to edges in E'. That is, $v \in V(E')$ if and only if there exists $e \in E'$ and $v \in e$. Recall from Subsect. 2.1 that $\mathrm{BCB}(G, k)$ is the convex hull of the characteristic vectors of $E' \subseteq E$ such that $(V(E'), E')$ is a complete bipartite subgraph of G and $|V(E') \cap V_1| = |V(E') \cap V_2| = k$.

We will first prove that for complete bipartite graphs the polytope of all balanced bi-cliques, with half of the nodes from each side, has super-polynomial extension complexity (Theorem 9).

Theorem 8. *Let $G = (V, E)$ be a graph with $|V| = n, |E| = m$ and n even. Then, there exists a bipartite graph G' such that G' has $m + \frac{n(n+2)}{8}$ nodes, and $\mathrm{BCB}(G', \frac{n(n-2)}{8})$ is an extension of $\mathrm{CLIQUE}\left(G, \frac{n}{2}\right)$.*

Proof. We will use the reduction from [10] that establishes NP-completeness of deciding whether an input graph contains a complete bipartite subgraph with k nodes on each side.

Let $k = \frac{n}{2}$ and $k' = \binom{k}{2} = \frac{n(n-2)}{8}$. So $k' - k = \frac{n(n-6)}{8}$. Now construct a graph $G' = (V', E')$ as follows: $V' = V \cup E \cup W, E' = \{\{e, w\} \mid e \in E, w \in W\} \cup \{\{e, v\} \mid e \in E, v \notin e\}$, where W is a set of $k' - k$ new elements. G' has $n + m + k' - k = m + \frac{n(n+2)}{8}$ nodes and it is a bipartite graph with E being one of the parts and $V \cup W$ the other part.

Now we show that $\mathrm{BCB}(G', k')$ contains a face which is an extension of $\mathrm{CLIQUE}(G, k)$.

Let $\mathbf{v}$ be a vertex of $\mathrm{BCB}(G', k')$. That is, $\mathbf{v}$ is the characteristic vector of a complete bipartite subgraph of G' with k' nodes of each side. For notational simplicity we will use $\mathbf{v}$ to denote both the vertex of the polytope as well as the corresponding subgraph of G'.

The subgraph $\mathbf{v}$ contains k' nodes from each part, in particular from the part $V \cup W$. Since W has only $k' - k$ nodes, $\mathbf{v}$ contains at least k nodes from V. This means that $\sum_{v \in V} \sum_{e \not\ni v} x_{\{e,v\}} \geqslant k \cdot k'$ is a valid inequality for $\mathrm{BCB}(G', k')$ defining face F which consists exactly of balanced complete bipartite subgraphs of G' that contain k nodes from V and $k' - k$ nodes from W. That is, $F := \mathrm{conv}\left(\{\mathbf{v} \in \mathrm{vert}(\mathrm{BCB}(G', k')) \mid \mathbf{v} \text{ has exactly } k \text{ nodes from } V\}\right)$ is a face of $\mathrm{BCB}(G', k')$.

To complete the proof, consider any vertex $\mathbf{v} \in F$. Let V_B be the set of nodes from V that are in the subgraph $\mathbf{v}$. The subgraph $\mathbf{v}$ also contains exactly k' nodes, say E_B, from E. Nodes in E_B are edges of G and these edges must be incident in G only to nodes in $V \setminus V_B$. Since, $|V_B| = k = |V|/2$ this means that $(V \setminus V_B, E_B)$ is a clique of size k in G. Therefore the linear map defined as $y_e = \frac{1}{k} \sum_{v \in V \setminus e} x_{\{e,v\}}$ for all $e \in E$ produces a vertex of $\mathrm{CLIQUE}(G, k)$.

Conversely, let $\mathbf{w}$ be a vertex of $\mathrm{CLIQUE}(G, k)$ with node set V_C of size k and edge set E_C. Then the subgraph of G' induced by $(V \setminus V_C) \cup E_C \cup W$ is a vertex of $\mathrm{BCB}(G', k')$ lying in F that projects to $\mathbf{w}$ under the projection described previously. $\qquad\square$

Theorem 9. *Let $m = \binom{n}{2}$ and $k = \frac{n(n-2)}{8}$, then $\mathrm{xc}(\mathrm{BCB}(K_{m,m}, k)) \geqslant 2^{\Omega(n)}$.*

Proof. Using K_n as G in Theorem 8 we construct bipartite G' with two parts of sizes m and $\frac{n(n+2)}{8}$ such that $\mathrm{BCB}(G', n(n-2)/8)$ contains an extension of $\mathrm{CLIQUE}(K_n, n/2)$ as a face. Therefore, $\mathrm{xc}(\mathrm{BCB}(G', n(n-2)/8)) \geqslant 2^n$. For $n \geqslant 2$ we have that $m = \binom{n}{2} \geqslant \frac{n(n+2)}{8}$. So consider the complete bipartite graph $K_{m,m}$. The polytope $\mathrm{BCB}(G', n(n-2)/8)$ is then a face of $\mathrm{BCB}(K_{m,m}, n(n-2)/8)$ obtained by setting $x_{\{e,v\}} = 0$ for all $e \notin E(G')$. This completes the proof. $\quad\square$

Now we argue that even though $\mathrm{BCB}(K_{m,m}, k)$ has large extension complexity, an extension for it can be built starting from the perfect matching polytope of bipartite graphs if we are able to perform partial permutations over tri-valued coordinates of the intermediate polytopes.

Building. $\mathrm{BCB}(K_{n,n}, k)$ from Bipartite matchings by partial permutations. Let $G = (V, E)$ be a graph. Define the polytope $\mathrm{EMP}(G, 2k)$ as the convex hull of vectors $(\mathbf{v}, \mathbf{m}, \mathbf{c}) \in \{0, 1\}^{|V|+|E|+|E|}$ such that $\mathbf{v}$ is the characteristic vector of a subset, say W, of nodes V with $|W| = 2k$, $\mathbf{m}$ is the characteristic vector of a matching where the set of matched nodes is exactly W, and $\mathbf{c}_{i,j} \leqslant |W \cap \{i, j\}|$ for all edges $\{i, j\} \in E$. Then, we have the following.

Lemma 3. $\mathrm{xc}(\mathrm{EMP}(G, 2k)) \leqslant \mathrm{xc}(\mathrm{CMP}(G, k)) + 6|E|$.

Proof. In CMP(G, k) we can add one coordinate x_v for each node $v \in V$ defined by the linear map $x_v = \sum_{e \ni v} x_e$. This creates the convex hull of all vectors $(\mathbf{v}, \mathbf{m})$ where $\mathbf{v}$ is a subset of $2k$ nodes that are matched in matching $\mathbf{m}$. Let us call this polytope $P_{E'}$ where $E' = \emptyset$ for now. Note that $\mathrm{xc}(P_\emptyset) = \mathrm{xc}(\mathrm{CMP}(G, k))$.

Now for each $e = \{i, j\} \in E \setminus E'$ we modify $P_{E'}$ to $P_{E' \cup \{e\}}$ as follows: $P_{E' \cup \{e\}} = \mathrm{conv}(\{(\mathbf{v}, \mathbf{m}, \mathbf{c}, \alpha_e, \beta_e, c_e) \mid (\mathbf{v}, \mathbf{m}, \mathbf{c}) \in \mathrm{vert}(P_{E'}), \alpha \in \{0, v_i\}, \beta \in \{0, v_j\}, c_e = \alpha_e + \beta_e\})$. This can be achieved by gluing the first coordinate of $Q = \mathrm{conv}(\{(0, 0), (1, 0), (1, 1)\})$ with the coordinate $\mathbf{x}_i$ of $P_{E'}$ and then again doing the same with a second copy of Q with the coordinate $\mathbf{x}_j$ of $P_{E'}$, and finally adding a coordinate $c_e = \alpha_e + \beta_e$. We then add e to set E' and repeat.

Since we are gluing polytopes over coordinates where every vertex has only binary values, using Lemma 2 we get that $\mathrm{xc}(P_{E' \cup \{e\}}) \leqslant \mathrm{xc}(P_{E'}) + 6$. Finally, we note that EMP$(G, 2k)$ is obtained by projecting out the coordinates α_e, β_e for all edges $e \in E$ from P_E. Thus $\mathrm{xc}(\mathrm{EMP}(G, 2k)) \leqslant \mathrm{xc}(P_E) \leqslant \mathrm{xc}(P_\emptyset) + 6|E| = \mathrm{xc}(\mathrm{CMP}(G, k)) + 6|E|$. □

Next we show how to modify EMP$(K_{n,n}, k)$ by copying some coordinates, applying partial permutation over those coordinates, and taking out of a face of the result to obtain an extension of BCB$(K_{n,n}, k)$.

Theorem 10. *Let n, k be natural numbers. There exists a polytope $P \subseteq \mathbb{R}^{3n^2 + 2n}$ and $I \subset [3n^2 + 2n]$ such that $\mathrm{xc}(P) = \mathrm{xc}(\mathrm{EMP}(K_{n,n}, 2k))$, and $\mathrm{xc}(\mathrm{BCB}(K_{n,n}, k)) \leqslant \mathrm{xc}(\mathrm{perm}_I(P))$.*

Proof. Recall that EMP$(G, 2k)$ is the convex hull of vectors $(\mathbf{v}, \mathbf{m}, \mathbf{c}) \in \{0, 1\}^{|V|} \times \{0, 1\}^{|E|} \times \{0, 1, 2\}^{|E|}$ such that $\mathbf{v}$ is the characteristic vector of a subset, say W, of nodes V with $|W| = 2k$, $\mathbf{m}$ is the characteristic vector of a matching where the set of matched nodes is exactly W, and $\mathbf{c}_{i,j} \leqslant |W \cap \{i, j\}|$ for all edges $\{i, j\} \in E$.

Let $P = \mathrm{conv}(\{(\mathbf{v}, \mathbf{m}, \mathbf{c}, \mathbf{z}) \mid (\mathbf{v}, \mathbf{m}, \mathbf{c}) \in \mathrm{vert}(\mathrm{EMP}(K_{n,n}, k)), \mathbf{z} = \mathbf{c}\})$. Observe that P is a copy of EMP$(K_{n,n}, k)$ obtained by copying the $\mathbf{c}$-coordinates. Set I to be the indices corresponding to the $\mathbf{z}$-coordinates. Now consider the face F of $\mathrm{perm}_I(P)$ defined by the requirements $\mathbf{z}_j = 2$ for $j \leqslant k^2$ and $\mathbf{z}_j = 0$ for all $j > k^2$. We claim that F is an extension of BCB$(K_{n,n}, k)$. This will complete the proof of the theorem.

We notice that $(\mathbf{v}, \mathbf{m}, \mathbf{c}, \mathbf{z})$ is a vertex of F if and only if it satisfies the following:

1. $\mathbf{v}$ is (the characteristic vector of) a subset, say U, of V of size $2k$,
2. $\mathbf{m}$ is (the characteristic vector of) a matching, say M, in $K_{n,n}$ of size k,
3. $\mathbf{c}_e = 2$ for exactly k^2 edges $e \in E$ and zero for all other edges. Furthermore, for e with $\mathbf{c}_e = 2$, both the endpoints are matched by M. And finally,
4. $\mathbf{z}$ is a sorted copy of $\mathbf{c}$.

To see this, let $(\mathbf{v}, \mathbf{m}, \mathbf{c}, \mathbf{z})$ be a vertex of F. It clearly satisfies properties (1) and (2) because all vertices of P satisfy these properties. Since it is a vertex of F, $\mathbf{z}$ has first k^2 coordinates 2 and the rest zero. Finally, since $(\mathbf{v}, \mathbf{m}, \mathbf{c}, \mathbf{z})$ is a vertex of $\mathrm{perm}_I(P)$ there exists a vertex $(\mathbf{v}, \mathbf{m}, \mathbf{c}, \mathbf{z}')$ such that $\mathbf{z}$ is obtained by permuting entries of $\mathbf{z}'$ and $\mathbf{z}' = \mathbf{c}$.

For the reverse direction, let $(\mathbf{v}, \mathbf{m}, \mathbf{c}, \mathbf{z})$ be a vector that satisfies properties (1)-(4). $(\mathbf{v}, \mathbf{m}, \mathbf{c}, \mathbf{c})$ is clearly a vertex of P since P is obtained by making copy of $\mathbf{c}$-coordinates of $\mathrm{EMP}(K_{n,n}, k)$ and so $(\mathbf{v}, \mathbf{m}, \mathbf{c}, \mathbf{z})$ is a vertex of $\mathrm{perm}_I(P)$. Since it satisfies property (4) it lies in F.

With the above properties we are ready to finish arguing that F is an extension of $\mathrm{BCB}(K_{n,n}, k)$.

Let $\mathbf{c}$ be the characteristic vector of a complete bipartite subgraph of $K_{n,n}$ with k nodes on each side. Take U to be this set of nodes and M to be any matching of $K_{n,n}$ that matches exactly U. Let $\mathbf{v}$ be the characteristic vector of U and $\mathbf{m}$ that of M. Then $\mathbf{c}$ has exactly k^2 one entries. Take $\mathbf{z}$ to be a (decreasingly) sorted copy of $\mathbf{c}$. Then $(\mathbf{v}, \mathbf{m}, 2\mathbf{c}, 2\mathbf{z})$ is a point in F.

Conversely, if $(\mathbf{v}, \mathbf{m}, \mathbf{c}, \mathbf{z})$ is a point in F then $\frac{1}{2}\mathbf{c}$ is the characteristic vector of a set of k^2 edges that are only between a set U of $2k$ nodes matched by $\mathbf{m}$. So $\mathbf{c}$ scaled by $1/2$ is the characteristic vector of the complete bipartite subgraph induced by the node set U.   □

Theorem 2, Lemma 3, Theorem 9, and Theorem 10 together yield the following.

Theorem 11. *For each natural number n, there exists a polytope $P \subseteq \mathbb{R}^{3n^2+2n}$ where each vertex is in the set $\{0, 1, 2\}^{3n^2+2n}$ and $I \subseteq [2n^2 + 2n]$ such that $\mathrm{xc}(P) \leqslant \mathcal{O}(n^2)$ but $\mathrm{xc}_I(P) \geqslant 2^{\Omega(n)}$.*

3.4   Sorting Polytopes

Notice that the proofs in the previous subsection worked by taking all permutations over a subset of coordinates and then choosing vertices where these coordinates were sorted. This immediately implies the following theorem for which we only provide a sketch of proof due to space constraints.

Theorem 12. *There exist polytopes P such that $\mathrm{xc}(\mathrm{sort}(P))$ is exponentially larger than $\mathrm{xc}(P)$.*

Proof (Sketch). By suitably adding a (different) value to each coordinate, we can make sure that only the first k coordinates change values in $\mathrm{sort}(P)$ for any choice of k, effectively allowing the proof of Theorem 11 to be simulated.

4   Concluding Remarks

Firstly, we would like to address the fact that the polytopes arising in our theorems are not full-dimensional. This can be easily remedied with asymptotically similar extension complexities by repeatedly forming pyramids until the polytope is full-dimensional and using the fact that $\mathrm{xc}(Q) = \mathrm{xc}(P) + 1$ if Q is a pyramid with base P.

As mentioned earlier, one primary motivation for considering permutations of a subset of coordinates was that if these operations did not significantly worsen the extension complexity, we would have a way to adding parity constraints to arbitrary polytopes. Applying this to the matching polytopes of bounded genus graphs, we would thus be able to construct extended formulations for the cut polytopes of bounded genus graphs. Cut polytope of toroidal graphs arise in Physics and so such a result would be quite useful. Since the matching polytope is a 0/1 polytope, we do not necessarily want to apply partial permutations to arbitrary polytopes – which as shown in this paper is hopeless anyway. Therefore, we leave the reader with the following open problem.

Question 1. Let $P \subseteq \mathbb{R}^d$ be a 0/1 polytope. That is, all vertices of P are in $\{0, 1\}^d$. Let $I \subseteq [d]$ be a subset of coordinates. Is $\mathrm{xc}(\mathrm{perm}_I(P))$ bounded above by a polynomial in $\mathrm{xc}(P)$?

References

1. Avis, D., Tiwary, H.R.: On the extension complexity of combinatorial polytopes. Math. Program. **153**(1), 95–115 (2015)
2. Balas, E.: Disjunctive programming: properties of the convex hull of feasible points. Disc. Appl. Math. **89**(1–3), 3–44 (1998)
3. Conforti, M., Cornuéjols, G., Zambelli, G.: Extended formulations in combinatorial optimization. Ann. OR **204**(1), 97–143 (2013)
4. Conforti, M., Pashkovich, K.: The projected faces property and polyhedral relations. In: Mathematical Programming, pp. 1–12 (2015). ISSN: 0025–5610
5. Fiorini, S., Massar, S., Pokutta, S., Tiwary, H.R., de Wolf, R.: Exponential lower bounds for polytopes in combinatorial optimization. J. ACM **62**(2), 17 (2015)
6. Garey, M.R., Johnson, D.S.: Computers and Intractability: A Guide to the Theory of NP-Completeness (1979)
7. Göös, M., Jain, R., Watson, T.: Extension complexity of independent set polytopes. SIAM J. Comput. **47**(1), 241–269 (2018)
8. Grötschel, M., Lovász, L., Schrijver, A.: Geometric Algorithms and Combinatorial Optimization, volume 2 of Algorithms and Combinatorics. Springer, Heidelberg (1988). https://doi.org/10.1007/978-3-642-78240-4. ISBN: 978-3-642-97883-8
9. Grünbaum, B., Kaibel, V., Klee, V., Ziegler, G.M.: Convex polytopes. Springer, New York (2003). https://doi.org/10.1007/978-1-4613-0019-9. ISBN: 9780387004242
10. Johnson, D.S.: The np-completeness column: an ongoing guide. J. Algor. **8**(3), 438–448 (1987). ISSN 0196–6774
11. Kaibel, V.: Extended formulations in combinatorial optimization. Optima **85**, 2–7 (2011)
12. Kaibel, V., Pashkovich, K.: Constructing extended formulations from reflection relations. In: 15th International Conference, IPCO 2011, pp. 287–300 (2011)
13. Kolman, P., Koutecký, M., Tiwary, H.R.: Extension complexity, MSO logic, and treewidth. Disc. Math. Theor. Comput. Sci. **22**(4) (2020)
14. Loebl, M.: Discrete mathematics in statistical physics - introductory lectures. In: Advanced Lectures in Mathematics. Vieweg (2009). ISBN: 978-3-528-03219-7

15. Margot, F.: Composition de polytopes combinatoires: une approche par projection. PhD thesis, EPFL, Switzerland (1994)
16. Rothvoß, T.: The matching polytope has exponential extension complexity. J. ACM **64**(6), 41:1–41:19 (2017)
17. Ziegler, G.M.: Lectures on polytopes, volume 152 of Graduate Texts in Mathematics, pp. ix+370. Springer-Verlag, Heidelberg (1995)

Non-linear Optimization

Piecewise Linearization of Bivariate Nonlinear Functions: Minimizing the Number of Pieces Under a Bounded Approximation Error

Aloïs Duguet$^{(\boxtimes)}$ and Sandra Ulrich Ngueveu

LAAS-CNRS, Université de Toulouse, CNRS, INP, Toulouse, France
`alois.duguet@laas.fr` , `ngueveu@laas.fr`

Abstract. This work focuses on the approximation of bivariate functions into piecewise linear ones with a minimal number of pieces and under a bounded approximation error. Applications include the approximation of mixed integer nonlinear optimization problems into mixed integer linear ones that are in general easier to solve. A framework to build dedicated linearization algorithms is introduced, and a comparison to the state of the art heuristics shows their efficiency.

Keywords: Piecewise linear approximation · Bivariate nonlinear functions · Mixed integer nonlinear programming · Heuristics

1 Problem Description and State of the Art

Let $(\mathcal{P})$ be the optimization problem of approximating a nonlinear function f of two variables by a piecewise linear (PWL) function g subject to approximation error constraints on domain $\mathbb{D}$ represented by functions l and u satisfying $l(x,y) \leq f(x,y) \leq u(x,y)$ for all $(x,y) \in \mathbb{D}$:

$$(\mathcal{P}) \begin{cases} min & n & (1) \\ subject\ to & l(x,y) \leq g(x,y) \leq u(x,y) \quad \forall (x,y) \in \mathbb{D} \subset \mathbb{R}^2 & (2) \\ & g \text{ is a PWL function with } n \text{ pieces} & (3) \end{cases}$$

Constraints (2) are pointwise approximation constraints, which gives an infinite number of constraints because $\mathbb{D}$ is a continuous domain. Piecewise linear approximation are more commonly minimizing an approximation error with a fixed number of pieces [9], but our objective is to minimize the number of pieces of g so that an MILP formulation of g introduces less binary variables [10]. It can be useful for the approximation of a Mixed Integer Nonlinear Programming problem (MINLP) with a Mixed Integer Linear Programming (MILP) by substituting each nonlinear functions by a PWL one. Moreover, it was shown in [4]

© The Author(s), under exclusive license to Springer Nature Switzerland AG 2022
I. Ljubić et al. (Eds.): ISCO 2022, LNCS 13526, pp. 117–129, 2022.
https://doi.org/10.1007/978-3-031-18530-4_9

that in some cases, MINLP can be solved by applying only techniques from MILP.

Heuristics and exact methods exist to approximate univariate nonlinear functions ($\mathbb{D} \subset \mathbb{R}$) [1,6,8]. We are interested in bivariate functions ($\mathbb{D} \subset \mathbb{R}^2$) and to the best of our knowledge, only two papers address this case, with heuristics only:

- The authors of [7] propose two heuristics to solve problem ($\mathcal{P}$) with continuous PWL functions. The first heuristic is based on an iterative subdivision of the domain $\mathbb{D}$ into triangles (2-simplexes) until for each subdomain a linear function satisfying the approximation error has been found. The verification that a given linear function fits the subdomain is made by solving a nonlinear programming problem (NLP). The second heuristic can be used if the contribution of the two variables in the function can be separated in two univariate functions (linearly or nonlinearly). In this case, an algorithm finds the two optimal continuous univariate PWL functions and combine them to build a single two-variable PWL function.
- In [5] an iterative process attempts to find a continuous PWL function written as a Difference of Convex Continuous PWL functions (DC CPWL) that satisfies the approximation error. The idea is to iteratively solve an MILP relaxation of ($\mathcal{P}$) and then to find lazy constraints to add to the relaxation until a solution found is feasible for ($\mathcal{P}$). The relaxation consists in replacing the infinite number of constraints (2) with a finite number of them.

After the introduction of definitions used throughout the paper in Sect. 2, the three key ideas of a framework for piecewise linearization are detailed in Sect. 3. It is followed by explanations on the instantiation of crucial parts of this framework to create different heuristics in Sect. 4, and finally, numerical experiments comparing the state of the art to our best heuristics are shown in Sect. 5.

2   Definitions

The vocabulary used throughout this work is presented below. They are in part extensions to $\mathbb{D} \subset \mathbb{R}^2$ of definitions from [1] for $\mathbb{D} \subset \mathbb{R}$.

Definition 1 (Polytope). *A polytope $\mathcal{P}$ is the convex hull of some points $X_i \in \mathbb{R}^m$: $\mathcal{P} = \{x \in \mathbb{R}^m, x = \sum_i \lambda_i X_i, \sum_i \lambda_i = 1, \lambda_i \geq 0 \ \forall i\}$.*

Throughout the paper, a *piece* will refer to a polytope that composes the graph of a PWL function, and $[\![1,n]\!] := \{1, ..., n\}$.

Definition 2 (PWL function). *Let $\mathbb{D}$ be a compact set of $\mathbb{R}^2$. A function $g : \mathbb{D} \mapsto \mathbb{R}$ is a PWL function with n pieces if and only if there exists $\{a_i\}_{i \in [\![1,n]\!]} \subset \mathbb{R}^2$, $\{b_i\}_{i \in [\![1,n]\!]} \subset \mathbb{R}$ and a family of polytopes $\{D_i\}_{i \in [\![1,n]\!]} \subset \mathbb{R}^2$ such that $\mathbb{D} = \cup_{i \in [\![1,n]\!]} \mathbb{D}_i$, and for $i \neq j$ the polytopes $\mathbb{D}_i$ and $\mathbb{D}_j$ can only intersect on their boundary and g is defined by $g(x,y) = \min\{a_i \cdot (x,y)^T + b_i | (x,y) \in \mathbb{D}_i \ \forall i \in [\![1,n]\!]\}$, with $\cdot$ denoting the standard scalar product.*

Precautions were taken to allow g to be *not necessarily continuous* at the boundary of a polytope $\mathbb{D}_i$. To prevent $g(x, y)$ to have multiple definitions because $\{D_i\}_{i \in [\![1,n]\!]}$ can intersect on their boundary, $g(x, y)$ is chosen as the minimum over all possible definitions, so that it is a lower semicontinuous function.

Definition 3 (Corridor). *Let* $\mathbb{D}$ *be a compact set of* $\mathbb{R}^2$. *Let* $u, l : \mathbb{D} \mapsto \mathbb{R}$ *be two continuous functions verifying* $u(x, y) > l(x, y), \forall (x, y) \in \mathbb{D}$. *Define the set* $\mathcal{C} = \{(x, y, z) \in \mathbb{R}^3 | (x, y) \in \mathbb{D}, l(x, y) \leq z \leq u(x, y)\}$ *as the corridor between* u *and* l. *If* $\mathbb{D} \subset \mathbb{R}^2$, *we call the area of* $\mathbb{D}$ *the domain area of* $\mathcal{C}$.

A similar definition can be made for $\mathbb{D}$ an interval $[a, b]$, in which case we call $b - a$ the *domain length* of $\mathcal{C}$.

Definition 4 (Corridor domain). *Let* $\mathcal{C}$ *be a corridor,* $\mathcal{C} \subset \mathbb{R}^3$. *The domain of corridor* $\mathcal{C}$ *noted* $D(\mathcal{C})$ *is the projection of* $\mathcal{C}$ *on its two first coordinates, which is also the domain on which* u *and* l *need to be defined.*

Definition 5 (Piece within a corridor). *A polytope* $\mathcal{P} \subset \mathbb{R}^3$ *is within a corridor* $\mathcal{C}$ *if and only if there exists a linear function* $g : \mathbb{D} \subset D(\mathcal{C})$, *such that* $\mathcal{P} = \{(x, y, g(x, y)), (x, y) \in \mathbb{D}\}$ *and* $\mathcal{P} \subset \mathcal{C}$.

Definition 6 (Fitting). *A PWL function* g *fits a corridor* $\mathcal{C}$ *if and only if the pieces of* g *(polytopes* $\{P\}_{i \in [\![1,n]\!]}$ *of the graph of* g*) are within* $\mathcal{C}$ *and* g *is defined on the entire domain* $D(\mathcal{C})$.

Definition 7 (PWL corridor). *A corridor* $\mathcal{C}$ *is called a PWL corridor if and only if* u *and* l *defining* $\mathcal{C}$ *are both PWL functions.*

Definition 8 (Inner corridor). *Let* $\mathcal{C}_0$ *be a corridor between* u_0 *and* l_0. *Let* $\mathcal{C}$ *be a corridor between* u *and* l. *We call* $\mathcal{C}$ *an inner corridor of* $\mathcal{C}_0$ *if and only if* $D(\mathcal{C}) = D(\mathcal{C}_0)$ *and* $l_0(x, y) \leq l(x, y) < u(x, y) \leq u_0(x, y)$.

Definition 9 ($\mathbb{R}^2$-corridor fitting problem). *The* $\mathbb{R}^2$*-corridor fitting problem consists in finding a PWL function* g *of two variables fitting a corridor* $\mathcal{C}$ *such that its number of pieces is minimized.*

$(\mathcal{P})$ is equivalent to an $\mathbb{R}^2$-corridor fitting problem with corridor $\mathcal{C}$ between u and l. Moreover, by extending the previous definitions to dimension $m \geq 1$, it is possible to define the $\mathbb{R}^m$-corridor fitting problem that refers to the problem with $D(\mathcal{C}) \subset \mathbb{R}^m$.

Definition 10 (Truncated-corridor). *Let* $\mathcal{C}_1, \mathcal{C}_2 \in \mathbb{R}^2$ *be two corridors both defined by functions* u, $l : \mathbb{R} \mapsto \mathbb{R}$ *on the intervals* $[a, b_1]$ *and* $[a, b_2]$. *We call* $\mathcal{C}_2$ *a truncated-corridor of* $\mathcal{C}_1$ *if and only if* $[a, b_2] \subset [a, b_1]$ *($b_2 \leq b_1$).*

Definition 11 (Maximal linear segment). *A maximal linear segment in a corridor* $\mathcal{C}$ *is a linear segment within* $\mathcal{C}$ *that induces a truncated-corridor of maximal domain length.*

Definition 12 (Truncated-corridor in direction d**).** *Let* C_1 *and* C_2 *be two corridors defined by the same functions* u *and* l *with compact corridor domains in* $\mathbb{R}^2$. *Let* $d \in \mathbb{R}^2 \setminus \{0\}$. *We call* C_2 *a truncated-corridor of* C_1 *in direction d if and only if there exists* $\sigma \in \mathbb{R}$ *for which* $D(C_2) = D(C_1) \cap \{(x,y) \in \mathbb{R}^2, (x,y) \cdot d \leq \sigma\}$, *i.e.* $D(C_2)$ *is the intersection of* $D(C_1)$ *with a half-plane.*

Definition 13 (Maximal piece in direction d**).** *A maximal piece in direction* $d \in \mathbb{R}^2 \setminus \{0\}$ *of a corridor* C *is a polytope within* C *that induces a truncated-corridor of* C *in direction d that is of maximal domain area.*

3 A Framework for Solving the $\mathbb{R}^2$-Corridor Fitting Problem

We present in this section a framework to create efficient algorithms for the $\mathbb{R}^2$-corridor fitting problem. The instantiation chosen for different parts of the framework are described in Sect. 4 as well as some details on the implementation.

Three key ideas are followed in the framework. The first two are used to avoid drawbacks encountered in [7], whereas the third one is meant to render a subproblem more tractable:

– **Key idea 1:** Pieces should be chosen among general convex polygons instead of triangles to constrain less the pieces chosen, possibly decreasing the number of pieces
– **Key idea 2:** Choose pieces that are good (ideally optimal) solutions of a maximal piece in direction d problem, to aim for a domain covered by a piece that is "as large as possible" because in [7] the size of triangles is fixed which increases the number of pieces necessary
– **Key idea 3:** Compute a good feasible solution of the maximal piece in direction d problem with a series of LP problems obtained after substituting C with a PWL inner corridor of C

The remainder of this section builds upon these principles.

3.1 Key Idea 1: Management of the Corridor Domain

The corridor domain should be tiled with shapes as general as possible provided that they can be formulated in an MILP. Such shapes are polygons, but we further restrict those shapes to convex polygons because formulating a non-convex polygon in an MILP introduces additional binary variables. It is expected that allowing convex polygons instead of only triangles as done in [7] will lead to a lower number of pieces.

The procedure that manages pieces and the remaining corridor domain is described in Algorithm 1. At each iteration, one piece is computed for each vertex of $D(C)$ by function *compute_piece*, and a function *score* selects the "most" suitable to obtain a PWL function with few pieces, see Sect. 4.1. Function *update_domain*(C, p) removes the part of $D(C)$ on which p is defined. This function also divides the new reduced corridor C in two corridors if polygon $D(C)$ only has angles at the vertices greater than 90 to avoid a bad behaviour.

Algorithm 1. Finding a PWL function fitting a corridor $\mathcal{C}$ with a low number of pieces

```
1: function PWL_2D_FITTING(C)
2:     P ← ∅                                          ▷ list of chosen pieces
3:     Q ← {C}            ▷ list of corridors with convex domains not already tiled
4:     while Q ≠ ∅ do
5:         C = pop(Q)
6:         candidate_pieces ← ∅
7:         for v vertex of D(C) do
8:             d ← choose_progress_direction(C, v)
9:             candidate_pieces ← candidate_pieces ∪ {compute_piece(C, v, d)}
10:        end for
11:        p ← argmax_{p∈candidate_pieces} score(p)
12:        P ← P ∪ {p}
13:        Q = update_domain(C, p)
14:    end while
15:    return P
16: end function
```

3.2 Key Idea 2: The Maximal Piece in Direction d Problem

We chose to find a new piece by covering an area starting from point v and extending as far as possible in direction d. This direction d points to the interior of $D(\mathcal{C})$ when starting from v and is computed via function *choose_progress_direction* of Algorithm 1. The hypothesis of starting from a vertex instead of any point of the border of the polygon $D(\mathcal{C})$ is made. Computing the piece consists in solving a maximal piece in direction d problem (MP_d):

$$(MP_d) \begin{cases} \max \ \sigma \\ \text{s.t.} \ \ \alpha x + \beta y + \gamma \in \mathcal{C}_\sigma^d \ \ \forall (x,y) \in D(\mathcal{C}_\sigma^d) \\ \alpha, \beta, \gamma, \sigma \in \mathbb{R} \end{cases} \tag{4}$$

where $D(\mathcal{C}_\sigma^d) = D(\mathcal{C}) \cap \{(x,y) \in \mathbb{R}^2 | (x,y)^T \cdot d \leq \sigma\}$ is the domain of $\mathcal{C}_\sigma^d$, the truncated-corridor of $\mathcal{C}$ in direction d. (MP_d) is a generalized semi-infinite programming problem (GSIP). Indeed, the number of pointwise constraints is infinite and depends on variable σ, while the number of variables is 4: a real variable σ for the half-plane intersection as well as 3 real variables (α, β, γ) to describe the linear function $g(x,y) = \alpha x + \beta y + \gamma$.

3.3 Key Idea 3: Computing a Feasible Solution of a Maximal Piece in Direction d Problem

As we want a computationally cheap solution to the GSIP (MP_d) because of the high number of times such a problem has to be solved, a feasible solution of (MP_d) is computed via a series of LP problems, as described below, using a PWL inner corridor of the original corridor $\mathcal{C}$, because it allows to replace the infinite number of nonlinear constraints by a finite number of linear constraints while ensuring feasibility.

Let corridor $\mathcal{C}_{PWL\sigma}^{d}$ be a PWL inner corridor of $\mathcal{C}_{\sigma}^{d}$ with associated functions $\tilde{u}$ and $\tilde{l}$ for readability; note $(D_{\tilde{u}}^{i})_{i \in I}$ and $(D_{\tilde{l}}^{j})_{i \in J}$ the subdomains of corridor $\mathcal{C}_{PWL\sigma}^{d}$ on which $\tilde{u}$ and $\tilde{l}$ are linear, indexed by I and J respectively. Then, (MP_d) is a relaxation of the following problem because $\mathcal{C}$ is replaced by an inner corridor.

$$(MP_d') \begin{cases} \max \ \sigma \\ s.t. \\ \alpha x + \beta y + \gamma - \tilde{u}_i(x,y) \leq 0 \ \forall (x,y) \in D(\mathcal{C}_{\sigma}^{d}) \cap D_{\tilde{u}}^{i}, \forall i \in I \\ \alpha x + \beta y + \gamma - \tilde{l}_j(x,y) \geq 0 \ \forall (x,y) \in D(\mathcal{C}_{\sigma}^{d}) \cap D_{\tilde{l}}^{j}, \forall j \in J \\ \alpha, \beta, \gamma, \sigma \in \mathbb{R} \end{cases} \quad (5)$$

Note that on each $D_{\tilde{u}}^{i}$, $g(x,y) - \tilde{u}(x,y)$ is a linear function, thus it suffices to check $g(X) - \tilde{u}(x,y) \leq 0$ for each vertex of convex polygonal domain $D_{\tilde{u}}^{i}$ to ensure constraint $g(x,y) - \tilde{u}(x,y) \leq 0$ on $D_{\tilde{u}}^{i}$. A similar reasoning leads to the same result for constraints involving $\tilde{l}$. Thus, the following problem is equivalent to (MP_d') but has the advantage of having only a finite number of linear constraints.

$$(MP_d'') \begin{cases} \max \ \sigma \\ s.t. \\ \alpha x + \beta y + \gamma - \tilde{u}_i(x,y) \leq 0 \ \text{ for each vertex } (x,y) \text{ of } D(\mathcal{C}_{\sigma}^{d}) \cap D_{\tilde{u}}^{i}, \forall i \in I \\ \alpha x + \beta y + \gamma - \tilde{l}_j(x,y) \geq 0 \ \text{ for each vertex } (x,y) \text{ of } D(\mathcal{C}_{\sigma}^{d}) \cap D_{\tilde{l}}^{j}, \forall j \in J \\ \alpha, \beta, \gamma, \sigma \in \mathbb{R} \end{cases}$$

$$(6)$$

Finally, problem (MP_d'') has constraints involving polygon intersections depending nonlinearly on variable σ, thus it is not an LP problem. Parameterizing (MP_d'') with σ, the following LP feasibility problem is obtained which can be repeatedly solved until a satisfactory σ value has been found.

$$(MP_{d,\sigma}'') \begin{cases} \alpha x + \beta y + \gamma - \tilde{u}_i(x,y) \leq 0 \ \text{ for each vertex } (x,y) \text{ of } D(\mathcal{C}_{\sigma}^{d}) \cap D_{\tilde{u}}^{i}, \forall i \in I \\ \alpha x + \beta y + \gamma - \tilde{l}_j(x,y) \geq 0 \ \text{ for each vertex } (x,y) \text{ of } D(\mathcal{C}_{\sigma}^{d}) \cap D_{\tilde{l}}^{j}, \forall j \in J \\ \alpha, \beta, \gamma \in \mathbb{R} \end{cases}$$

$$(7)$$

4   Framework Key Points Instantiation

In this section, choices made on key points of the framework of Sect. 3 are described: the function *score* of Algorithm 1 in Sect. 4.1, the choice of a direction d in Sect. 4.2 and the computation of a PWL inner corridor in Sect. 4.3. $\mathcal{C}$ refer to a corridor between u and l in the remainder of the section.

4.1   Scoring the Quality of Pieces

In Algorithm 1, a function *score* ranks the quality of candidate pieces and the piece with highest score is kept. Two scoring functions have been implemented:

- *Area* measures the area of the domain covered by piece p
- *Partial Derivatives Total Variation* (PaD) approximates the sum of the *total variation* of each partial derivative of u and l on the domain covered by p

The total variation of a function g is a measure of how much that function varies on its domain $\mathcal{D}$. The total variation of g on $\mathcal{D}$ is equal to $\int_{\mathcal{D}} \|\nabla g(x)\|_2 dx$. The interest of PaD needs the introduction of the *pointwise height* of a corridor.

Definition 14 (pointwise height). *We call $\mathcal{C}_{PH}(x,y) = u(x,y) - l(x,y)$ the pointwise height of corridor $\mathcal{C}$ at point (x,y).*

Remark 1. The most commonly used type of approximation error for a function f is the absolute error. It induces a corridor $\mathcal{C}$ such that $l(x,y) = f(x,y) - \delta$ and $u(x,y) = f(x,y) + \delta$ with $\delta > 0$, that has constant pointwise height.

Area is a straightforward and simple idea to evaluate the piece quality, it will serve as a reference to evaluate other scoring functions. PaD is thought to be an adaptation for two-variable functions of Theorem 1 of [3]. Indeed, it states that for a corridor $\mathcal{C}$ of constant pointwise height 2δ with $D(\mathcal{C}) = [a,b]$, when $\delta \to 0$, the minimum number of pieces $s(\delta)$ satisfies the asymptotic approximation $s(\delta) \sim \frac{1}{4\delta} \int_a^b \sqrt{|u''(x,y)|} dx$. Note first that $u''(x,y) = l''(x,y)$ because it is a corridor with constant pointwise height, and second that the integral computed is the total variation of function u' (and l'). It is thus expected that PaD performs better than Area for small values of pointwise height. For large values, the total variation should be less relevant to estimate the difficulty of fitting a piece, thus PaD could be less efficient.

4.2   Choose a Progress Direction

Line 8 of Algorithm 1 selects progress direction d knowing starting vertex v. Two options were tested for this choice.

- the direction going along the bisector of the two edges of $D(\mathcal{C})$ having v as starting point, denoted *bd* for Bisector Direction
- Compute two maximal linear segments starting from v and following each edge of $D(\mathcal{C})$ having v as endpoint. The direction orthogonal to the line joining the two ends of the maximal linear segments is chosen as the progress direction d. It is denoted *med* for Mean progress along Edges Direction

The first is a naive option, while the second is meant to take into account the "difficulty" of progressing along the two extremal directions given by the two edges starting at v.

4.3   Inner Approximation of a Corridor

In the function *compute_piece*, the computation of a PWL inner corridor $\mathcal{C}_{PWL}$ of $\mathcal{C}$ boils down to computing PWL functions $\tilde{u}$ and $\tilde{l}$ verifying $l(x,y) \leq \tilde{l}(x,y) \leq \tilde{u}(x,y) \leq u(x,y)$, for all $(x,y) \in D(\mathcal{C})$.

To compute $\tilde{u}$ (a similar method works for $\tilde{l}$), a basic idea is to divide $D(\mathcal{C})$ into rectangular pieces of same sizes, and then to compute a third coordinate $\tilde{u}(x, y)$ to each vertex $v = (x, y)$ of each rectangular piece such that $\tilde{u}(x, y) \leq u(x, y)$. Interval analysis on the gradient of u suffices to compute the values of $\tilde{u}(x, y)$ such that $\tilde{u}$ is an underestimation of u, as explained in Proposition 1.

Proposition 1. *Let* $u \in C^1$ *defined on* $\mathcal{D} = [a, b] \times [c, d] \subset \mathbb{R}^2$. *Let* $l_x = b - a$ *and* $l_y = d - c$. *Let* ∇u *be the gradient of* u. *Let* $[D_x^{low}, D_x^{high}] \times [D_y^{low}, D_y^{high}]$ *be such that* $\nabla u(x, y) \in [D_x^{low}, D_x^{high}] \times [D_y^{low}, D_y^{high}]$ *for all* $(x, y) \in \mathcal{D}$. *Let* $(M_x, M_y) = (\frac{a+b}{2}, \frac{c+d}{2})$. *Define:*

$$u_{(a,c)}^- := u(M_x, M_y) - D_x^{high} \cdot l_x - D_y^{high} \cdot l_y \tag{8}$$

$$u_{(b,c)}^- := u(M_x, M_y) + D_x^{low} \cdot l_x - D_y^{high} \cdot l_y \tag{9}$$

$$u_{(b,d)}^- := u(M_x, M_y) + D_x^{low} \cdot l_x + D_y^{low} \cdot l_y \tag{10}$$

$$u_{(a,d)}^- := u(M_x, M_y) - D_x^{high} \cdot l_x + D_y^{low} \cdot l_y \tag{11}$$

If a linear function f *satisfies:*

$$f(a, c) \leq u_{(a,c)}^-, \ f(b, c) \leq u_{(b,c)}^-, \ f(b, d) \leq u_{(b,d)}^- \text{ and } f(a, d) \leq u_{(a,d)}^- \tag{12}$$

Then $f(x, y) \leq u(x, y)$ *for all* $(x, y) \in \mathcal{D}$.

Proof. Let f be a linear function satisfying the 4 inequalities (12). Let $M = (M_x, M_y, f(M_x, M_y))$ be the point on the surface defined by u corresponding to the middle of $\mathcal{D}$. Let $(x_0, y_0) \in \mathcal{D}$. We have:

$$u(M_x, M_y) - D_x^{high} \cdot (M_x - x_0) - D_y^{high} \cdot (M_y - y_0) \leq u(x_0, y_0) \quad \text{if } x_0 \leq M_x, y_0 \leq M_y$$

$$u(M_x, M_y) + D_x^{low} \cdot (x_0 - M_x) - D_y^{high} \cdot (M_y - y_0) \leq u(x_0, y_0) \quad \text{if } x_0 \geq M_x, y_0 \leq M_y$$

$$u(M_x, M_y) + D_x^{low} \cdot (x_0 - M_x) + D_y^{low} \cdot (y_0 - M_y) \leq u(x_0, y_0) \quad \text{if } x_0 \geq M_x, y_0 \geq M_y$$

$$u(M_x, M_y) - D_x^{high} \cdot (M_x - x_0) + D_y^{low} \cdot (y_0 - M_y) \leq u(x_0, y_0) \quad \text{if } x_0 \leq M_x, y_0 \geq M_y$$

because $[D_x^{low}, D_x^{high}] \times [D_y^{low}, D_y^{high}]$ are bounds of ∇u on domain $\mathcal{D}$. Now, define a PWL function f_{PWL} with four rectangle pieces with vertices positionned at $\{(a, c), (b, c), (b, d), (a, d), (\frac{a+b}{2}, \frac{c+d}{2})\}$ and height the left-hand sides of (12) as well as $u(M_x, M_y)$ respectively. In particular, $f_{PWL} \leq u$ on $\mathcal{D}$. In addition, direct computations show that $f(x, y) \leq f_{PWL}(x, y)$ for $(x, y) \in \{(a, c), (b, c), (b, d), (a, d), (\frac{a+b}{2}, \frac{c+d}{2})\}$. Finally, as f and f_{PWL} are linear on the four pieces domain of f_{PWL}, we have $f \leq f_{PWL} \leq u$ on $\mathcal{D}$. $\qed$

To build a PWL inner corridor exploiting Proposition 1 in our algorithm, a method called *efficiency refinement* is used. It is described after the introduction of the bounding efficiency η.

Definition 15 (bounding efficiency η**).** *Let* $\mathcal{C}$ *be a corridor with* $D(\mathcal{C})$ *a polygonal domain of* $\mathbb{R}^2$. *Let* $\mathcal{C}_{PWL}$ *be a PWL inner corridor of* $\mathcal{C}$. *We say that* $\mathcal{C}_{PWL}$ *achieves a bounding efficiency for* $\mathcal{C}$ *of* $\eta \in [0, 1]$ *if the pointwise height (PH) ratio* $\frac{(\mathcal{C}_{PWL})_{PH}(x,y)}{\mathcal{C}_{PH}(x,y)}$ *is greater or equal to* η *for each* $(x, y) \in D(\mathcal{C})$.

Proposition 2. *Let $\mathcal{C}$ be a corridor with $D(\mathcal{C})$ a polygonal domain of $\mathbb{R}^2$ and let $\eta \in [0,1]$. If $\mathcal{C}$ has constant pointwise height, then a PWL inner corridor $\mathcal{C}_{PWL}$ has a bounding efficiency for $\mathcal{C}$ of η if the pointwise height ratio $\frac{(\mathcal{C}_{PWL})_{PH}(x,y)}{\mathcal{C}_{PH}(x,y)}$ is greater or equal to η for each (x,y) vertex of a piece domain of $\tilde{u}$ or $\tilde{l}$.*

Proof. $\mathcal{C}$ has constant pointwise height. Thus for each piece of $\mathcal{C}_{PWL}$, the minimum pointwise height ratio is on an extreme point of the piece domain, that is to say on a vertex of the piece domain, which is lower bounded by η by hypothesis. □

The efficiency refinement procedure builds a PWL inner corridor $\mathcal{C}_{PWL}$ of $\mathcal{C}$ achieving a bounding efficiency of η if it has a constant pointwise height, but without this property, it only checks that the pointwise height ratio at the vertices of each piece is η.

After having found a rectangle $\mathcal{D}$ containing $D(\mathcal{C})$, it creates an initial PWL corridor likely unvalid ($l \not\leq u$) with only one rectangular piece on $\mathcal{D}$, and then iteratively refines the pieces that do not satisfy a bounding efficiency of η at the 4 vertices into 4 new pieces until each piece satisfies the efficiency, which implies the validity of the PWL inner corridor as well.

Parameter η needs to be adjusted depending on the quality of PWL inner corridor $\mathcal{C}_{PWL}$ wanted. To produce a really good approximation of corridor $\mathcal{C}$, η near 1 shall be used, but the number of pieces forming $\mathcal{C}_{PWL}$ increases consequently, thus increasing the computation time of $(MP''_{d,\beta})$ to be solved later on.

5 Numerical Experiments

The performance of our framework is compared to the state of the art. Our best heuristic (found with experiments in [2]) called DN99 uses scoring function PaD, starting vertex set to med and $\eta = 0.99$. Also, to illustrate the effects of parameter η, results of a heuristic which uses the same parameters as DN99 but with $\eta = 0.95$, are shown. This second heuristic is called DN95.

Heuristics from the state of the art are described in articles [1,5,7]. Recall that [7] proposes two heuristics: one based on triangulation (RK2D), and one based on a decomposition of the bivariate function into a sum of one variable functions to approximate separately (RK1D). Heuristic LinA2D is based on the same decomposition as RK1D, but uses library LinA from [1] to compute the approximation of univariate functions. Obviously LinA2D outperforms RK1D since LinA computes optimal non necessarily continuous univariate PWL functions instead of optimal continuous univariate PWL functions for RK1D. The heuristic of [5] is denoted *KL2D*.

Those 5 heuristics are compared on the benchmark instances of [7], which consist of 45 instances obtained from 9 functions and 5 different absolute approximation errors for each function. An absolute approximation error of δ for function f means that the corresponding corridor $\mathcal{C}$ is between functions u and l

satisfying $l(x,y) = f(x,y) - \delta$ and $u(x,y) = f(x,y) + \delta$. The first two functions of the benchmark are linearly separable and refered to as L1,L2. The remaining seven are not linearly separable and thus refered to as N1,...,N7. The expression and domain of each function are described in Table 1.

For our heuristics as well as LinA2D, JuMP (v. 0.21.8) is used as modeling language, Gurobi (v. 9.1) is the (MI)LP solver and a CPU of 4.4 GHz using a single core and 32 GB RAM. Moreover, heuristic RK2D uses the modeling language GAMS (v. 23.6), the global optimization solver LindoGlobal (v. 23.6.5), a CPU intel i7 with a single core, 2.93 GHz and 12 GB RAM. Finally, KL2D uses Pyomo (v. 5.6.4) as modeling language, CPLEX (v. 12.8) and 10 threads to solve MILPs, COUENNE (v.0.5.8) and 1 thread to solve NLPs, a CPU with 3.6 GHz and 32 GB RAM. As for the time limit, LinA2D and RK2D do not have one, KL2D allows 3600 s seconds for each MILP problem, while our heuristics allow 3600 s seconds for each LP problem. The differences in computation time between heuristics are in several orders of magnitude, therefore these differences are mainly due to the algorithms differences and are only marginally impacted by the differences in hardware.

Table 1. Expression and domain of benchmark functions

Ref	Expression	Domain	Ref	Expression	Domain
L1	$x^2 - y^2$	$[0.5, 7.5] \times [0.5, 3.5]$	N3	$x\sin(y)$	$[1.0, 4.0] \times [0.05, 3.1]$
L2	$x^2 + y^2$	$[0.5, 7.5] \times [0.5, 3.5]$	N4	$\frac{\sin(x)}{x}y^2$	$[1.0, 3.0] \times [1.0, 2.0]$
			N5	$x\sin(x)\sin(y)$	$[0.05, 3.1] \times [0.05, 3.1]$
N1	xy	$[2.0, 8.0] \times [2.0, 4.0]$	N6	$(x^2 - y^2)^2$	$[1.0, 2.0] \times [1.0, 2.0]$
N2	$xe^{-x^2-y^2}$	$[0.5, 2.0] \times [0.5, 2.0]$	N7	$e^{-10(x^2-y^2)^2}$	$[1.0, 2.0] \times [1.0, 2.0]$

Results are shown in Table 2. For each of the five heuristics we present the number of pieces n obtained by the heuristic as well as its computation time in seconds. Bold integer highlight the best solution found for each instance, i.e. the ones with the minimum number of pieces. "TO" means that the heuristic has stopped because of a time out, thus without any valid solution. In this case, "-" means that no solutions were found.

In terms of minimum number of pieces, DN99 performs the best with 28 best solutions out of 45 instances, followed by KL2D with 19 out of 45, DN95 with 15 out of 45, LinA2D with 8 out of 45 and finally RK2D with 1 out of 45. As for the computation time, the hardware and software used for the different heuristics are of different quality, thus it will not be a precise tool to compare the time needed for each heuristics. However, it can be said that LinA2D takes less than 1 s to execute, RK2D and DN95 take seconds to minutes, while KL2D and DN99 take seconds to hours.

A more in-depth analysis shows that LinA2D is the best only for instances with linearly separable functions (L1 and L2), but it does really poorly on the

Table 2. Comparison with state of the art heuristics

Ref.	δ	DN95		DN99		RK2D		LinA2D		KL2D	
		n	time	n	time	n	time	n	time	n	time
L1	1.5	8	20.6	7	44.7	16	30.8	**6**	0.0	**6**	87.1
	1.0	10	28.2	9	67.6	20	84.4	8	0.0	**6**	148.9
	0.5	22	54.5	20	175.1	48	150.4	**15**	0.0	–	TO
	0.25	46	110.9	42	258.2	80	272.6	**21**	0.0	–	TO
	0.1	113	356.8	106	762.9	224	380.6	**60**	0.0	–	TO
L2	1.5	8	29.6	**6**	39.1	24	26.8	**6**	0.0	–	TO
	1.0	13	39.7	11	101.2	28	7.4	8	0.0	**6**	1495.5
	0.5	26	47.6	27	117.9	84	38.0	**15**	0.0	–	TO
	0.25	59	114.0	53	279.7	121	35.8	**21**	0.0	–	TO
	0.1	146	287.2	136	642.5	351	171.7	**60**	0.0	–	TO
N1	1.0	4	9.5	3	21.2	4	0.8	6	0.0	4	18.5
	0.5	10	22.4	**6**	52.4	12	72.4	8	0.0	**6**	416.8
	0.25	16	70.2	14	116.1	20	4.7	18	0.0	**12**	14762.5
	0.1	32	115.8	**28**	324.5	59	59.3	45	0.0	–	TO
	0.05	68	293.2	**56**	564.6	94	45.3	91	0.0	–	TO
N2	0.1	**1**	4.3	**1**	4.1	2	0.3	4	0.0	**1**	6.0
	0.05	**2**	14.6	**2**	47.8	6	18.7	9	0.0	**2**	13.4
	0.03	**4**	24.6	**4**	57.0	10	12.7	12	0.0	**4**	39.0
	0.01	**11**	87.9	**11**	450.4	31	54.6	30	0.0	–	TO
	0.001	105	1041.9	**96**	3832.1	350	652.6	238	0.0	–	TO
N3	1.0	**3**	10.4	**3**	17.1	5	1.0	12	0.0	–	TO
	0.5	**4**	10.0	**4**	37.4	8	13.1	16	0.0	–	TO
	0.25	7	19.5	**6**	42.2	16	30.0	22	0.0	8	342.6
	0.1	18	57.5	**16**	306.1	44	74.6	68	0.0	–	TO
	0.05	34	193.0	**30**	681.2	85	141.9	120	0.0	–	TO
N4	0.5	**2**	15.5	**2**	23.2	**2**	1.8	3	0.0	**2**	13.3
	0.25	**3**	9.2	4	57.8	4	1.0	8	0.0	4	19.6
	0.1	7	41.2	**6**	230.0	9	25.8	21	0.0	**6**	66.9
	0.05	14	161.7	**11**	632.6	23	14.4	27	0.0	**12**	7246.5
	0.03	21	272.7	**19**	1066.4	40	161.4	44	0.0	–	TO
N5	1.0	2	8.3	**1**	5.0	6	7.7	20	0.0	**1**	6.3
	0.5	8	33.5	5	96.5	6	1.3	42	0.0	4	86.9
	0.25	12	49.9	14	242.7	21	30.8	72	0.0	5	2091.2
	0.1	**26**	103.5	**26**	646.9	96	73.0	168	0.0	–	TO
	0.05	56	313.4	**51**	954.9	274	305.5	340	0.0	–	TO
N6	1.0	**3**	11.7	**3**	51.4	6	22.8	9	0.3	3	23.3
	0.5	5	22.7	5	78.6	9	15.6	9	0.0	4	127.2
	0.25	9	42.7	8	188.2	12	22.9	16	0.0	**6**	305.5
	0.1	**16**	136.7	**16**	803.9	40	202.8	49	0.0	–	TO
	0.05	34	516.9	**28**	1986.8	87	174.1	81	0.0	–	TO
N7	1.0	**1**	8.1	**1**	23.8	2	0.8	4	0.0	**1**	2.9
	0.5	**2**	32.8	**2**	234.8	4	66.6	4	0.0	**2**	9.5
	0.25	**4**	145.7	**4**	1370.6	6	4.4	9	0.0	**4**	35.5
	0.1	6	462.4	7	6210.7	84	231.5	16	0.0	5	377.7
	0.05	**10**	2050.0	–	TO	86	57.8	36	0.0	–	TO

other functions. KL2D is the best 19 times out of the 24 instances where it terminates with a solution, which shows a clear limit to the size of the instance it can tackle, due to the increasing size of MILPs it solves. The effect of the change of parameter η from 95% to 99% in heuristics DN95 and DN99 is on average a decrease of 9.1% of the number of pieces of the solutions, at the cost of an increase of 301% of the computation time.

6   Conclusion

We introduced a framework to create linearization algorithms based on the solution of the $\mathbb{R}^2$-corridor fitting problem. Convex polygons were used to tile the domain instead of triangles, a scoring function ranking candidate pieces was developed and a good feasible solution of a GSIP was computed via a series of LP feasibility problems. Finally, numerical experiments showed that our heuristics outperforms the state of the art on not linearly separable functions. Further work could attempt to diminish the relatively high computation time of the heuristics, search for better instanciation of the different parts of the framework or apply those heuristics to approximate real-world MINLPs to show their practical usage.

References

1. Codsi, J., Ngueveu, S.U., Gendron, B.: Lina: a faster approach to piecewise linear approximations using corridors and its application to mixed-integer optimization. Technical report (2021)
2. Duguet, A.: Appendix of "piecewise linearization of bivariate nonlinear functions: minimizing the number of pieces under a bounded approximation error": determining the best heuristic (2022). https://homepages.laas.fr/sungueve/2dpwl.html
3. Frenzen, C., Sasao, T., Butler, J.T.: On the number of segments needed in a piecewise linear approximation. J. Comput. Appl. Math. **234**(2), 437–446 (2010). https://doi.org/10.1016/j.cam.2009.12.035
4. Geißler, B., Martin, A., Morsi, A., Schewe, L.: Using piecewise linear functions for solving MINLPs. In: Mixed Integer Nonlinear Programming. The IMA Volumes in Mathematics and its Applications, vol. 154, pp. 287–314. Springer, Heidelberg (2012). https://doi.org/10.1007/978-1-4614-1927-3_10
5. Kazda, K., Li, X.: Nonconvex multivariate piecewise-linear fitting using the difference-of-convex representation. Comput. Chem. Eng. **150**, 107310 (2021). https://doi.org/10.1016/j.compchemeng.2021.107310
6. Ngueveu, S.U.: Piecewise linear bounding of univariate nonlinear functions and resulting mixed integer linear programming-based solution methods. Eur. J. Oper. Res. **275**(3), 1058–1071 (2019). https://doi.org/10.1016/j.ejor.2018.11.021
7. Rebennack, S., Kallrath, J.: Continuous piecewise linear delta-approximations for bivariate and multivariate functions. J. Optim. Theory Appl. **167**(1), 102–117 (2014). https://doi.org/10.1007/s10957-014-0688-2
8. Rebennack, S., Krasko, V.: Piecewise linear function fitting via mixed-integer linear programming. INFORMS J. Comput. **32**(2), 507–530 (2020). https://doi.org/10.1287/ijoc.2019.0890

9. Toriello, A., Vielma, J.P.: Fitting piecewise linear continuous functions. Eur. J. Oper. Res. **219**(1), 86–95 (2012). https://doi.org/10.1016/j.ejor.2011.12.030
10. Vielma, J.P., Ahmed, S., Nemhauser, G.: Mixed-integer models for nonseparable piecewise-linear optimization: Unifying framework and extensions. Oper. Res. **58**(2), 303–315 (2010). https://doi.org/10.1287/opre.1090.0721

An Outer-Approximation Algorithm for Maximum-Entropy Sampling

Marcia Fampa[1]([✉]) and Jon Lee[2]

[1] Universidade Federal do Rio de Janeiro, Rio de Janeiro, Brazil
fampa@cos.ufrj.br
[2] University of Michigan, Ann Arbor, MI, USA
jonxlee@umich.edu

Abstract. We apply the well-known MINLO outer-approximation algorithm (OA) to the maximum-entropy sampling problem (MESP), using the linx and NLP convex relaxations for MESP. We enhance our approach using disjunctive cuts.

Keywords: Maximum-entropy sampling · Mixed-integer nonlinear optimization · Outer approximation · Disjunctive cuts · No-good

1 Introduction

The *maximum-entropy sampling problem* is

$$\max\left\{\mathrm{ldet}C[S,S] \ : \ |S| = s\right\}. \qquad \text{(MESP)}$$
$$= \max\left\{\mathrm{ldet}C[S(x),S(x)] \ : \ \mathbf{e}^\top x = s, \ x \in \{0,1\}^n\right\},$$

where C is a positive-semidefinite symmetric matrix (having rank at least s) with rows/columns indexed by $N := \{1,2,\ldots,n\}$, $C[S,S]$ denotes the principal submatrix of C with rows/columns indexed by S, ldet denote log determinant, s is an integer satisfying $1 < s < n$, and $S(x)$ denotes the support of x. When C is the covariance matrix of multivariate Gaussian random-vector Y_N, then MESP corresponds to maximizing the "differential entropy" of a size-s subvector Y_S (see [24]). MESP has broad applicability in experimental design, and it has been applied in re-designing environmental-monitoring networks. Matrices C from that application can be prepared and validated (as being covariance matrices of a Gaussian random vector) from readily-obtainable data (see [2] and [1]).

MESP was introduced in [24], and shown to be NP-Hard in [18]. An eigenvalue-based upper-bounding method and branch-and-bound scheme was developed in [18]. Extensions of this upper-bounding method were given in: [6,9,17,21]. The eigenvalue based upper-bounding method was extended to the

M. Fampa was supported in part by CNPq grants 305444/2019-0 and 434683/2018-3. J. Lee was supported in part by AFOSR grant FA9550-19-1-0175 and ONR grant N00014-21-1-2135.

constrained version of MESP in [19]. Upper-bounding methods based on convex relaxation were given in [3–5,22,23], and then further combined and developed in [10,11].

All of these approaches were considered and developed in the context of the original branching scheme of [18]. We do note that [12,13] developed an OA (outer approximation) scheme connected with the so-called "BQP bound" for entropy. Helmberg (private communication) suggested (essentially) the BQP bound in 1995 (see [14,20]) to Anstreicher and Lee, but no one developed it at all until, independently and in different ways, [3] and [12,13]. Later, [10] leveraged and further developed the BQP bound.

Our goal is to take one of the most successful upper-bounding methods based on convex relaxation, Anstreicher's "linx bound" (see [4]), and instead of using it within the original branching scheme of [18] (as [4] and most the other bounding methods before it did), we marry it with the successful OA scheme of mixed-integer nonlinear optimization (MINLO), and other state-of-the-art techniques (e.g., disjunctive cuts). In fact, we found that this overall approach is enhanced using "variable fixing" based on the linx bound but also based on the so-called "NLP bound" for MESP (see [5]).

In Sect. 2, we describe OA for our setting, without yet specifying our relaxation. In Sect. 3, we describe the (scaled) linx relaxation (see [4]) and the (scaled) NLP relaxation (see [5]). We used both relaxations for variable fixing, but we base our OA solely on linx. Previously, linx had only been used in a branch-and-bound setting; ours is the first effort to build an OA based on linx. In Sect. 4, we use disjunctive cuts as a means to enhance our OA. In Sect. 5, we present the results of preliminary computational experiments on a benchmark problem from the literature. In Sect. 6, we describe some next steps.

Notation: • denotes matrix inner product, and ∘ denotes Hadamard product.

2   Outer Approximation

Our starting point is the OA algorithm for MINLO. But our needs are very specific. Our problem of interest has the form:

$$z := \max\left\{ f(x) \ : \ Ax \geq b, \ \mathbf{e}^\top x = s, \ x \in \{0,1\}^n \right\}, \tag{P}$$

where $f : \mathbb{R}^n \to \mathbb{R}$ is a smooth concave function; we use $Ax \geq b$ to accumulate cuts. Notably, we have only 0/1 variables, and our only nonlinearity is f. A convenient reference that covers such a situation is [15] (note that they work with minimization, their y variables are the integer ones, and they have continuous variables additionally, which they call x). While the framework of [15] does cover our situation, there appears to be no successful experience in the literature of successfully applying OA in the context of such a problem as P.

At a high level, at each iteration we solve a MILO "main problem". The main problem at iteration ℓ has the form,

$$z(\mathcal{T}^\ell) := \max \ \eta$$
$$\text{subject to:}$$
$$\eta \leq f(\tilde{x}) + \nabla f(\tilde{x})^\top (x - \tilde{x}), \ \text{for } \tilde{x} \in \mathcal{T}^\ell, \qquad (M^\ell)$$
$$A^\ell x \geq b^\ell,$$
$$\mathbf{e}^\top x = s, \ x \in \{0,1\}^n.$$

The set of linearization points $\mathcal{T}^\ell$ grows at each iteration, and also the $A^\ell x \geq b^\ell$ constraints accumulate inequalities. Therefore $z(\mathcal{T}^\ell)$ is nonincreasing as the algorithm progresses. The $z(\mathcal{T}^\ell)$ form a nonincreasing sequence, and *in spirit* these values are upper bounds on z. But to be precise, we will maintain a lower bound LB, which is the incumbent objective value—the best solution that we have seen so far for P. And actually $UB := \max\{z(\mathcal{T}^\ell), LB\}$ is an upper bound on z, because even though we will accumulate inequalities in $A^\ell x \geq b^\ell$ to exclude solutions that we have already seen, the best of those is captured in LB.

The OA algorithm for P. At iteration ℓ, we solve the MILO problem M^ℓ. If M^ℓ is infeasible, then we stop the algorithm with the determination that $z = LB$. If not, we let the optimal solution of M^ℓ be x^ℓ. We update $LB \leftarrow \max\{LB, f(x^\ell)\}$. We stop if $UB - LB < \tau$, an input optimality tolerance. If not, we update our set of linearization points $\mathcal{T}^{\ell+1} \leftarrow \mathcal{T}^\ell \cup \{x^\ell\}$. Next, we update $A^\ell x \geq b^\ell$ to form $A^{\ell+1} x \geq b^{\ell+1}$ by appending the no-good cut

$$\sum_{j \in N \setminus S^\ell} x_j \geq 1, \qquad (1)$$

where S^ℓ denotes the support of x^ℓ. The inequality excludes (only) x^ℓ from being visited again—even if it turns out that x^ℓ is optimal, its value was already captured when we updated LB.

An Enhancement. First, we can potentially improve LB at iteration ℓ by running a best-improvement local-search heuristic, starting from S^ℓ, using the true objective function $f(x(S))$, and using the neighbors $S \leftarrow S - i + j$, for $i \in S$, $j \in N \setminus S$. The local search visits a sequence of sets $S^\ell =: S_0^\ell, S_1^\ell, \ldots, S_p^\ell =: S_*^\ell$, starting at S^ℓ, finally reaching some locally-optimal S_*^ℓ. For each set S_k^ℓ visited along the way, $k \leq p$, we can do the following:

– update the linearization set: $\mathcal{T}^{\ell+1} \leftarrow \mathcal{T}^{\ell+1} \cup \{x(S_k^\ell)\}$, and
– update $A^{\ell+1} x \geq b^{\ell+1}$, for $2 \leq s \leq n-2$, by appending the no-good cut

$$\sum_{j \in N \setminus S_k^\ell} x_j \geq 2 \qquad (2)$$

(or equivalently, $\sum_{j \in S_k^\ell} x_j \leq s - 2$), which rules out precisely S_k^ℓ and its immediate neighbors (which we know cannot be improving on S_k^ℓ, because we are performing a best-improvement local search).

Finally, we update the lower bound: $LB \leftarrow \max\{LB, f(x(S_*^\ell))\}$; note that $f(x(S_*^\ell))$ is the best function value observed (over all solutions visited and their neighbors) in the current application of local search.

With more work, we can improve on the cuts (2), for $3 \leq s \leq n - 3$; we can choose a set $S^+ \subseteq N$, with $|S^+| = s+1$. Next, we search over all $S = S^+ - e$, for all $e \in S^+$, and all neighbors of such S. Taking the best of all of these solutions, we update LB, and then we have the no-good cut

$$\sum_{j \in N \setminus S^+} x_j \geq 2 \tag{3}$$

(or equivalently, $\sum_{j \in S^+} x_j \leq s-2$), which is valid for any solution with objective value greater than LB. This inequality is stronger than each $\sum_{j \in N \setminus S} x_j \geq 2$, for $S = S^+ - e$ with $e \in S^+$. We can see the validity directly, or by

$$\sum_{\substack{S = S^+ - e : \\ e \in S^+}} \frac{1}{s}\left(\sum_{j \in S} x_j \leq s - 2\right) \iff \sum_{j \in S^+} x_j \leq (s-1) - \frac{2}{s},$$

and the right-hand side rounds down to $s - 2$ for $s \geq 3$.

3   Convex Relaxations for MESP

There are several upper-bounding methods for MESP based on convex relaxation, and these methods can be effectively combined in various ways (see [10,11]). For our present investigation, we concentrate on two of the best ones.

For $\gamma > 0$, we have the *(scaled) linx bound*

$$\max \tfrac{1}{2} \left(\mathrm{ldet}(\gamma C \mathrm{Diag}(x) C + \mathrm{Diag}(e - x)) - s \log \gamma\right)$$
$$\text{subject to:} \tag{linx}$$
$$e^\top x = s, \; 0 \leq x \leq e.$$

The convex-programming linx bound was introduced in [4], which also includes a methodology for fixing some variables at 0/1, based on convex duality and a lower bound for MESP; also see [10] regarding optimizing the choice of γ.

For $\gamma > 0$, we have the *(scaled) NLP bound*

$$\max \; \mathrm{ldet}\left(\gamma \mathrm{Diag}(x^{\frac{p}{2}}) C \mathrm{Diag}(x^{\frac{p}{2}}) + \mathrm{Diag}((\gamma d)^x - \gamma d \circ x^p)\right) - s \log(\gamma)$$
$$\text{subject to:} \tag{NLP}$$
$$e^\top x = s, \; 0 \leq x \leq e,$$

where $\gamma > 0, d_i > 0$, and $p_i \geq 1$, and $u^w := (u_1^{w_1}, u_2^{w_2}, \ldots, u_n^{w_n})$, for all $u, w \in \mathbb{R}^n$.

The NLP bound was introduced in [5], which also presents methodology for choosing the parameters so as to make the objective function concave, and a methodology for fixing some variables at 0/1, based on convex duality and a lower bound for MESP.

4 Disjunctive Cuts

A powerful and well-known methodology for convexification is disjunctive programming; see [7]. In this section we work out a formulation of a "cut-generating nonlinear program" which we use to enhance our OA algorithm for MESP.

We rewrite the linx relaxation for MESP as

$$\max \eta$$
$$\text{subject to:}$$
$$\eta \le \tfrac{1}{2}(\operatorname{ldet} W - s\log(\gamma)), \tag{4}$$
$$L(x) = W,$$
$$\mathbf{e}^\top x = s, \ 0 \le x \le \mathbf{e},$$

where $L(x) := \gamma C \operatorname{Diag}(x) C + \operatorname{Diag}(\mathbf{e} - x)$. Note that we can write $L(x) = L_0 + L_1 x_1 + L_2 x_2 + \cdots + L_n x_n$, where

$$L_0 = I \quad \text{and} \quad L_k[i,j] = \left\{ \begin{array}{ll} C_{ik}C_{kj}, & \text{for } i \ne k \text{ or } j \ne k, \\ C_{ii}^2 - 1, & \text{for } i = j = k, \end{array} \right\} \text{ for } i,j,k \in N.$$

A finite disjunction by linear inequalities in $\mathbb{R}^n$ is described as $\vee_{k=1}^q (D_k x \ge d_k)$, where $D_k \in \mathbb{R}^{m_k \times n}$ and $d_k \in \mathbb{R}^{m_k}$. The disjunction is valid for P if every feasible solution of P satisfies $D_k x \ge d_k$ *for some* k. For example, a valid *elementary disjunction* chooses some j $(1 \le j \le n)$, and is of the form $(-x_j \ge 0) \vee (x_j \ge 1)$. In general, for the disjunction $\vee_{k=1}^q (D_k x \ge d_k)$, we define Q to be the convex closure of $\cup_{k=1}^q \{(\eta; x) \in \mathbb{R}^{n+1} \mid \eta \le \tfrac{1}{2}(\operatorname{ldet} L(x) - s\log(\gamma)), \ \mathbf{e}^\top x = s, \ 0 \le x \le \mathbf{e}, \ D_k x \le d_k\}$. Then, for a given $\alpha \in \mathbb{R}^n$ and each $k = 1, \ldots, q$, we consider

$$\beta_k^p := \min \ -\eta + \alpha^\top x \qquad\qquad \text{dual variables}$$
$$\text{subject to:}$$
$$\eta \le \tfrac{1}{2}(\operatorname{ldet} W - s\log(\gamma)), \qquad (z_k)$$
$$L(x) = W, \qquad (\Theta_k)$$
$$\mathbf{e}^\top x = s, \qquad (\nu_k) \tag{5}$$
$$x \ge 0, \qquad (w_k)$$
$$x \le \mathbf{e}, \qquad (u_k)$$
$$D_k x \le d_k. \qquad (\delta_k)$$

The Lagrangian dual problem of (5) can be formulated as

$$\beta_k^d := \max \ \alpha^\top x - \tfrac{1}{2}(\operatorname{ldet} W - s\log(\gamma)) + \Theta_k \bullet (W - L(x))$$
$$+ \nu_k(s - \mathbf{e}^\top x) - w_k^\top x + u_k^\top (x - \mathbf{e}) + \delta_k^\top (D_k x - d_k)$$
$$\text{subject to:} \tag{6}$$
$$-\tfrac{1}{2}W^{-1} + \Theta_k = 0,$$
$$\alpha_i - \Theta_k \bullet L_i - \nu_k - w_{k_i} + u_{k_i} + \delta_k^\top D_k[\cdot, i] = 0, \qquad i \in N,$$
$$w_k \ge 0, u_k \ge 0, \delta_k \ge 0.$$

Theorem 1. The following equation is valid for all feasible solutions of (6):

$$\alpha^\top x + \Theta_k \bullet (W - L(x)) + \nu_k(s - \mathbf{e}^\top x) - w_k^\top x + u_k^\top (x - \mathbf{e}) + \delta_k^\top (D_k x - d_k)$$
$$= \tfrac{n}{2} - \operatorname{Tr}(\Theta_k) - \mathbf{e}^\top u_k - d_k^\top \delta_k + s\nu_k.$$

Proof. From the group of constraints in (6), for all $i \in N$, we have

$$\alpha^\top x - \Theta_k \bullet \sum_{i=1}^n L_i x_i - \nu_k \mathbf{e}^\top x - w_k^\top x + u_k^\top x + \delta_k^\top D_k x = 0 \Rightarrow$$
$$\alpha^\top x + \Theta_k \bullet (W - L(x)) + \nu_k (s - \mathbf{e}^\top x) - w_k^\top x + u_k^\top (x - \mathbf{e}) + \delta_k^\top (D_k x - d_k) =$$
$$\Theta_k \bullet W - \Theta_k \bullet L_0 + \nu_k s - u_k^\top \mathbf{e} - \delta_k^\top d_k.$$

From the first constraint in (6) we have that $W = (2\Theta_k)^{-1} = \Theta_k^{-1}/2$, so the result follows.   □

Then, we can rewrite the dual problems (6), for $k = 1, \ldots, q$, as

$$\beta_k^d := \max \tfrac{1}{2}(\operatorname{ldet}(2\Theta_k) + s\log(\gamma)) - \operatorname{Tr}(\Theta_k) - \mathbf{e}^\top u_k - d_k^\top \delta_k + s\nu_k + \tfrac{n}{2}$$
$$\text{subject to:} \tag{7}$$
$$\alpha - \operatorname{diag}(\gamma C\Theta_k C - \Theta_k) - \nu_k \mathbf{e} + u_k + D_k^\top \delta_k \geq 0,$$
$$\Theta_k \succ 0, \ u_k \geq 0, \ \delta_k \geq 0.$$

(7) is a convex optimization problem. (5) is also convex and Slater's condition holds for it. Therefore, strong duality holds for these problems and $\beta_k^d = \beta_k^p =: \beta_k$ for all k. Moreover $\beta := \min_k \{\beta_k\}$ is the minimum value of $-\eta + \alpha' x$ for all $(\eta, x) \in Q$, so $-\eta + \alpha' x \geq \beta$ is a valid inequality for Q. Finally, β can be computed by

$$\max \beta$$
$$\text{subject to:}$$
$$\tfrac{1}{2}(\operatorname{ldet}(2\Theta_k) + s\log(\gamma)) - \operatorname{Tr}(\Theta_k) - \mathbf{e}^\top u_k - d_k^\top \delta_k + s\nu_k + \tfrac{n}{2} \geq \beta, \ k = 1, \ldots, q,$$
$$\alpha - \operatorname{diag}(\gamma C\Theta_k C - \Theta_k) - \nu_k \mathbf{e} + u_k + D_k^\top \delta_k \geq 0, \qquad k = 1, \ldots, q,$$
$$\Theta_k \succ 0, \ u_k \geq 0, \ \delta_k \geq 0, \qquad k = 1, \ldots, q.$$

Next, we formulate the cut-generation nonlinear program for a given solution $(\hat{\eta}, \hat{x})$ of (4), as

$$\min -\hat{\eta} + \alpha^\top \hat{x} - \beta \tag{CGNLP}$$
$$\text{subject to:}$$
$$\frac{1}{2}(\operatorname{ldet}(2\Theta_k) + s\log(\gamma)) - \operatorname{Tr}(\Theta_k) - \mathbf{e}^\top u_k - d_k^\top \delta_k + s\nu_k + \frac{n}{2} \geq \beta, \quad k = 1, \ldots, q,$$
$$\alpha - \operatorname{diag}(\gamma C\Theta_k C - \Theta_k) - \nu_k \mathbf{e} + u_k + D_k^\top \delta_k \geq 0, \qquad k = 1, \ldots, q,$$
$$\Theta_k \succ 0, \ u_k \geq 0, \ \delta_k \geq 0, \qquad k = 1, \ldots, q.$$

where the variables are $\alpha \in \mathbb{R}^n$, $\beta \in \mathbb{R}$, $u_k \in \mathbb{R}_+^n$, $\delta_k \in \mathbb{R}_+^{m_k}$, $\nu_k \in \mathbb{R}$, $\Theta_k \in \mathbb{S}_{++}^n$, for $k = 1, \ldots, q$. If the objective value of CGNLP is negative, then we obtain a valid inequality $-\eta + \alpha^\top x \geq \beta$ for Q, which is violated by $(\hat{\eta}, \hat{x})$.

Theorem 2. Let $(\hat{\alpha}, \hat{\beta}, \hat{u}_k, \hat{\delta}_k, \hat{\nu}_k, \hat{\Theta}_k)$ be a feasible for CGNLP with $\det \hat{\Theta}_k \leq 1/(2^n \exp(n)\gamma^s)$, then $\tau(\hat{\alpha}, \hat{\beta}, \hat{u}_k, \hat{\delta}_k, \hat{\nu}_k, \hat{\Theta}_k)$ is feasible for CGNLP, for all $\tau > 1$.

Proof. For the second group of constraints of CGNLP, the result is easily verified because the constraints are linear. The nonnegativity constraints clearly hold as well. So we focus on the first group of constraints. It suffices to show that

$$\mathrm{ldet}(2\tau\hat{\Theta}_k) + s\log(\gamma) + n \geq \tau(\mathrm{ldet}(2\hat{\Theta}_k) + s\log(\gamma) + n). \tag{8}$$

From the hypothesis, we have that $\mathrm{ldet}2\hat{\Theta}_k + s\log(\gamma) + n \leq 0 < n$. Then,

$$\log(\tau) \leq \tau - 1 \Rightarrow$$
$$\log(\tau)(\mathrm{ldet}(2\hat{\Theta}_k) + s\log(\gamma) + n) \geq (\tau - 1)(\mathrm{ldet}(2\hat{\Theta}_k) + s\log(\gamma) + n) \Rightarrow$$
$$n\log(\tau) \geq (\tau - 1)(\mathrm{ldet}(2\hat{\Theta}_k) + s\log(\gamma) + n) \Rightarrow$$
$$n\log(\tau) + \mathrm{ldet}(2\hat{\Theta}_k) + s\log(\gamma) + n \geq \tau(\mathrm{ldet}(2\hat{\Theta}_k) + s\log(\gamma) + n),$$

and (8) follows.   □

Theorem 2 implies that when $\det\Theta_k$ is small for a solution of CGNLP having negative objective value, then CGNLP is unbounded. So we conclude, as is typical for constraint-generating programs in the context of disjunctive programming, that we should add a normalization constraint to CGNLP to bound its objective function from below. In our computational experiments we used the normalization: $\|\alpha\|_1 \leq 1$. It might be, in our nonlinear setting, that our normalization could exclude some violated cuts from consideration; but from a practical point of view, considering numerical stability, it is valuable to restrict the coefficients of disjunctive cuts to reasonable values.

5 Experiments

We coded our algorithm in `Matlab` R2016b, and we ran our experiments under Windows 10, on an Intel Core i5-8265 @ 1.60 GHz processor with 8 GB of RAM. We solved the cut-generation nonlinear program CGNLP with `Mosek`, v9.2, the main problem M^ℓ with Gurobi, v9.0.3, and the convex relaxation linx with the conical-optimization software `SDPT3`, v4, (see [25]), within the `Yalmip` (v20200116) `Matlab` framework. As the convex relaxation NLP cannot be solved by `SDPT3`, we have coded an interior-point algorithm to compute the NLP bound, also in `Matlab`. The solution procedure is the same as described in [5, Sec. 3].

We consider a benchmark covariance matrix obtained from J. Zidek (University of British Columbia), coming from an application to re-designing an environmental monitoring network; see [16,17]. This matrix have been used extensively in testing and developing algorithms for MESP; see [3–6,9,17–19,21].

We summarize our complete algorithmic framework in Algorithm 1, where we first apply an iterative procedure to fix variables in MESP based on a lower bound obtained with a greedy/interchange heuristic described in [18], and on the upper bounds given by linx and NLP (lines 2–13). We note that the parameters used on both relaxations are computed for the original problem and then recomputed each time the problem is reduced by the variable-fixing procedure.

To optimize the parameter γ in linx we do a one-dimensional search, exploiting the fact that the linx bound is convex in the logarithm of γ (see [10]). Concerning the parameters γ, d and p in NLP, three different strategies are presented in [5] for choosing them in order to have NLP proven to be a convex program. In our numerical experiments, we have chosen these parameters based on the so-called "NLP-Trace" strategy, where d minimizes the trace of $\mathrm{Diag}(d) - C$, subject to $\mathrm{Diag}(d) - C$ being positive semidefinite. Once d is chosen, the scaling parameter γ is selected in the interval $[1/d_{\max}, 1/d_{\min}]$. We note that, unlike what we have for linx, there is no theory to pick the best γ for NLP. Once d and γ are fixed, p can be determined to generate the best possible bound (see [5] for more details on the selection of the NLP parameters).

The next procedure in Algorithm 1 iteratively adds disjunctive cuts to the linx relaxation of the possibly reduced problem, by solving CGNLP, until no cut violated by the current solution of linx is found or the maximum number of cuts is reached (lines 18–26).

Finally, we apply the OA algorithm to MESP considering its formulation given by linx, but constraining x to be binary (lines 27–35). The set of linearization points T^1 used for the first main problem solved (M^1), includes the solutions of the heuristic and of the NLP and linx relaxations already solved, the points visited when our local-search procedure is applied to the heuristic solution, points constructed with the dual variables of CGNLP, and some convex combinations of the linx solution with the heuristic solution (lines 16,25).

The no-good cuts (2) are derived for the feasible solutions for MESP obtained during the execution of Algorithm 1, as well as for the points visited when applying the local-search procedure from them (lines 17,33) (in some experiments, we also derive the stronger cuts (3) for the locally-optimal points obtained by the local search). As the cuts are generated, they are included in M^ℓ. We note that these cuts do not lead the algorithm to miss a possible optimal solution because the best current solution found is saved during its execution. After solving M^ℓ at each iteration of the OA algorithm, we also solve linx with the no-good cuts (line 34), and include its solution in $T^{\ell+1}$. The points visited when applying the local-search procedure from the solution of M^ℓ, and some convex combinations of the linx solution with the solution of M^ℓ are also included in $T^{\ell+1}$ (line 35).

It is well-known that the successful application of OA is highly related to the set of linearization points used and the set of valid cuts included in the M^ℓ formulation during the execution of the algorithm. We highlight three procedures in Algorithm 1, where we invest significant computational effort in these two features of OA. These procedures are specifically identified below, as well as the labels that represent them in our tables.

- "disj": represents the procedure in lines 18–26 of Algorithm 1, where we iteratively solve CGNLP and linx (with the disjunctive cuts generated by CGNLP). At each iteration, we select the variable to do the elementary disjunction in CGNLP as the most fractional variable in the solution of linx. The main purpose of this procedure is to generate linearization points for M^1 that we hope to be close to the optimal solution of MESP, leading to a good

---

Algorithm 1: Algorithmic framework.
Input: $C \in \mathbb{S}_+^n$, s, ϵ, *max.num.disj.cuts*.
Output: x^*

1 $\mathcal{T}^1 := \emptyset$, *no.good.cuts*$^1 = \emptyset$, *disj.cuts* $= \emptyset$, $\ell := 1$;
2 $\bar{x} := \text{heuristic}(C, s)$;
3 $LB := \text{ldet}C[S(\bar{x}), S(\bar{x})]$;
4 *fix* := *true*;
5 **while** *fix* **do**
6 $\quad$ *fix* := *false*;
7 $\quad$ $[UB_{\text{NLP}}, x_{\text{NLP}}^*] := \text{NLP}(C, s)$;
8 $\quad$ $[UB_{\text{linx}}, x_{\text{linx}}^*] := \text{linx}(C, s)$;
9 $\quad$ **if** *possible to fix variables* **then**
10 $\quad\quad$ Update C, and s;
11 $\quad\quad$ $\bar{x} := \text{heuristic}(C, s)$;
12 $\quad\quad$ $LB := \text{ldet}C[S(\bar{x}), S(\bar{x})]$;
13 $\quad\quad$ *fix* := *true*;

14 $[\bar{x}^+, \mathcal{N}(\bar{x})] := \text{local-search}(C, s, \bar{x})$;
15 $LB := \text{ldet}C[S(\bar{x}^+), S(\bar{x}^+)]$;
16 $\mathcal{T}^1 := \mathcal{T}^1 \cup \{x_{\text{NLP}}^*, \bar{x}^+\} \cup \{\lambda x_{\text{linx}}^* + (1-\lambda)\bar{x} : \lambda = 0, .25, .5, .75, 1\} \cup \mathcal{N}(\bar{x})$;
17 *no.good.cuts* := *no.good.cuts* $\cup$ no-good-cuts($\{\bar{x}, \bar{x}^+\} \cup \mathcal{N}(\bar{x})$);
18 *violation* := *true*;
19 **while** *violation* && $|disj.cuts| < max.num.disj.cuts$ **do**
20 $\quad$ *violation* := *false*;
21 $\quad$ $[\tilde{x}, d.cut] := \text{CGNLP}(C, s, x_{\text{linx}}^*)$;
22 $\quad$ **if** *d.cut is violated by* x_{linx}^* **then**
23 $\quad\quad$ *disj.cuts* := *disj.cuts* $\cup \{d.cut\}$;
24 $\quad\quad$ $[UB_{\text{linx}}, x_{\text{linx}}^*] := \text{linx.cuts}(C, s, \gamma, disj.cuts)$;
25 $\quad\quad$ $\mathcal{T}^1 := \mathcal{T}^1 \cup \{x_{\text{linx}}^*, \tilde{x}\}$;
26 $\quad\quad$ *violation* := *true*;

27 $UB := \min\{UB_{\text{linx}}, UB_{\text{NLP}}\}$;
28 **while** $UB - LB > \epsilon$ **do**
29 $\quad$ $[UB, \hat{x}] := M^\ell(C, s, \mathcal{T}^\ell, no.good.cuts^\ell)$;
30 $\quad$ $[x^*, \mathcal{N}(\hat{x})] := \text{local.search}(C, s, \hat{x})$;
31 $\quad$ **if** $LB < \text{ldet}C[S(x^*), S(x^*)]$ **then**
32 $\quad\quad$ $LB := \text{ldet}C[S(x^*), S(x^*)]$;
33 $\quad$ *no.good.cuts*$^{\ell+1}$:= *no.good.cuts*$^\ell \cup$ no.good.cuts($\{x^*, \hat{x}\} \cup \mathcal{N}(\hat{x})$);
34 $\quad$ $x_{\text{linx}}^* := \text{linx.cuts}(C, s, \gamma, no.good.cuts^\ell)$;
35 $\quad$ $\mathcal{T}^{\ell+1} := \mathcal{T}^\ell \cup \{x^*\} \cup \{\lambda x_{\text{linx}}^* + (1-\lambda)\hat{x} : \lambda = 0, .5, 1\} \cup \mathcal{N}(\hat{x})$;
36 $\quad$ $\ell := \ell + 1$;

---

outer-approximation of the problem in the vicinity of its optimal solution. The linearization points included in $\mathcal{T}^1$ at each iteration are the solution of linx and points constructed with the dual variables associated to CGNLP (see [8], for example).

- "(3)": represents the procedure that improves on the no-good cut (2) corresponding to the locally-optimal point obtained by our local-search procedure, replacing it with the cut (3). When this procedure is executed, this stronger cut is included in the formulation of M^ℓ (line 29) and also of linx (line 34).
- "lOA": refers to the solution of linx in line 34 of Algorithm 1. The purpose of solving linx with the no-good cuts at each iteration of the OA algorithm is to generate more linearization points for M^ℓ. After solving it, convex combinations of its solution and the solution of M^ℓ are added to the set of linearization points (line 35).

In order to verify the impact of the three procedures described above, we have implemented eight different versions of Algorithm 1, using or not each one of them. The two possibilities are identified in our tables, respectively with (y) or (n) after the name of the procedure.

We summarize our results in Tables 1 and 2. In the first four columns of Table 1, we show the ability of the fixing procedure shown in lines 2–13, to reduce the problem ($\bar{n}$ and $\bar{s}$ are the updated values of n and s after the procedure is applied). We see that the procedure is very effective for these instances. We were able to reduce the size of the covariance matrix in up to 49%, which was significant to the successful application of OA. In the last two columns of Table 1, we show the number of violated disjunctive cuts that were added to linx during the execution of Algorithm 1 when considering disj(y), and the consequent percentage decrease in the absolute gap between the solutions of linx and the heuristic. We have set the parameter $max.num.disj.cuts := 30$. We see that for all instances, except for the one with $s = 44$, the disjunctive cut was able to cut the solution of linx every time it was computed. Furthermore, the cuts reduce the gap between linx and the heuristic solutions by 8.3% on average.

In the columns 5–9 of Table 1, we show the number of iterations of the OA algorithm when using the different implementations of Algorithm 1. The results show an expected trade-off between the execution of the procedures and the number of iterations and we can verify that the three procedures are effective in reducing it. Considering the use of each procedure independently, we find that lOA, (3) and disj are able to reduce the number of OA iterations by 37%, 1.5%, and 15% respectively. By combining them we conclude that no procedure dominates another as it is always possible to reduce the number of iterations even further by adding one procedure to another. Comparing procedure (yyy) to (nnn), we see a 40.5% reduction. The more effective procedure lOA shows the importance of obtaining good linearization points at each iteration of OA, which we accomplish by considering solutions for linx with constraints that cut solutions for MESP already known. The same objective of generating good linearization points is achieved with disj, which calculates linearization points for M^1 by iteratively solving CGNLP and linx. We note that the reduction in the number of iterations led by the use of the stronger cut (3) can increase if we strengthen each cut (2) generated and not just the one corresponding to the locally-optimal solution obtained by the local search each time it is applied. However, in Table 2 we show the most significant elapsed times spent in Algorithm 1, and we conclude that with our current implementation, the time to compute these stronger cuts does not compensate. On the other side, we see that the time to execute lOA is compensated by the decrease in the number of iterations. Finally, the solution of CGNLP is still too time consuming, but in this case, we believe there is room for improvement with the use of a more appropriate solver.

6   Next Steps

There is room for a lot of tuning in our OA approach to MESP to improve its performance, with the ultimate goal being to perform better than branch-and-bound. For example, our methodology for choosing linearization points can be

Table 1. Fixing reduction/OA iterations/Effect of disjunction cuts

				disj(n) (3)(n) lOA(n)	disj(n) (3)(n) lOA(y)	disj(n) (3)(y) lOA(n)	disj(n) (3)(y) lOA(y)	disj(y) (3)(n) lOA(n)	disj(y) (3)(n) lOA(y)	disj(y) (3)(y) lOA(n)	disj(y) (3)(y) lOA(y)	disj. cuts added	perc. gap decr.
n	s	$\bar{n}$	$\bar{s}$										
63	37	57	31	1230	919	1181	943	1099	756	1099	970	30	5.8
63	38	52	27	1663	936	1452	985	1157	1093	1316	862	30	5.9
63	39	51	27	1342	752	1332	657	1257	816	1019	749	30	5.9
63	40	48	25	1165	747	972	782	988	587	1052	596	30	6.7
63	41	46	24	782	641	1074	665	901	492	932	521	30	8.4
63	42	41	20	934	679	1152	532	1032	616	938	675	30	7.6
63	43	41	21	1025	743	1077	487	827	565	755	612	30	9.6
63	44	40	21	1004	540	869	590	807	648	865	671	13	9.1
63	45	40	22	894	486	896	537	642	477	837	509	30	5.5
63	46	37	20	948	503	818	683	629	538	662	396	30	8.9
63	47	34	18	701	473	747	464	573	460	542	440	30	7.6
63	48	32	17	519	327	501	300	536	339	413	353	30	10.8
63	49	32	18	479	270	381	277	317	277	397	248	30	10.9
63	50	32	19	312	173	348	156	277	175	286	201	30	13.5
Average:				928.4	584.9	914.3	575.6	788.7	559.9	793.8	557.4	28.8	8.3

Table 2. Average elapsed time (seconds)

	disj(n) (3)(n) lOA(n)	disj(n) (3)(n) lOA(y)	disj(n) (3)(y) lOA(n)	disj(n) (3)(y) lOA(y)	disj(y) (3)(n) lOA(n)	disj(y) (3)(n) lOA(y)	disj(y) (3)(y) lOA(n)	disj(y) (3)(y) lOA(y)
linx	1.1	441.7	1.1	300.6	333.9	608.2	342.7	601.3
M^{ℓ}	558.0	248.9	535.2	157.2	302.3	152.6	318.5	143.3
local search	510.6	339.6	1658.3	730.0.9	378.6	220.6	1335.5	710.7
CGNLP	0.0	0.0	0.0	0.0	709.6	641.4	715.1	626.3
Total time	1069.7	1030.2	2194.6	1187.8	1724.4	1622.8	2711.9	2081.6

improved and would also need to be rethought to handle the constrained version of MESP. Additionally, the quality of our disjunctive cuts may well improve by normalizing differently for CGNLP or choosing different disjunctions.

There are many extensions of our approach that appear promising. [11] demonstrated that variable fixing based on the so-called "factorization bound" can be fruitfully combined with variable fixing based on the linx bound. Furthermore, for many instances, the factorization bound gives the best bound, and so it maybe be useful to do OA based on it too. Also, there is a "bound mixing" methodology (see [10]) which may be useful for our OA.

References

1. Al-Thani, H., Lee, J.: MESgenCov (2020). github.com/hessakh/MESgenCov
2. Al-Thani, H., Lee, J.: An R package for generating covariance matrices for maximum-entropy sampling from precipitation chemistry data. SN Oper. Res. Forum **1**, Article 17, 21 (2020)
3. Anstreicher, K.M.: Maximum-entropy sampling and the Boolean quadric polytope. J. Glob. Optim. **72**, 603–618 (2018)
4. Anstreicher, K.M.: Efficient solution of maximum-entropy sampling problems. Oper. Res. **68**, 1826–1835 (2020)
5. Anstreicher, K.M., Fampa, M., Lee, J., Williams, J.: Using continuous nonlinear relaxations to solve constrained maximum-entropy sampling problems. Math. Program. Ser. A **85**, 221–240 (1999)
6. Anstreicher, K.M., Lee, J.: A masked spectral bound for maximum-entropy sampling. In: Di Bucchianico, A., Läuter, H., Wynn, H.P. (eds.) mODa 7-Advances in Model-Oriented Design and Analysis. Contributions to Statistics, pp. 1–12. Physica, Heidelberg (2004). https://doi.org/10.1007/978-3-7908-2693-7_1
7. Balas, E.: Disjunctive programming. Springer, Heidelberg (2018). https://doi.org/10.1007/978-3-030-00148-3
8. Bonami, P., Linderoth, J., Lodi, A.: Disjunctive cuts for mixed integer nonlinear programming problems. In: Mahjoub, A. (ed.) Progress in Combinatorial Optimization, pp. 521–544. John Wiley & Sons Inc. (2011)
9. Burer, S., Lee, J.: Solving maximum-entropy sampling problems using factored masks. Math. Program. Ser. B **109**, 263–281 (2007)
10. Chen, Z., Fampa, M., Lambert, A., Lee, J.: Mixing convex-optimization bounds for maximum-entropy sampling. Math. Prog. Ser. B **188**, 539–568 (2021)
11. Chen, Z., Fampa, M., Lee, J.: On computing with some convex relaxations for the maximum-entropy sampling problem (2022). arxiv.org/abs/2112.14291
12. Choi, H.L., How, J., Barton, P.: An outer-approximation algorithm for generalized maximum entropy sampling. In: Proceedings of the ACC 2008, pp. 1818–1823. IEEE (2008)
13. Choi, H.L., How, J., Barton, P.: An outer-approximation approach for information-maximizing sensor selection. Optim. Lett. **7**, 745–764 (2013)
14. Fedorov, V., Lee, J.: Design of experiments in statistics. In: Wolkowicz, H., Saigal, R., Vandenberghe, L. (eds.) Handbook of Semidefinite Programming. International Series in Operations Research & Management Science, vol. 27, pp. 511–532. Springer, Boston, MA (2000). https://doi.org/10.1007/978-1-4615-4381-7_17
15. Fletcher, R., Leyffer, S.: Solving mixed integer nonlinear programs by outer approximation. Math. Program. **66**, 327–349 (1994)

16. Guttorp, P., Le, N.D., Sampson, P.D., Zidek, J.V.: Using entropy in the redesign of an environmental monitoring network. In: Patil, G., Rao, C., Ross, N. (eds.) Multivariate Environmental Statistics, vol. 6, pp. 175–202. North-Holland (1993)
17. Hoffman, A., Lee, J., Williams, J. (2001). New upper bounds for maximum-entropy sampling. In: Atkinson, A.C., Hackl, P., Mäller, W.G. (eds.) mODa 6–Advances in Model-Oriented Design and Analysis. Contributions to Statistics, pp. 143–153. Physica, Heidelberg. https://doi.org/10.1007/978-3-642-57576-1_16
18. Ko, C.W., Lee, J., Queyranne, M.: An exact algorithm for maximum entropy sampling. Oper. Res. 43, 684–691 (1995)
19. Lee, J.: Constrained maximum-entropy sampling. Oper. Res. 46, 655–664 (1998)
20. Lee, J.: Maximum entropy sampling. In: El-Shaarawi, A., Piegorsch, W. (eds.) Encyclopedia of Environmetrics, 2nd ed., pp. 1570–1574. Wiley (2012)
21. Lee, J., Williams, J.: A linear integer programming bound for maximum-entropy sampling. Math. Program. Ser. B 94, 247–256 (2003)
22. Li, Y., Xie, W.: Best principal submatrix selection for the maximum entropy sampling problem: scalable algorithms and performance guarantees (2020). preprint at: https://arxiv.org/abs/2001.08537
23. Nikolov, A.: Randomized rounding for the largest simplex problem. In: Proceedings of STOC 2015, pp. 861–870. ACM, New York (2015)
24. Shewry, M.C., Wynn, H.P.: Maximum entropy sampling. J. Appl. Stat. 46, 165–170 (1987)
25. Toh, K.C., Todd, M.J., Tütüncü, R.H.: SDPT3: a Matlab software package for semidefinite programming, v. 1.3. Optim. Meth. Soft. 11(1–4), 545–581 (1999)

Mitigating Anomalies in Parallel Branch-and-Bound Based Algorithms for Mixed-Integer Nonlinear Optimization

Prashant Palkar[1]([✉]) and Ashutosh Mahajan[2]

[1] Institute of Mathematics, University of Augsburg, 86159 Augsburg, Germany
prashant.palkar@math.uni-augsburg.de
[2] Industrial Engineering and Operations Research, IIT Bombay,
Powai, Mumbai 400076, India
amahajan@iitb.ac.in
http://www.ieor.iitb.ac.in/ppalkar

Abstract. We address detrimental anomalies in parallel versions of two state-of-the-art algorithms for convex mixed-integer nonlinear programs (MINLPs): nonlinear branch-and-bound (NLP-BB) and the LP/NLP based branch-and-bound (QG). A detrimental anomaly is when a parallel algorithm performs worse than its sequential counterpart. Unambiguous node selection functions have been developed in the past to avoid these anomalies. We extend this notion of unambiguity to subroutines for branching and generating linearization cuts. We implement the proposed unambiguous branching and cut generation strategies alongside unambiguous node selection in NLP-BB and QG in the open-source MINLP solver Minotaur. Our computational experiments on convex instances from the MINLPLib library show that detrimental anomalies can be reduced to a great extent in practical algorithms. We also compare these algorithms with opportunistic parallel versions. Our results highlight that opportunistic versions perform better in terms of wall clock times, while the deterministic versions avoid detrimental anomalies with theoretically established guarantees and also provide reproducible results, a feature that is desirable while developing parallel algorithms. We recommend settings in Minotaur that yield opportunistic or deterministic runs for the parallel NLP-BB and QG algorithms.

Keywords: Mixed-integer nonlinear programming · Parallel branch-and-bound · Anomalies · LP/NLP based branch-and-bound

1 Introduction

Mixed-Integer Nonlinear Programs (MINLPs) are discrete optimization problems that involve integer-constrained decision variables and nonlinear functions. An MINLP can be mathematically expressed as

$$\underset{x}{\text{minimize}}\, c^T x \qquad\qquad\qquad (P)$$

$$\text{subject to}\ \ g(x) \leq b,$$

$$x \in \mathcal{X}, x_j \in \mathbb{Z} \quad \forall j \in \mathcal{I},$$

I. Ljubić et al. (Eds.): ISCO 2022, LNCS 13526, pp. 143–156, 2022.
https://doi.org/10.1007/978-3-031-18530-4_11

where the set $\mathcal{I}$ contains the indices of integer-constrained variables, the constraint functions $g : \mathbb{R}^n \to \mathbb{R}^m$ are assumed to be nonlinear and twice continuously differentiable, and the set $\mathcal{X}$ is non-empty and compact. MINLPs arise in many important real-life applications [2,4,16], however, they are difficult to solve as their special cases such as MILPs belong to the class of NP-hard problems [8].

State-of-the-art algorithms for MINLPs are based on branch-and-bound (BB) framework. It partitions the search space recursively into smaller and usually disjoint regions until a solution is found, or no further partitioning is possible. BB starts by solving a relaxation that is easier to solve and whose solution provides a valid lower bound on the optimal value (say z^*) of (P). Then the search-space is branched to create smaller subproblems. If a solution to any of the subproblems is feasible for (P), its objective value provides an upper bound on z^*. This partitioning continues until the lower bound and the upper bound on z^* coincide. This setup can be analyzed as a tree-search where the tree-nodes denote the subproblems and the edges denote the branches that divide a subproblem.

The BB framework is naturally suitable for parallel processing due to the presence of independent subproblems. The performance of a parallel BB algorithm depends on the way it has been implemented on a software platform. Different MILP and MINLP solvers use distinctive data structures, classes, subsolvers, etc., which make each implementation/solver unique in its own way. Several studies have addressed the practical aspects of parallel BB algorithms [3,6,7,14,18]. Theoretical analysis of parallel BB algorithms has also been performed earlier [9–11,13]. Some of these studies focus on the unpredictable performance of parallel branch-and-bound algorithms with respect to the number of processors used. These phenomena are referred to as "anomalies". We focus on reducing "detrimental anomalies" that may arise due to two subroutines in BB for convex MINLPs: selecting a branching candidate, and adding linearization inequalities at a node. While the issue of node-selection has been studied in the past [10,11], these two aspects have not been addressed to the best of our knowledge. We concentrate on two well-known BB based algorithms for convex MINLPs: NLP-BB and LP/NLP based BB, also called QG (an acronym for Quesada and Grosmann [15] who proposed this algorithm).

The outline of the paper is as follows. Section 3 presents the opportunistic parallel extensions of NLP-BB and QG in Minotaur[1] [12]. In Sect. 4, we define unambiguous branching functions and show how a parallel NLP-BB algorithm without detrimental anomalies can be implemented using unambiguous algorithmic components (referred to as "nondetrimental" NLP-BB). Section 5 presents unambiguous functions for generating linearization cuts and a nondetrimental QG algorithm. Section 6 shows the computational performance of the opportunistic and the nondetrimental algorithms and Sect. 7 presents the conclusions.

---

[1] Available at http://github.com/minotaur-solver/minotaur.

2   Anomalies in Parallel Algorithms

Existence of anomalies in parallel tree-search based algorithms is shown in [9] and sufficient conditions to avoid detrimental anomalies caused by ambiguous node selection are proposed in [10]. These results are based on the concept of an "iteration" in a parallel BB algorithm, which we now define.

Definition 1 (Iteration). *An iteration is one cycle of all operations such as node-presolving, node-processing, branching, adding cuts, checking stopping conditions, inserting new nodes in the memory, etc., that a set of processors perform simultaneously.*

Instead of the wall clock time, the number of iterations taken by a parallel algorithm is suited for our analysis because it does not vary based on the hardware, software implementation and factors like the computational load on a system at a point in time. Let k denote the number of processors, and $T(k, 0)$ be the number of iterations taken, where 0 indicates that we seek an exact optimal solution (see [10] for an analysis when solutions with a predefined tolerance are acceptable). We make the following assumptions.

Assumption 1. *The processors operate "synchronously" i.e., at most k open nodes are selected and solved simultaneously in an iteration.*

Assumption 1 restricts a processor from starting a new cycle of operations until all the other processors have also finished their part in the iteration. Due to this synchronization, the processors that finish earlier incur a waiting time. This waiting for synchronization makes the procedure reproducible, but possibly at a cost of longer running time.

Assumption 2. *All the subsolvers used within the algorithm are deterministic.*

The term "deterministic" in Assumption 2 means that the same solution would be obtained using the subsolvers if the same initial conditions are provided to them. Typically, an LP or an NLP subsolver is used in most MINLP algorithms. Assumption 2 holds for certain LP and NLP solvers subject to the use of sufficiently small tolerance values.

Definition 2 [10]. *A behavior exhibited by a parallel tree-search algorithm using k processors is a detrimental anomaly if $T(k, 0) > T(1, 0)$.*

A depiction of a detrimental anomaly is shown in Figs. 1–2 where a parallel algorithm ($k = 2$) performs worse than the sequential algorithm (depicted as $k = 1$). Table 1 lists the nodes that are processed in each iteration by each processor for both the cases. The number of iterations required when two processors are used (5 iterations) is more than that for the sequential algorithm (3 iterations). As mentioned in [10], one of the main reasons for detrimental anomalies is the ambiguous selection of nodes in the sequential and the parallel versions. One can observe in Table 1 that while the sequential algorithm processes node labeled 4

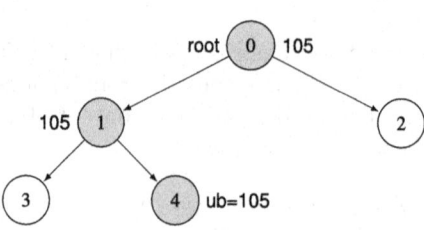

Fig. 1. A sequential branch-and-bound tree ($k = 1$). The algorithm processes nodes $0, 1, 4$ respectively and then terminates.

Table 1. Nodes solved in different iterations of the sequential ($k = 1$) and the parallel algorithm ($k = 2$).

	Node processed		
	$k = 1$	$k = 2$	
Iter	thread0	thread0	thread1
1	0	0	–
2	1	1	2
3	4	9	10
4	–	11	12
5	–	13	14

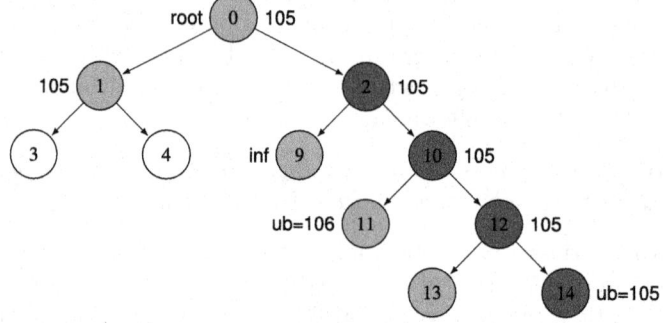

Fig. 2. A BB tree explored using two threads ($k = 2$). Two nodes are solved in parallel in each iteration except in the first iteration. thread0 solves the yellow colored nodes and thread1 solves the red ones. The algorithm terminates after node 14 is processed.

in Iteration 3, the parallel algorithm selects nodes 9 and 10 (and not 4) and ends up processing six other nodes (in three more iterations).

Sufficient conditions on node selection functions to avoid detrimental anomalies are as follows.

Definition 3 [10]. *Given a set of open nodes $\mathcal{P}$, a heuristic node selection function $h(\cdot)$ is referred to as unambiguous if it satisfies the following two properties.*

1. $h(P_i) \neq h(P_j)$, for any $P_i, P_j \in \mathcal{P}$, $P_i \neq P_j$
2. $h(P_i) \leq h(P_j)$, if P_j is a descendant of P_i.

All nodes encountered while moving down the tree starting from a node are referred to as the "descendants" of that node. Similarly, the nodes encountered while moving up the tree starting from a node are called its "ancestor" nodes. The function h maps nodes in $\mathcal{P}$ to real values. A node with a lower heuristic function value has a higher priority for getting selected.

Theorem 1 [10]. *If h is unambiguous, then $T(k, 0) \leq T(1, 0)$.*

Although Figs. 1–2 demonstrate how ambiguity in node selection cause detrimental anomalies, ambiguity in other components of BB can also give rise to detrimental anomalies. In fact, the tree might evolve differently in presence of an ambiguous algorithmic component such as branching, cut generation or primal heuristics when using different number of processors. State-of-the-art branching rules decide upon a branching disjunction using the scores calculated based on the information obtained from the nodes already processed in the BB tree. Considering the BB trees shown in Figs. 1–2, Table 2 demonstrates how an ambiguous branching decision occurs if all available information is used for branching at the node labeled 4. The column "br. info." mentions the set of nodes processed. It can be seen in Iteration 3 that the sequential ($k = 1$) and parallel ($k = 2$) algorithms have access to different information sets and are likely to select a different branching disjunction at the node labeled 4, which will result in different child nodes, hence different BB trees.

Table 2. Ambiguous branching resulting in generation of dissimilar nodes

	$k = 1$		$k = 2$			
	thread0		thread0		thread1	
Iteration	node	br. info.	node	br. info.	node	br. info.
1	0	–	0	–	–	–
2	1	{0}	1	{0}	2	{0}
3	4	{0,1}	4	{0,1,2}	3	{0,1,2}

In [13, Assump. (A1)], the authors assume that the branching scheme at a node P_i depends only on the information obtained along the path from P_i to the root node. However, their analysis does not include unambiguous branching functions. In this work, we explicitly define functions for unambiguous branching and unambiguous cut generation. We compare the performance of parallel nondetrimental algorithms obtained using these functions with parallel "opportunistic" algorithms in Minotaur.

3   Opportunistic Parallel Branch-and-Bound in Minotaur

The parallel implementation of NLP-BB in Minotaur uses classes available in its MINLP framework. A single pool of open nodes ($\mathcal{P}$) is maintained by the class `TreeManager` to be processed simultaneously by k threads (one corresponding to each of the k processors) until $\mathcal{P}$ is empty. In the beginning, the first thread solves the root relaxation and creates two child subproblems (if not pruned). An idle thread then requests a node (if any) from `TreeManager` for processing. Each thread sends the child nodes generated after branching to the `TreeManager` that are added to $\mathcal{P}$. Node-level parallel extensions of NLP-BB and QG have been implemented in Minotaur [17, Sec. 4.1 and Sec. 4.3] that use OpenMP `for` loops that induce synchronization at the end of the loop. This section mentions new node-level parallel extensions of the NLP-BB and the QG algorithms that process multiple tree-nodes simultaneously in a more "opportunistic" way compared to the shared-memory parallel algorithms presented in [17, Sec. 4.1 and Sec. 4.3]. Here, we use the term opportunistic in two respects. First, the threads attempt to get a new open node from $\mathcal{P}$ as soon as they finish a cycle without waiting for the other threads. Second, this algorithm is "not deterministic" in terms of reproducibility of results.

Parallel NLP-BB. We refer to the opportunistic parallel extension of NLP-BB in Minotaur as *mcbnbOpp* (here, the prefix *mc* indicates multi-core). This algorithm completely avoids the synchronization of threads after each iteration unlike the algorithm in [17, Sec. 4.1] that has an implicit synchronization at the end of OpenMP `for` loop. This means that if a thread finishes processing a node earlier than the other threads, it will not wait.

Parallel QG. The parallelization scheme of opportunistic parallel extension of QG (*mcqgOpp*) is similar to that of *mcbnbOpp*. The two major dissimilarities from *mcbnbOpp* are that LPs (instead of NLPs) are solved at nodes, and linearization cuts are generated at nodes that yield integer solutions by solving an NLP in which integer variables are fixed.

4   Reducing Detrimental Anomalies in Parallel NLP-BB

If none of the algorithmic components in NLP-BB induce ambiguity, the overall algorithm would avoid detrimental anomalies. First, we focus on branching and define unambiguous branching functions.

4.1   Unambiguous Branching Functions

Typically, a node is constructed by adding a set of branching constraints to its parent. We define unambiguity (as in Definition 3) for branching functions for creating simple variable disjunctions. This definition can be easily extended to other branching functions. Variable disjunctions usually select a branching variable from a set $\mathcal{I}_C := \{j \in \mathcal{I} : x_j^* \notin \mathbb{Z}\}$ of candidates at a node with an optimal solution x^*.

Definition 4. *Consider a node $P_i \in \mathcal{P}$ that has been processed and that is not pruned. Let x^* be the optimal solution of P_i and $\mathcal{I}_C := \{j \in \mathcal{I} : x_j^* \notin \mathbb{Z}\}$. A branching variable selection function $\nu(\cdot)$ over $\mathcal{I}_C$ at P_i is referred to as unambiguous if*

- *$\nu(\cdot)$ uses information obtained only from P_i and its ancestors,*
- *$\nu(j) \neq \nu(k)$ for $j, k \in \mathcal{I}_C, j \neq k$.*

Without loss of generality, the candidate with the highest function value can be used for branching. We refer to a branching scheme as unambiguous if it uses an unambiguous branching function.

The assumption that branching functions must be unambiguous is implicit in the examples and results in [9,10]. However, for sophisticated branching schemes used in state-of-the-art solvers, unambiguity cannot be assumed. Hence, we formally prove that if an unambiguous branching function is used to branch at a node, then "equivalent" child nodes would be created in both the sequential and the parallel BB tree. We say that two nodes are equivalent if they represent the same subproblem. We state this result for simple variable branching. Let ϕ_1 and ϕ_k denote the BB tree explored by the sequential algorithm and the parallel algorithm (using k processors), respectively. Consider a node P_s where $s \in \mathbb{N}$ represents a label (a unique identifier of a node in a BB tree). For the two child nodes of P_s, we assign a label $2s$ to the left child node (generated using the $\leq$ disjunction) and $2s + 1$ to the other child node. We refer to this labeling mechanism as the $2s\_2s + 1$ rule. The root node is assigned a label 1.

Proposition 1. *Let the nodes in ϕ_1 and ϕ_k be generated using an unambiguous branching function and labeled using the $2s\_2s+1$ rule. Let $P_s^1 \in \phi_1$ and $P_s^k \in \phi_k$ be two nodes with the same label s. Then P_s^1 and P_s^k are equivalent.*

Proof. The equivalence of P_s^1 and P_s^k can be shown using equivalence of their corresponding ancestor nodes upto the root node. The $2s\_2s + 1$ rule implies that the labels of the ancestor nodes on the unique paths from the root node (P_1) to P_s^1 and P_s^k, respectively, are equal. Since, P_1 is common to both ϕ_1 and ϕ_k, the branching disjunctions at the root node are equivalent by Assumption 2 and Definition 4. Hence, the ancestor nodes P_2^1 and P_2^k or P_3^1 and P_3^k are equivalent. By Definition 4, $\nu(\cdot)$ generates the branching disjunctions at these nodes using scores obtained from only the current node and its ancestors, which are equivalent. Similarly, the equivalence of P_s^1 and P_s^k is shown. $\square$

Since k is arbitrary in Proposition 1, nodes with the same label are equivalent irrespective of the number of threads used in the algorithm when an unambiguous branching rule is used. It can be easily verified that some well-known branching strategies are naturally based on unambiguous branching functions. For example, the lexicographic branching scheme (*lex*) that selects the variable with the smallest subscript in $\mathcal{I}_C$ satisfies Definition 4. Also, strong branching (*str*), which involves partly or fully solving an LP/NLP subproblem satisfies Definition 4 if a deterministic LP/NLP subsolver is used and if the objective values obtained are all distinct. However, while *lex* is known to result in large BB trees, *str* is considered expensive due to the requirement of solving many LPs/NLPs. Thus, we present an unambiguous variant of a practically effective branching scheme called the reliability branching [1].

4.2 Unambiguous Reliability Branching Scheme

Reliability branching (*rel*) is a hybrid scheme that attempts to combine the benefits of *str* and the pseudocost branching scheme; see [1] for a detailed description of these schemes. *rel* updates the pseudocost scores of variables when they are used for branching as the BB tree evolves. However, as shown in Table 2, using information from all the processed nodes could result in ambiguous branching. Hence, we present a scheme that uses limited information available in the BB tree in a way that avoids ambiguities.

ancestRel Branching

An unambiguous version of *rel* can be obtained by using pseudocosts only from a node P_i and its ancestors.

Since *ancestRel* uses some node-processing information generated in the tree, it possibly provides better branching decisions compared to simple unambiguous branching schemes like *lex*.

A variable x_j with the maximum branching score is chosen for branching. Any ties between candidates with the same score are broken lexicographically.

Corollary 1. *The ancestRel branching strategy with the lexicographic tie-breaking rule is unambiguous.*

Proof. Since, the pseudocosts from only the ancestor nodes are used to calculate scores, the selected branching candidate is independent of the number of processors used to generate this tree by Proposition 1. Also, the set of branching candidates, $\mathcal{I}_C$, at a node P_i is not ambiguous because the NLP subsolver used is deterministic, and the lexicographic tie-breaking rule ensures that a unique branching candidate is chosen. Hence, the conditions of Definition 4 are satisfied by *ancestRel*. □

Compared to *sharedRel* branching rule in [17] that is ambiguous, the implementation of *ancestRel* incurs a storage overhead because at each node, it stores a list of variables used for branching at its ancestors, their pseudocosts, the number of times they are branched, and the iteration when they are last strong-branched. Parameters like the number of variables on which strong branching must be applied and the limit on the number of NLP/LP iterations for strong branching are as per default settings of Minotaur.

4.3   A Hybrid Unambiguous Node Selection Strategy

Node selection strategies like depth-first, width-first or best-first along with tie-breaking rules have been shown to be unambiguous [10,13]. State-of-the-art solvers generally use hybrid node selection mechanisms to mitigate the drawbacks and combine the advantages of the above mentioned search strategies. We show that one such hybrid strategy called the best-then-dive strategy coupled with a tie-breaking rule is unambiguous. This strategy first selects a node with the best lower bound, and then keeps diving (processing one of the two immediate child nodes) until a node is pruned.

Let P_i be a node that has been processed and branched, and has an optimal value $\hat{z}$. Assume that P_j is the child node preferred for diving, we assign the following heuristic function values to the child nodes of P_i:

$$h(P_j) = \begin{cases} -\infty, & \text{if } P_j \text{ is the preferred child of } P_i, \\ \hat{z}, & \text{otherwise.} \end{cases}$$

In case a node gets pruned, an open node with the lowest lower bound value is selected. Very often, multiple such nodes exist, for which an unambiguous tie-breaking mechanism is required.

Proposition 2. *The best-then-dive node selection strategy with the $2s\_2s + 1$ tie-breaking rule is unambiguous.*

We omit the proof as it is easy to show that both the conditions of Definition 3 are met by this strategy.

4.4 Nondetrimental NLP-BB

We denote by *mcbnbDeter*, the parallel extension of NLP-BB in Minotaur that uses the following unambiguous components.

- best-then-dive node selection strategy with the $2s\_2s + 1$ tie-breaking rule
- *ancestRel* branching strategy with the lexicographic tie-breaking rule
- a deterministic NLP solver.

In addition to the above, we require synchronization of threads at few stages in the algorithm, for example, to avoid passing ambiguous initial conditions to the subsolvers, as well as unambiguity of other algorithmic components. We also disabled "guided diving" in Minotaur, which sometimes causes ambiguity depending upon when the best solution is obtained during an iteration.

Theorem 2. *The algorithm mcbnbDeter satisfies $T(k,0) \leq T(1,0)$ for $k > 1$.*

Proof. The result follows from [10, Theorem 1] and the unambiguity of the node selection function (Proposition 2) and the branching function (Corollary 1), and due to the deterministic NLP solver (Assumption 2). □

5 Reducing Detrimental Anomalies in Parallel QG

In this section, we formally define the notion of unambiguity for cutting planes, an integral component of branch-and-cut based algorithms.

Definition 5. *A vector valued function $\pi(\cdot)$ for generating linearization cuts at a node P_i is referred to as unambiguous if $\pi(\cdot)$ depends only on the information obtained from P_i and its ancestors.*

We consider the basic linearization cuts in QG, which are obtained using an integer optimal solution x^* obtained at a node P_i. An NLP is solved in which the integer variables are fixed to their values in x^*. Using a point $\hat{x}$ returned from the NLP solver, a linearization cut is obtained. The point $\hat{x}$ is either an integer feasible solution to (P) or a point from a feasibility problem [2, Sec. 3.2.1] that minimizes some measure of constraint violation at this node. In traditional QG, these linearizations are applied to all open nodes in $\mathcal{P}$. However, we store and apply these cuts only to the descendants of P_i to avoid generation of ambiguous relaxations or nodes when using different number of threads. We refer to the strategy that uses this cut generating function as *cutGenQG*.

Proposition 3. *The cut generation strategy cutGenQG is unambiguous.*

Proof. Since the integer feasible LP solution x^* at a node is obtained from a deterministic LP solver, the resulting fixed NLP will not be ambiguous. This ensures that the point $\hat{x}$ returned from a deterministic NLP solver, and hence, the cuts generated at P_i are not ambiguous. All descendants of P_i exhibit similar behavior, hence, *cutGenQG* satisfies the conditions of Definition 5.   □

Next, we denote by *mcqgDeter*, the parallel extension of QG in Minotaur that uses the following unambiguous components.

- best-then-dive node selection strategy with the $2s\_2s + 1$ tie-breaking rule
- *ancestRel* branching strategy with the lexicographic tie-breaking rule
- *cutGenQG* cutting plane strategy
- a deterministic NLP solver and a deterministic LP solver

Theorem 3. *The algorithm mcqgDeter satisfies $T(k,0) \leq T(1,0)$ for $k > 1$.*

Other linearization inequalities proposed in [17] for QG can also be included in an unambiguous way in *mcqgDeter*. However, by restricting the application of these cuts only to descendant nodes, the relaxations at other nodes would be weaker than those in the traditional QG and might result in a larger BB tree.

Reproducibility of Results The use of unambiguous components in mcbnbDeter and mcqgDeter results in a deterministic behavior of these algorithms. Reproducibility in parallel algorithms is desired for performance analysis, debugging during code development, etc. In Minotaur, we provide appropriate options that synchronize various subroutines, and ensures unambiguity.

6   Computational Results

We have carried out our computational experiments on a server with two 64-bit Intel(R) Xeon(R) E5-2670 v2, 2.50 GHz CPUs with 10 cores each and sharing 128 GB RAM. Our schemes are implemented in Minotaur[2]. The algorithms are written in C++ and complied with GCC-4.9.2. We use OpenMP-4.0 provided by

---

[2] Available at http://github.com/minotaur-solver/minotaur.

GCC for our parallel constructs. IPOPT-3.12 with MA97 linear-systems solver is used for solving NLPs and CPLEX-12.8 for solving LPs. We have disabled hyperthreading to highlight the effect of explicit parallelism. We have used 374 convex instances from MINLPLib [5] (we refer to them as testset TS). A limit of one hour on the wall clock time has been used for all our experiments.

First, we briefly mention the performance of opportunistic algorithms. *mcbnbOppor16* could solve 29 additional instances compared to *mcbnbOppor1* and reduced the time taken by more than 60%. This performance is better compared to the opportunistic schemes reported earlier in [17].

The "scalability graphs" [17] for *mcbnbOppor* are shown in Fig. 3. The plot for *mcbnbOppor1* is a base step function for which the peak value (about 0.71 here) indicates the fraction of instances solved using *mcbnbOppor1*. The ordinate corresponding to a value at, say 2^{-1}, indicates the fraction of instances that are solved by a multithreaded variant by a factor of two or more as compared to *mcbnbOppor1*. For example, *mcbnbOppor16* solves 40% of the instances at least twice as fast as *mcbnbOppor1*. The rightmost values on the plots show the fraction of instances that could be solved within the time limit.

mcbnbDeter can be run in Minotaur using the following options: *–brancher ancestRel –tb_rule 2s_2s+1 –mcbnb_deter_mode 1*. The scalability graphs in terms of wall clock time are shown in Fig. 4. Overall, the performance of *mcbnbOppor* is better than *mcbnbDeter* in terms of wall clock time because the former exploits parallelism in an opportunistic way. However, *mcbnbDeter* variants can provide a guarantee to not be worse than *mcbnbDeter1* and also be reproducible.

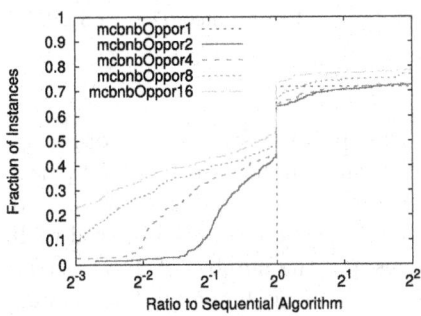

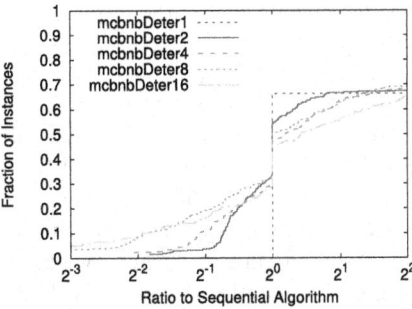

Fig. 3. Scalability graphs of wall clock times taken by *mcbnbOppor* on TS.

Fig. 4. Scalability graphs of wall clock times for *mcbnbDeter* without guided diving.

We are able to avoid detrimental anomalies in *mcbnbDeter* using the above mentioned options in Minotaur in terms of the number of iterations except for seven instances. Disabling guiding diving (using `--guided_dive 0`) eliminates detrimental anomalies in three of these instances. The remaining instances exhibit anomalies due to ambiguity induced by floating point precision in Minotaur and the subsolvers. Figure 4 and 7 demonstrate good scalability of *mcbnbDeter* both in terms of wall clock times and the number of iterations, respectively.

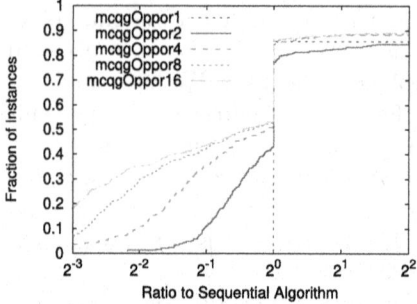

Fig. 5. Scalability graphs of wall clock times taken by *mcqgOppor* on *TS*.

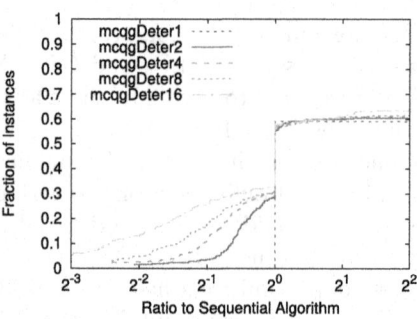

Fig. 6. Scalability graphs of wall clock times taken by *mcqgDeter* on *TS*.

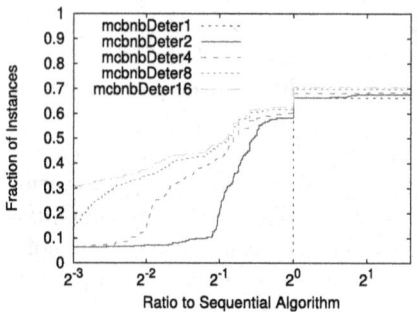

Fig. 7. Scalability graphs of no. of iterations for *mcbnbDeter* without guided diving.

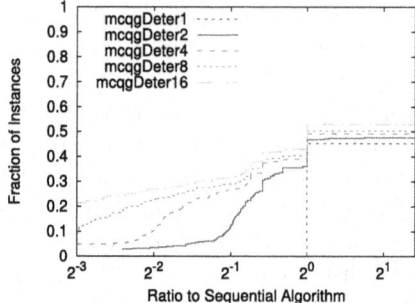

Fig. 8. Scalability graphs of number of iterations taken by *mcqgDeter* on *TS*.

Figure 5 shows the scalability graphs for the opportunistic variants of *mcqg*. *mcqgOppor16* could solve about 40% of the instances in half the time compared to *mcqgOppor1*. For difficult instances, improvement up to 88% is obtained, and overall, 16 additional instances could be solved. *mcqgDeter* also scales well with the number of threads in both wall clock times and the number of iterations as shown in Fig. 6 and 8, respectively. For 41 instances, *mcqgOppor* exhibits anomalous behavior in terms of wall clock time. We note that this list is longer compared to that corresponding to *mcbnbOppor*. On the other hand, five instances solved by multithreaded *mcqgDeter* took more iterations than that by *mcqgDeter1*. Again, this is because of the ambiguities induced by floating point precision in Minotaur.

7   Conclusions and Future Directions

It is important to study anomalies in parallel branch-and-bound algorithms to design better strategies in practice that can enhance scalability of parallel algorithms. We addressed detrimental anomalies in two convex MINLP algorithms,

NLP-BB and QG by extending the notion of unambiguity to functions for variable branching and generating cuts. We also showed that these theoretical ideas can translate to practically effective algorithmic components such as hybrid node selection strategies like best-then-dive (instead of pure strategies like depth-first, best-first, etc.), branching rules like *ancestRel* (instead of lexicographic branching rule), etc. Our computational experiments show that detrimental anomalies can be eliminated to a great extent practically. Opportunistic versions perform better in terms of wall clock times than the deterministic versions on average, because deterministic versions tend to synchronize more and incur some extra intervals of waiting time. However, deterministic versions can avoid detrimental anomalies with guarantees and also provide reproducible results.

The analysis presented so far depends on the number of iterations. It remains to be explored how unambiguity and speed can be achieved simultaneously. Also, unambiguity can be extended to other MINLP algorithms like the MILP based outer approximation. Another immediate extension of our work is to avoid general k_1-k_2 anomalies in MINLP algorithms where $k_2 > k_1 > 1$ is the number of processors used.

References

1. Achterberg, T., Koch, T., Martin, A.: Branching rules revisited. Oper. Res. Lett. **33**(1), 42–54 (2005)
2. Belotti, P., Kirches, C., Leyffer, S., Linderoth, J., Luedtke, J., Mahajan, A.: Mixed-integer nonlinear optimization. Acta Numer **22**, 1–131 (2013)
3. Berthold, T., Farmer, J., Heinz, S., Perregaard, M.: Parallelization of the FICO Xpress-Optimizer. Optim. Methods Softw. **33**(3), 518–529 (2018)
4. Boukouvala, F., Misener, R., Floudas, C.A.: Global optimization advances in mixed-integer nonlinear programming, MINLP, and constrained derivative-free optimization. CDFO. Eur. J. Oper. Res. **252**(3), 701–727 (2016)
5. Bussieck, M.R., Drud, A.S., Meeraus, A.: MINLPLib - a collection of test models for mixed-integer nonlinear programming. INFORMS J. Comput. **15**(1), 114–119 (2003)
6. Crainic, T.G., Le Cun, B., Roucairol, C.: Parallel branch-and-bound algorithms. Parallel Comb. Optim. **1**, 1–28 (2006)
7. Hart, W.E., Phillips, C.A., Eckstein, J.: PEBBL: An object-oriented framework for scalable parallel branch and bound. Technical report, Sandia National Laboratories (SNLNM), Albuquerque, NM (United States) (2013)
8. Kannan, R., Monma, C.L.: On the computational complexity of integer programming problems. In: Henn, R., Korte, B., Oettli, W. (eds.) Optimization and Operations Research. LNEMS, vol. 157, pp. 161–172. Springer, Berlin, Heidelberg (1978). https://doi.org/10.1007/978-3-642-95322-4_17
9. Lai, T.H., Sahni, S.: Anomalies in parallel branch-and-bound algorithms. Commun. ACM **27**(6), 594–602 (1984)
10. Li, G.J., Wah, B.W.: Coping with anomalies in parallel branch-and-bound algorithms. IEEE Trans. Comput. **100**(6), 568–573 (1986)
11. Li, G.J., Wah, B.W.: Computational efficiency of parallel combinatorial OR-tree searches. IEEE Trans. Softw. Eng. **16**(1), 13–31 (1990)

12. Mahajan, A., Leyffer, S., Linderoth, J., Luedtke, J., Munson, T.: Minotaur: a mixed-integer nonlinear optimization toolkit. Math. Program. Comput. **13**, 1–38 (2020)
13. Mans, B., Roucairol, C.: Performances of parallel branch and bound algorithms with best-first search. Discret. Appl. Math. **66**(1), 57–74 (1996)
14. Menouer, T.: Solving combinatorial problems using a parallel framework. J. Parallel Distrib. Comput. **112**, 140–153 (2018)
15. Quesada, I., Grossmann, I.E.: An LP/NLP based branch and bound algorithm for convex MINLP optimization problems. Comput. Chem. Eng. **16**(10–11), 937–947 (1992)
16. Sahinidis, N.V.: Mixed-integer nonlinear programming 2018. Optim. Eng. **20**(2), 301–306 (2019). https://doi.org/10.1007/s11081-019-09438-1
17. Sharma, M., Palkar, P., Mahajan, A.: Linearization and parallelization schemes for convex mixed-integer nonlinear optimization. Comput. Optim. Appl. **81**, 1–56 (2022)
18. Shinano, Y., Heinz, S., Vigerske, S., Winkler, M.: FiberSCIP - a shared memory parallelization of SCIP. INFORMS J. Comput. **30**(1), 11–30 (2017)

Game Theory

Exact Price of Anarchy for Weighted Congestion Games with Two Players

Joran van den Bosse, Marc Uetz, and Matthias Walter[✉]

Department of Applied Mathematics, University of Twente,
Enschede, The Netherlands
j.v.vandenbosse@alumnus.utwente.nl,
{m.uetz,m.walter}@utwente.nl

Abstract. This paper gives a complete analysis of worst-case equilibria for various versions of weighted congestion games with two players and affine cost functions. The results are exact price of anarchy bounds, which are parametric in the weights of the two players, and establish exactly how the attributes of the game enter into the quality of equilibria. Interestingly, some of the worst-cases are attained when the players' weights only differ slightly. Our findings also show that sequential play improves the price of anarchy in all cases, however, this effect vanishes with an increasing difference in the players' weights. Methodologically, we obtain exact price of anarchy bounds by a duality-based proof mechanism, based on a compact linear programming formulation that computes worst-case instances. This mechanism yields duality-based optimality certificates, which can eventually be turned into purely algebraic proofs.

1 Introduction

This paper studies the quality of equilibria for games with two players that compete for a set of resources, when the cost (or congestion) on each resource is given by an affine function that depends on the total load of that resource. The games that we consider fall into a class of games known as weighted Rosenthal congestion games [20,28]. As we address several variants of such games, before discussing our actual contribution, let us first define these different settings.

Problem Definition. There are two players denoted $i = 1, 2$, and the possible *actions* of player i are given by admissible subsets $\mathcal{A}_i \subseteq 2^R$ of a finite set of resources R. This includes so-called *atomic network routing games*, where the resources are edges E in a directed graph $G = (V, E)$, and each player wants to establish a path between source and target vertices $s_i, t_i \in V(G)$, so that the admissible actions of player i are the edge sets of the directed (s_i, t_i)-paths. A congestion game is called *symmetric* if $\mathcal{A}_1 = \mathcal{A}_2$. For network routing games, that means that all players have the same source and the same target.

We address two different game theoretic settings. In *simultaneous* games, both players choose their actions simultaneously, and the admissible actions $\mathcal{A}_i$

© The Author(s), under exclusive license to Springer Nature Switzerland AG 2022
I. Ljubić et al. (Eds.): ISCO 2022, LNCS 13526, pp. 159–171, 2022.
https://doi.org/10.1007/978-3-031-18530-4_12

of a player i are exactly the player's strategies in the corresponding strategic form bimatrix game. We also consider *sequential* extensive form games where the players choose actions after each other, w.l.o.g. first player 1, then player 2. Here, the first player's strategies are the actions $\mathcal{A}_1$, and a strategy for the second player consists of one action from $\mathcal{A}_2$ for *each* possible strategy (= action) of the first player. In both cases, the game results in an outcome where both players have chosen one action, denoted *action profile* $A = (A_1, A_2)$, where $A_i \in \mathcal{A}_i$.

The players have to pay for each resource $r \in R$ that they choose, and the cost of a resource $r \in R$ depends on its load, which again depends on the set of players using it. In the unweighted version of the problem, the load x_r of a resource r equals the *number* of players using it, and the game is a potential game [22,28]. In the weighted version of the problem, each player i has a *player-specific weight* w_i, and the load x_r of a resource $r \in R$ equals the total weight of the players that have chosen it. So if $A = (A_1, A_2)$ is an action profile, the load of a resource $r \in R$ equals $x_r(A) = \sum_{i | r \in A_i} w_i$. When $w_1 = w_2$, this corresponds to the unweighted version (possibly after scaling).

Throughout the paper, we assume that the cost function for each resource r is an affine function in its load x_r, with non-negative coefficients. That is, for each resource $r \in R$, there are non-negative reals $\alpha_r \geq 0$ and $\beta_r \geq 0$, and with x_r being the total load of resource r, the cost of resource r equals $\alpha_r + \beta_r x_r$.

For weighted congestion games, two different cost functions appear in the literature, under different names. We here follow the nomenclature as used in [15]. The first class of cost functions is *uniform costs* [10,11,15,16,21,26], where the cost of a resource can be thought of as a delay or latency (as, e.g., in traffic), which is the same for all players choosing that resource, so that each player pays the costs for the loads of all chosen resources. That means that an action profile $A = (A_1, A_2)$ yields as cost for player i

$$C_i^{\mathrm{uni}}(A) := \sum_{r \in A_i} \left(\alpha_r + \beta_r \sum_{j: r \in A_j} w_j \right). \tag{1}$$

The second class of cost functions is *proportional costs* [13,15], where the cost of a resource can be thought of as as per-unit cost, and each player pays proportionally to the load that the player imposes on a resource. Then, an action profile $A = (A_1, A_2)$ yields as cost for player i

$$C_i^{\mathrm{prop}}(A) := w_i \sum_{r \in A_i} \left(\alpha_r + \beta_r \sum_{j: r \in A_j} w_j \right). \tag{2}$$

In the following we also use $C_i(\cdot)$ to denote either of the two cost functions. Note that the cost functions agree in the (unweighted) case when $w_1 = w_2 = 1$.

Equilibria. For simultaneous games, a strategy profile $(A_1^{\mathrm{equi}}, A_2^{\mathrm{equi}})$ is a (pure) Nash equilibrium [23,24] if none of the players can improve unilaterally, i.e.,

$$C_1(A_1^{\mathrm{equi}}, A_2^{\mathrm{equi}}) \leq C_1(A_1, A_2^{\mathrm{equi}}) \qquad \forall A_1 \in \mathcal{A}_1, \tag{3a}$$
$$C_2(A_1^{\mathrm{equi}}, A_2^{\mathrm{equi}}) \leq C_2(A_1^{\mathrm{equi}}, A_2) \qquad \forall A_2 \in \mathcal{A}_2. \tag{3b}$$

Since we assume that the cost functions per resource are affine, the existence of pure Nash equilibria is guaranteed [14].

For sequential games, let us assume w.l.o.g. that the players choose their actions in the order 1, and then 2. The first player then has strategy set $\mathcal{A}_1$. A strategy of the second player can be written as a tuple of length $|\mathcal{A}_1|$, specifying the response of player 2 to any possible strategy of player 1.

Since we consider full information games, in equilibrium both players can choose a (pure) strategy that minimizes the player's cost. For player 2 that means, if A_1 is chosen by player 1, to choose any A_2^{equi} so that

$$C_2(A_1, A_2^{\text{equi}}) \leq C_2(A_1, A_2) \text{ for all } A_2 \in \mathcal{A}_2.$$

If we denote by $A_2^{\text{equi}}(A_1)$, $A_1 \in \mathcal{A}_1$, such an equilibrium strategy for player 2, player 1 chooses any strategy A_1^{equi} so as to minimize her cost, that is,

$$C_1(A_1^{\text{equi}}, A_2^{\text{equi}}(A_1^{\text{equi}})) \leq C_1(A_1, A_2^{\text{equi}}(A_1)) \text{ for all } A_1 \in \mathcal{A}_1.$$

All pairs of strategies $(A_1^{\text{equi}}, A_2^{\text{equi}}(A_1^{\text{equi}}))$ that fulfill both conditions are precisely the (pure) subgame-perfect equilibria [30] for the sequential, extensive form game. For the full information games considered here, (pure) subgame perfect strategies exist and can easily be computed by the "procedure" as sketched above, known as backward induction [27]. A subgame perfect equilibrium yields as outcome an action profile $(A_1^{\text{equi}}, A_2^{\text{equi}})$. It is well known that, even for network routing games, subgame perfect outcomes need not be a Nash equilibrium in the corresponding simultaneous game [8], which is one of the difficulties in analyzing subgame-perfect equilibria.

Price of Anarchy. It is well known that if players choose their actions $A = (A_1, A_2)$ selfishly while only considering their own cost functions $C_i(A)$, there may be outcomes that are stable with respect to either Nash or subgame-perfect equilibrium, but that solution need not minimize the cost of both players together $C(A) := C_1(A) + C_2(A)$. The price of anarchy [18] measures these negative effects of selfish behaviour. It is defined as the maximum cost of an equilibrium outcome, relative to the cost of the so-called *social optimum*, which is an action profile that minimizes total cost. Formally, for a given instance I of a weighted congestion game, if we define $\mathcal{A}^{\text{equi}}(I)$ as the set of action profiles that may result as outcome from some equilibrium, and $A^{\text{opt}}(I)$ as a social optimum, so any profile that minimizes total costs $C(\cdot)$, the price of anarchy is

$$\text{PoA}(I) := \max_{A \in \mathcal{A}^{\text{equi}}(I)} \frac{C(A)}{C(A^{\text{opt}}(I))}.$$

For sequential games, the maximum is also taken over all possible orders of the players. The price of anarchy for a class of games is the supremum of $\text{PoA}(I)$ over all instances I of that class. For sequential games, the price of anarchy has also been called the *sequential price of anarchy* [25]; it has been analyzed in several settings [3, 9, 17, 25].

Table 1. Known results and improvements for lower bounds (lb) and upper bounds (ub) on the price of anarchy for weighted *simultaneous* congestion games with two players. The first columns indicate restriction (✓) to network or symmetric games or no restriction (✗), as well as the cost function. In the first column, "✗, ✓" indicates that the upper bound holds in general, while the lower bound is attained even for network routing games. Bounds are underlined if the gap between lower and upper bound is closed. Empty cells indicate that there was no old bound or that there is no new bound.

Net	Sym	Cost	Old lb	New lb	New ub	Old ub	Result
✗, ✓	✓		<u>1.6</u> [8]			<u>1.6</u> [7]	
✗, ✓	✓	uni	1.6 [8]	<u>2</u>	<u>2</u>		Theorem 4, Corollary 3
✗, ✓	✓	prop	1.6 [8]	≈ 1.6096	≈ 1.6096	≈ 2.618 [1]	Corollary 4
✗, ✓	✗		<u>2</u> [folklore]			<u>2</u> [7]	
✗, ✓	✗	uni	2 [folklore]	≈ 2.155	≈ 2.155		Corollary 1
✗, ✓	✗	prop	2 [folklore]	≈ 2.0411	≈ 2.0411	≈ 2.618 [1]	Corollary 2

Contribution and Known Results. This paper gives the exact price of anarchy results for several classes of *weighted* two-player congestion games with respect to Nash equilibria for simultaneous games, and subgame-perfect equilibria for sequential games. We further distinguish between arbitrary congestion games and network routing games, uniform and proportional cost functions, as well as symmetric and asymmetric games. This gives rise to 16 different classes of two-player games. An overview of previously known bounds as well as new ones presented in this paper can be found in Tables 1 and 2. The tables also contain the previously known results for the special case of unweighted congestion games (if known).

Table 2. Known results and improvements on the price of anarchy for weighted *sequential* congestion games with two players.

Net	Sym	Cost	Old lb	New lb	New ub	Old ub	Result
✗, ✓	✓		<u>1.4</u> [8]			<u>1.4</u> [8]	
✗, ✓	✓	uni	1.4 [8]		2		Corollary 5
✗, ✓	✓	prop	1.4 [8]		1.5		Corollary 6
✗	✗		<u>1.5</u> [17]			<u>1.5</u> [17]	
✓	✗		1	<u>1.5</u>		<u>1.5</u> [17]	
✗, ✓	✗	uni	1	<u>2</u>	<u>2</u>		Corollary 5
✗, ✓	✗	prop	1	<u>1.5</u>	<u>1.5</u>		Corollary 6

We not only give the worst-case bounds, but also prove the exact price of anarchy bounds parametric in the weight ratio w_1/w_2 of the two players, which provides additional insight. Figure 1 depicts these findings.

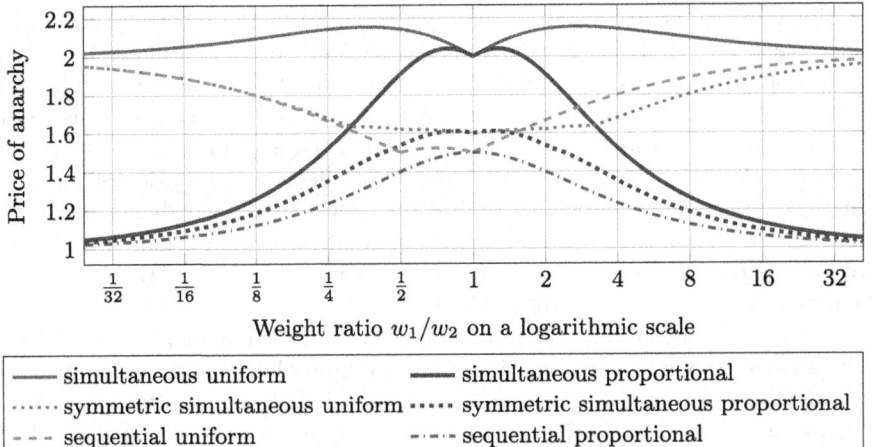

Fig. 1. Price of anarchy for different types of weighted affine two-player congestion games depending on the weight ratio w_1/w_2. No graphs are shown for sequential games with symmetric strategy spaces, as tight lower bound instances are not known.

Our results close a gap in the existing literature on the analysis of equilibria for affine congestion games. Let us therefore briefly discuss what is known.

The price of anarchy for *unweighted* affine congestion games with an arbitrary number of players is known to be bounded from above by 5/2 [1,7]. This bound is tight for $n \geq 3$ players [7]. For unweighted two-player games, the tight bound is 2 [7]. The 5/2 upper bound improves to $(5n-2)/(2n+1)$ for *symmetric* games with n players [7], which is tight even in the class of network routing games [8]. For non-symmetric, unweighted and affine congestion games, the bound 5/2 is tight (for $n \to \infty$) even for *singleton* congestion games, where each player chooses a single resource only, i.e., $|A_i| = 1$ for all $A_i \in \mathcal{A}_i$ [6]. For unweighted and affine singleton congestion games that are symmetric, so when each player can choose every single resource, the price of anarchy equals 4/3 [19]. This singleton model can also be interpreted as a network routing model with parallel source-sink arcs, and is also dubbed the *parallel link* model, see e.g. [29].

For *weighted* affine congestion games, the price of anarchy with an arbitrary number of players is slightly larger than 5/2; it equals $1 + \phi$ with $\phi = (1 + \sqrt{5})/2 \approx 1.618$ being the golden ratio [1]. This is again tight for $n \geq 3$ players [2]. Here, we are not aware of improved bounds for special cases or with two players.

For the *sequential* version of *unweighted*, affine congestion games, it is known that the (sequential) price of anarchy equals 1.5 for two players [17], it equals $1039/488 \approx 2.13$ for $n = 3$ players [17], and for $n = 4$ players it equals $28679925/10823887 \approx 2.65$ [4]. Moreover, for symmetric network routing games and two players, it equals 1.4 [8], while it can be as large as $\Omega(\sqrt{n})$ for an arbitrary number n of players [8].

The above summary suggests that the potential loss of efficiency caused by selfish players in affine congestion games is by now pretty well understood. Yet the two-player case is a fundamental base case, and it does not seem to be well understood when the two players are not identical, i.e., the weighted case. Our results exhibit how the attributes of the game, such as symmetry of strategies, proportionality of costs, or sequentiality, have effects on the price of anarchy, which we believe are interesting and not obvious. It is conceivable that comparable results can be expected for congestion games with more than two players. However, extending our approach to, say, the case with three players, although theoretically possible, seems to be tedious.

As to the technical approach and contribution of this paper, all our results take as starting point a compact linear programming based idea to compute the cost functions of worst-case instances that extends the one used for unweighted congestion games in [17]. The linear program computes the parameters of all resource cost functions for the finite worst case instance, which can be done because it can be proved that such finite worst-case instances must exist [9,17]. However for weighted games this can only be done for any *fixed* pair of weights w_1, w_2. Doing this for several weight ratios w_1/w_2, however, one can produce "educated guesses" for the dependence of the cost parameters on the weights, and the same can be done for the corresponding optimal dual solution. The so derived expressions for primal and dual solutions can finally be turned into algebraic proofs for a matching upper bound on the price of anarchy. This primal-dual procedure is somewhat mechanical, yet non-trivial. Note that this approach is reminiscent of the primal-dual technique as suggested in [2], however here we work with a different linear programming formulation.

Outline. We present our results on the price of anarchy for simultaneous, symmetric simultaneous and sequential games in Sect. 2. Our methodology is explained in Sect. 3 and a few concluding remarks are made in Sect. 4.

2   Results

The main results of this paper are best summarized and illustrated in Fig. 1, which displays the exact price of anarchy bounds in dependence on the weight ratio w_1/w_2 for six different cases. This section supports this illustration: The subsequent Theorems 1–7 give the exact price of anarchy bounds in dependence on the players' weights w_1 and w_2. Corollaries are obtained by taking the supremum over all feasible weight pairs w_1, w_2, where "weight ratios" refers to the ratios w_1/w_2 or w_2/w_1, respectively. Corresponding worst-case instances as well as proofs can be found in the full version of our paper [5].

Theorem 1 (uniform, simultaneous, specific weights). *Let $w_1, w_2 \geq 0$ with $w_1 + w_2 > 0$. The price of anarchy for simultaneous uniformly weighted two-player congestion/network routing games with weights w_1, w_2 is equal to*

$$1 + \frac{2w_1 w_2 + \max(w_1, w_2)^2}{w_1^2 + w_1 w_2 + w_2^2}.$$

Corollary 1 (uniform, simultaneous). *The price of anarchy for simultaneous uniformly weighted two-player congestion/network routing games is equal to $1 + 2/\sqrt{3} \approx 2.155$, which is attained for weight ratios equal to $1 + \sqrt{3} \approx 2.732$.*

Theorem 2 (proportional, simultaneous, specific weights). *Let $w_1, w_2 \geq 0$ with $w_1 + w_2 > 0$. The price of anarchy for simultaneous proportionally weighted two-player congestion/network routing games with weights w_1, w_2 is equal to*

$$1 + w_1 w_2 \frac{w_1 + w_2 + \max(w_1, w_2)}{w_1^3 + w_2^3 + w_1 w_2 \min(w_1, w_2)}.$$

Corollary 2 (proportional, simultaneous). *The price of anarchy for simultaneous proportionally weighted two-player congestion/network routing games is approximately 2.0411, which is attained for weight ratios ≈ 1.2704.*

For the next case we have an irrational weight ratio at which the behavior changes. We denote it by $\tau := \frac{1}{3}(2\sqrt[3]{62 - 3\sqrt{183}} + \sqrt[3]{62 + 3\sqrt{183}}) \approx 3.1527$.

Theorem 3 (uniform, symmetric, simultaneous, specific weights). *Let $w_1, w_2 \geq 0$ with $w_1 + w_2 > 0$. The price of anarchy for symmetric simultaneous uniformly weighted two-player congestion games with weights w_1, w_2 is equal to*

(a) $\frac{3w_1^3 + 9w_1^2 w_2 + 9 w_1 w_2^2 + 3w_2^2}{2w_1^3 + 5w_1^2 w_2 + 5 w_1 w_2^2 + 2w_2^3 + w_1 w_2 \min(w_1, w_2)}$ *if* $\frac{1}{2} w_2 \leq w_1 \leq 2w_2$,

(b) $\frac{2w_1^3 + 5w_1^2 w_2 + 5 w_1 w_2^2 + 2w_2^3 + 3 \max(w_1^2 w_2 + w_1^3, w_1 w_2^2 + w_2^3)}{2w_1^3 + 4w_1^2 w_2 + 4 w_1 w_2^2 + 2w_2^3}$ *if* $2w_2 \leq w_1 \leq \tau w_2$ *or if* $\frac{1}{\tau} w_2 \leq w_1 \leq \frac{1}{2} w_2$.

(c) $\frac{2w_1^2 + 2w_1 w_2 + 2w_2^2}{(w_1 + w_2)^2}$ *if* $w_1 \geq \tau w_2$ *or if* $w_1 \leq \frac{1}{\tau} w_2$.

Corollary 3 (uniform, symmetric, simultaneous, congestion). *The price of anarchy for symmetric simultaneous uniformly weighted two-player congestion games is equal to 2, attained for weight rations $\to \infty$.*

Our computed worst case instances for which the price of anarchy approaches its supremum are no network routing games. However, there is a family of network routing games for which the price of anarchy also approaches 2, although slower than that for congestion games.

Theorem 4 (uniform, symmetric simultaneous, network routing). *The price of anarchy for symmetric simultaneous uniformly weighted two-player network routing games is equal to 2.*

Also for proportional cost functions and symmetric games we have an irrational weight ratio at which the behavior changes. It is the unique real root σ of the polynomial $x^4 - 3x^2 - 3x - 1$ and is approximately $\sigma \approx 2.14790$.

Theorem 5 (proportional, symmetric, simultaneous, specific weights). *Let $w_1, w_2 \geq 0$ with $w_1 + w_2 > 0$. The price of anarchy for symmetric simultaneous proportionally weighted two-player congestion/network routing games with weights w_1, w_2 is equal to*

(a) $\dfrac{2w_1^4+6w_1^3w_2+8w_1^2w_2^2+6w_1w_2^3+2w_2^4}{2w_1^4+3w_1^3w_2+4w_1^2w_2^2+3w_1w_2^3+2w_2^4+w_1w_2\min(w_1^2,w_2^2)}$ $if\ \dfrac{w_2}{\sigma}\leq w_1\leq\sigma w_2$.

(b) $\dfrac{2w_1^3w_2+4w_1^2w_2^2+2w_1w_2^3+2\max(w_1^4+w_1^3w_2,w_1w_2^3+w_2^4)}{w_1^3w_2+2w_1^2w_2^2+2w_1w_2^3+w_2^4+2w_1w_2\min(w_1^2,w_2^2)+\max(w_1^4,w_2^4)}$ $if\ w_1\geq\sigma w_2\ or\ w_1\leq\dfrac{w_2}{\sigma}$.

Corollary 4 (proportional, symmetric, simultaneous). *The price of anarchy for symmetric simultaneous proportionally weighted two-player congestion/network routing games is ≈ 1.6096, attained for weight ratios ≈ 1.1940.*

Theorem 6 (uniform, sequential, specific weights). *Let $w_1, w_2 \geq 0$ with $w_1 + w_2 > 0$. The price of anarchy for sequential uniformly weighted two-player congestion/network routing games with weights w_1, w_2 is equal to*

(a) $1 + \frac{w_1}{w_1+w_2}$ *if $w_2 \leq w_1$,*

(b) $1 + \frac{2w_1w_2}{2w_1^2+w_1w_2+w_2^2}$ *if $w_1 \leq w_2 \leq 2w_1$ and*

(c) $1 + \frac{w_2}{2w_1+w_2}$ *if $w_2 \geq 2w_1$.*

Note that the uniform, sequential case is the only case where the exact price of anarchy is not symmetric around the weight ratio 1.

Corollary 5 (uniform, sequential). *The price of anarchy for sequential uniformly weighted two-player congestion games is equal to 2, attained for weight ratios $\to \infty$.*

Theorem 7 (proportional, sequential, specific weights). *Let $w_1, w_2 \geq 0$ with $w_1 + w_2 > 0$. The price of anarchy for sequential proportionally weighted two-player congestion/network routing games with weights w_1, w_2 is equal to*

$$1 + \frac{w_1w_2}{w_1^2 + w_2^2}.$$

Corollary 6 (proportional, sequential). *The price of anarchy for sequential proportionally weighted two-player congestion games is equal to 1.5, which is attained if and only if $w_1 = w_2$ holds.*

As symmetry cannot be handled by our approach for sequential games, a gap remains between the lower and upper bounds; see Table 2. We briefly discuss this issue in Sect. 3 preceding Theorem 8.

3   LP Based Proofs

First, observe that for fixed weights w_1 and w_2 the cost functions (1) and (2) are linear. The idea of computing the price of anarchy is greedy: we construct an LP for an instance of a game with a minimal set of actions $\mathcal{A}$ that are required for a worst-case instance. The LP has as variables the cost parameters α_r, β_r of the cost functions per resource $r \in R$. This is finite, since as in [17] we can argue that by pigeonhole principle at most $2^{|\mathcal{A}|}$ resources are needed for any such an instance: if two resources appear in precisely the same actions, then these could be combined into one (adding their costs), which yields an instance with fewer resources but the same price of anarchy. Given that the LP's cost functions per resource are yet undetermined, one can w.l.o.g. "label" actions as social optimum and equilibrium, respectively. These labels are $\mathcal{A} = \{A_1^{\text{opt}}, A_2^{\text{opt}}, A_1^{\text{equi}}, A_2^{\text{equi}}\}$. For sequential games we extend $\mathcal{A}$ by the label $A_2^{\text{equi}'}$ for an optimal action for player 2 after player 1 has played A_1^{opt}. The observation above justifies to view resources as subsets of labels, i.e., $R = 2^{\mathcal{A}}$. Therefore, if $r \in A$, we can also write $A \in r$, as we view r as the set of all actions $A \ni r$. In addition to the cost variables α_r, β_r for each $r \in R$ the LP has variables for the costs $C_i(A_1, A_2)$ for $i = 1, 2$ and for all labels $A_j \in \mathcal{A}_j$ that are admissible for player $j \in \{1, 2\}$. We now introduce the constraints of the LP that ensure correctness of the labels and then prove that optimal solutions indeed correspond to worst-case instances. We start with the basic (but incomplete) LP:

$$\max \ C_1(A_1^{\text{equi}}, A_2^{\text{equi}}) + C_2(A_1^{\text{equi}}, A_2^{\text{equi}}) \tag{4a}$$

$$\text{s.t.} \ C_1(A_1^{\text{opt}}, A_2^{\text{opt}}) + C_2(A_1^{\text{opt}}, A_2^{\text{opt}}) = 1 \tag{4b}$$

$$C_1(A_1, A_2) + C_2(A_1, A_2) \geq 1 \quad \forall A_1 \in \mathcal{A}_1, \ \forall A_2 \in \mathcal{A}_2 \tag{4c}$$

$$\alpha_r, \beta_r \geq 0 \quad \forall r \in R \tag{4d}$$

Note that the normalization of the social optimum to 1 via (4b) is without loss of generality since a value of 0 would also yield zero costs for each equilibrium. In case of uniform cost functions (1) we add

$$C_i(A_1, A_2) = \sum_{r \in R: \ A_i \in r} (\alpha_r + \beta_r \sum_{j: A_j \in r} w_j) \quad \forall A_1 \in \mathcal{A}_1, \ \forall A_2 \in \mathcal{A}_2, \ i = 1, 2, \tag{5}$$

while in case of proportional cost functions (2) we add

$$C_i(A_1, A_2) = w_i \sum_{r \in R: \ A_i \in r} (\alpha_r + \beta_r \sum_{j: A_j \in r} w_j) \quad \forall A_1 \in \mathcal{A}_1, \ \forall A_2 \in \mathcal{A}_2, \ i = 1, 2. \tag{6}$$

Simultaneous Games. For simultaneous games we need to add the Nash inequalities (3) to enforce that A_1^{equi} and A_2^{equi} form a Nash equilibrium. For general games we define $\mathcal{A}_1 := \{A_1^{\text{opt}}, A_1^{\text{equi}}\}$ and $\mathcal{A}_2 := \{A_2^{\text{opt}}, A_2^{\text{equi}}\}$, while for symmetric games we allow both players to use the union $\mathcal{A}_1 := \mathcal{A}_2 := \{A_1^{\text{opt}}, A_1^{\text{equi}}, A_2^{\text{opt}}, A_2^{\text{equi}}\}$ of these actions.

Sequential Games. The following constraints model that A_2^{equi} and $A_2^{\text{equi}'}$ are optimal actions for player 2 after player 1 has played A_1^{equi} and A_1^{opt}, respectively, as well as the requirement that A_1^{equi} is a subgame-perfect action for player 1:

$$C_2(A_1^{\text{equi}}, A_2^{\text{equi}}) \leq C_2(A_1^{\text{equi}}, A_2) \quad \forall A_2 \in \mathcal{A}_2, \tag{7a}$$

$$C_2(A_1^{\text{opt}}, A_2^{\text{equi}'}) \leq C_2(A_1^{\text{opt}}, A_2) \quad \forall A_2 \in \mathcal{A}_2, \tag{7b}$$

$$C_1(A_1^{\text{equi}}, A_2^{\text{equi}}) \leq C_1(A_1^{\text{opt}}, A_2^{\text{equi}'}). \tag{7c}$$

We thus define $\mathcal{A}_1 := \{A_1^{\text{opt}}, A_1^{\text{equi}}\}$ and $\mathcal{A}_2 := \{A_2^{\text{opt}}, A_2^{\text{equi}}, A_2^{\text{equi}'}\}$. Our approach does not work for symmetric sequential games. The reason is that if the actions from $\mathcal{A}_2$ are available to player 1 as well, then for each of these actions we would need to introduce new actions for the subgame-perfect action of player 2 in such a case, which in turn would be available to player 1, and so on.

We now establish correctness of all variants of this LP.

Theorem 8. *For fixed weights $w_1, w_2 \geq 0$ with $w_1 + w_2 > 0$ the LP (4) for the action sets $\mathcal{A}_1, \mathcal{A}_2$ specified above, and extended by either (5) or (6) and by either (3) or (7) computes the price of anarchy for affine (sequential or simultaneous or symmetric simultaneous) congestion games with these specific weights.*

Proof. Consider a class of games and an LP as defined in the theorem. It is straight forward to see that every primal solution of the LP actually represents a game and that the objective value is equal to its price of anarchy.

It remains to show that for *every* game there exists a primal solution of the LP such that its objective value is equal to the price of anarchy of that game. To this end, consider a game with resources $\bar{R}$, actions $\bar{\mathcal{A}}_1, \bar{\mathcal{A}}_2 \subseteq 2^{\bar{R}}$ for the two players as well as cost coefficients $\bar{\alpha}, \bar{\beta} \in \mathbb{R}^{\bar{R}}$. By scaling we can assume that the cost of the social optimum is equal to 1. We now create a mapping π from labels to the game's actions. To this end, let $\pi(A_1^{\text{opt}}) \in \bar{\mathcal{A}}_1$ and $\pi(A_2^{\text{opt}}) \in \bar{\mathcal{A}}_2$ be actions that constitute a social optimum. Moreover, we consider an equilibrium (either Nash or subgame-perfect) for which the price of anarchy of this game is attained and let $\pi(A_1^{\text{equi}}) \in \bar{\mathcal{A}}_1$ and $\pi(A_2^{\text{equi}}) \in \bar{\mathcal{A}}_2$ constitute such an equilibrium. In the sequential case, let $\pi(A_2^{\text{equi}'}) \in \bar{\mathcal{A}}_2$ be an action that is subgame-perfect for player 2 after player 1 has played $\pi(A_1^{\text{opt}})$.

The mapping π of labels $\mathcal{A}$ to actions $\pi(\mathcal{A}) \subseteq \bar{\mathcal{A}}$ identifies the "relevant" actions from $\bar{\mathcal{A}}$. For any resource $\bar{r} \in \bar{R}$, we can now associate with it the "incidence pattern" of the set of actions from $\pi(\mathcal{A})$ in which $\bar{r}$ is contained. This induces a reverse mapping χ that maps each resource $\bar{r} \in \bar{R}$ to the unique $r \in R$ that has the same incidences. Formally, $A \in \chi(\bar{r}) \iff \bar{r} \in \pi(A)$ must hold for all $A \in \mathcal{A}$. This allows us to aggregate the resources $\bar{R}$ accordingly via

$$\alpha_r := \sum_{\bar{r} \,:\, r=\chi(\bar{r})} \bar{\alpha}_{\bar{r}} \quad \text{and} \quad \beta_r := \sum_{\bar{r} \,:\, r=\chi(\bar{r})} \bar{\beta}_{\bar{r}}. \qquad (8)$$

By Eqs. (5) or (6), the values of the remaining variables are determined uniquely. Moreover, (8) ensures that for $i = 1, 2$ and for any profile $(A_1, A_2) \in \mathcal{A}_1 \times \mathcal{A}_2$, $C_i(A_1, A_2)$ is equal to the cost of player i if actions $\pi(A_1)$ and $\pi(A_2)$ are played in the game. In particular, constraints (4b), (4c) as well as either (3) or (7) are satisfied. Consequently, the objective value corresponds to the price of anarchy of this game since the social optimum equals 1. Hence the value of the optimal LP solution equals the price of anarchy for the given class of games. $\qquad \square$

Arbitrary Weights. In order to derive the price of anarchy for arbitrary weights, the LPs from Theorem 8 can be used as an auxiliary tool. First, an approximate version of Fig. 1 can be produced, from which we could guess intervals of weight ratios w_1/w_2 for which the same LP basis is optimal. Second, for each such interval, several weight pairs are chosen and optimal primal and dual solutions are computed. Using the LP solver SoPlex, we computed exact rational solutions (see [12]), which helped to derive educated guesses for algebraic expressions.

Even though this procedure is the core of our contribution, due to space constraints we could not include it here. A concrete example of such a derivation can be found in the full version of our paper [5]. Here, we still mention two tricks that were important. First, cancellations in the expressions can be avoided by choosing prime numbers for the weights. Second, we observed that often several optimal solutions exist, which makes it hard to make an educated guess of the weight-dependent expression for each variable. We were able to circumvent this difficulty by forcing some of the cost variables to 0 while ensuring that optimality of the solution is maintained.

4   Concluding Remarks

One of the findings of our paper is the fact that the worst-cases for simultaneous games are attained for players' weights that differ only slightly, and that sequential play reduces the price of anarchy irrespective of the players' weights. While for simultaneous games the symmetry with respect to players implies that the prices of anarchy for weight ratios λ and $1/\lambda$ are equal, no such implication holds for sequential games. Surprisingly, a different symmetry holds for sequential games with uniform costs, namely equality for weight ratios $\lambda/2$ and $1/\lambda$. As to methodology, the algebraic expressions for the price of anarchy are already quite complicated for two players, especially for symmetric simultaneous games. Hence, a similar analysis for the 3-player case seems to be out of reach.

References

1. Awerbuch, B., Azar, Y., Epstein, A.: The price of routing unsplittable flow. In: Proceedings of the Thirty-Seventh Annual ACM Symposium on Theory of Computing, STOC 2005, pp. 57–66. ACM, New York (2005)
2. Bilò, V.: A unifying tool for bounding the quality of non-cooperative solutions in weighted congestion games. In: Erlebach, T., Persiano, G. (eds.) WAOA 2012. LNCS, vol. 7846, pp. 215–228. Springer, Heidelberg (2013). https://doi.org/10.1007/978-3-642-38016-7_18
3. Bilò, V., Flammini, M., Monaco, G., Moscardelli, L.: Some anomalies of farsighted strategic behavior. Theor. Comput. Syst. **56**(1), 156–180 (2013). https://doi.org/10.1007/s00224-013-9529-1
4. van den Bosse, J.: Computing the sequential price of anarchy of affine congestion games using linear programming techniques. Master's thesis, University of Twente (2021). https://essay.utwente.nl/87784/
5. van den Bosse, J., Uetz, M., Walter, M.: Exact price of anarchy for weighted congestion games with two players (2022). arXiv:2203.01740
6. Caragiannis, I., Flammini, M., Kaklamanis, C., Kanellopoulos, P., Moscardelli, L.: Tight bounds for selfish and greedy load balancing. Algorithmica **61**(3), 606–637 (2010). https://doi.org/10.1007/s00453-010-9427-8

7. Christodoulou, G., Koutsoupias, E.: The price of anarchy of finite congestion games. In: Proceedings of the Thirty-Seventh Annual ACM Symposium on Theory of Computing, pp. 67–73. STOC '05, ACM, New York, NY, USA (2005)
8. Correa, J., de Jong, J., de Keijzer, B., Uetz, M.: The Inefficiency of Nash and Subgame Perfect Equilibria for Network Routing. Math. Oper. Res. **44**(4), 1286–1303 (2019)
9. de Jong, Jasper, Uetz, Marc: The sequential price of anarchy for atomic congestion games. In: Liu, Tie-Yan., Qi, Qi., Ye, Yinyu (eds.) WINE 2014. LNCS, vol. 8877, pp. 429–434. Springer, Cham (2014). https://doi.org/10.1007/978-3-319-13129-0_35
10. Fotakis, D., Kontogiannis, S., Spirakis, P.: Selfish unsplittable flows. Theoret. Comput. Sci. **348**(2), 226–239 (2005)
11. Gairing, M., Monien, B., Tiemann, K.: Routing (un-) splittable flow in games with player-specific linear latency functions. In: Bugliesi, M., Preneel, B., Sassone, V., Wegener, I. (eds.) Automata, Languages and Programming, pp. 501–512. Springer, Berlin Heidelberg, Berlin, Heidelberg (2006)
12. Gleixner, A., Steffy, D., Wolter, K.: Iterative refinement for linear programming. Technical report, 15–15, ZIB, Takustr. 7, 14195 Berlin (2015)
13. Goemans, M.X., Mirrokni, V., Vetta, A.: Sink equilibria and convergence. In: 46th Annual IEEE Symposium on Foundations of Computer Science (FOCS 2005), pp. 142–151 (2005)
14. Harks, T., Klimm, M.: On the existence of pure Nash equilibria in weighted congestion games. Math. Oper. Res. **37**(3), 419–436 (2012)
15. Harks, T., Klimm, M.: Congestion games with variable demands. Math. Oper. Res. **41**(1), 255–277 (2016)
16. Ieong, S., McGrew, R., Nudelman, E., Shoham, Y., Sun, Q.: Fast and compact: a simple class of congestion games. In: Proceedings of the 20th National Conference on Artificial Intelligence, AAAI 2005, vol. 2, pp. 489–494. AAAI Press (2005)
17. de Jong, J., Uetz, M.: The sequential price of anarchy for affine congestion games with few players. Oper. Res. Lett. **47**(2), 133–139 (2019)
18. Koutsoupias, E., Papadimitriou, C.: Worst-case equilibria. Comput. Sci. Rev. **3**(2), 65–69 (2009)
19. Lücking, T., Mavronicolas, M., Monien, B., Rode, M.: A new model for selfish routing. Theoret. Comput. Sci. **406**(3), 187–206 (2008)
20. Milchtaich, I.: Congestion games with player-specific payoff functions. Game. Econ. Behav. **13**(1), 111–124 (1996)
21. Milchtaich, I.: The equilibrium existence problem in finite network congestion games. In: Spirakis, P., Mavronicolas, M., Kontogiannis, S. (eds.) Internet Network Econ., pp. 87–98. Springer, Berlin Heidelberg, Berlin, Heidelberg (2006)
22. Monderer, D., Shapley, L.S.: Potential games. Game. Econ. Behav. **14**(1), 124–143 (1996)
23. Nash, J.: Equilibrium points in n-person games. Proc. Natl. Acad. Sci. **36**(1), 48–49 (1950)
24. Nash, J.: Non-cooperative games. Ann. Math. **54**(2), 286–295 (1951)
25. Paes Leme, R., Syrgkanis, V., Tardos, É.: The curse of simultaneity. In: ITCS 2012: Proceedings of the 3rd Innovations in Theoretical Computer Science Conference, pp. 60–67 (2012)
26. Panagopoulou, P.N., Spirakis, P.G.: Algorithms for pure Nash equilibria in weighted congestion games. ACM J. Exp. Algorithmics **11**, 2–7 (2006)
27. Peters, H.: Game Theory - A Multi-leveled Approach. Springer, Heidelberg (2008). https://doi.org/10.1007/978-3-540-69291-1

28. Rosenthal, R.W.: A class of games possessing pure-strategy Nash equilibria. Int. J. Game Theory **2**(1), 65–67 (1973)
29. Roughgarden, T.: The price of anarchy is independent of the network topology. J. Comput. Syst. Sci. **67**(2), 341–364 (2003)
30. Selten, R.: A simple model of imperfect competition, where 4 are few and 6 are many. Int. J. Game Theory **2**, 141–201 (1973). https://doi.org/10.1007/BF01737566

Nash Balanced Assignment Problem

Minh Hieu Nguyen, Mourad Baiou, and Viet Hung Nguyen[✉]

INP Clermont Auvergne, Univ Clermont Auvergne, Mines Saint-Etienne, CNRS,
UMR 6158 LIMOS, 1 Rue de la Chebarde, Aubiere Cedex, France
{minh_hieu.nguyen,mourad.baiou,viet_hung.nguyen}@uca.fr

Abstract. In this paper, we consider a variant of the classic *Assignment Problem* (AP), called the *Balanced Assignment Problem* (BAP) [2]. The BAP seeks to find an assignment solution which has the smallest value of *max-min distance*: the difference between the maximum assignment cost and the minimum one. However, by minimizing only the max-min distance, the total cost of the BAP solution is neglected and it may lead to an inefficient solution in terms of total cost. Hence, we propose a fair way based on Nash equilibrium [1,3,4] to inject the total cost into the objective function of the BAP for finding assignment solutions having a better trade-off between the two objectives: the first aims at minimizing the total cost and the second aims at minimizing the max-min distance. For this purpose, we introduce the concept of *Nash Fairness* (NF) solutions based on the definition of proportional-fair scheduling adapted in the context of the AP: a transfer of utilities between the total cost and the max-min distance is considered to be fair if the percentage increase in the total cost is smaller than the percentage decrease in the max-min distance and vice versa.

We first show the existence of a NF solution for the AP which is exactly the optimal solution minimizing the product of the total cost and the max-min distance. However, finding such a solution may be difficult as it requires to minimize a concave function. The main result of this paper is to show that finding all NF solutions can be done in polynomial time. For that, we propose a Newton-based iterative algorithm converging to NF solutions in polynomial time. It consists in optimizing a sequence of linear combinations of the two objective based on Weighted Sum Method [5]. Computational results on various instances of the AP are presented and commented.

Keywords: Combinatorial optimization · Balanced assignment problem · Proportional fairness · Proportional-fair scheduling · Weighted sum method

1 Introduction

The *Assignment Problem* (AP) is a fundamental combinatorial optimization problem. It can be formally defined as follows. Given a set n workers, a set of n jobs and a $n \times n$ cost matrix whose elements are positive representing

© The Author(s), under exclusive license to Springer Nature Switzerland AG 2022
I. Ljubić et al. (Eds.): ISCO 2022, LNCS 13526, pp. 172–186, 2022.
https://doi.org/10.1007/978-3-031-18530-4_13

the assignment of any worker to any job, the AP aims at finding an one-to-one worker-job assignment (i.e., a bipartite perfect matching) that minimizes certain objective functions.

In the classic AP, we seek to find an assignment solution minimizing the total cost. It is a well-known optimization problem that can be solved by Hungarian algorithm in $O(n^3)$ [7]. The *Balanced Assignment Problem* (BAP) is a variant of the classic AP where instead of minimizing the total cost, we minimize *the max-min distance* which is the difference between the maximum assignment cost and the minimum one in the assignment solution. In [2], the authors proposed an efficient threshold-based algorithm to solve the BAP in $O(n^4)$. However, by minimizing only the max-min distance, the total cost of the BAP solution is neglected and it may lead to a very inefficient solution in terms of total cost.

In this paper, to overcome the possible inefficiency of the solutions for the BAP, we propose a fair way based on Nash equilibrium to inject the total cost into the objective function of the BAP. Nash equilibrium is the most common optimality notion for sharing resources among users [1,3,4]. We are interested in assignment solutions for the AP achieving a Nash equilibrium between two players: the first aims at minimizing the total cost and the second aims at minimizing the max-min distance. For this purpose, we introduce the *Nash Fairness* (NF) solutions based on the definition of proportional-fair scheduling adapted in the context of the AP: a transfer of utilities between the total cost and the max-min distance is considered to be fair if the percentage increase in the total cost is smaller than the percentage decrease in the max-min distance and vice versa.

In fact, we have introduced the concept of NF solutions for the Balanced Traveling Salesman Problem (BTSP) in a recent paper [10]. In [10], we proposed an algorithm converging to particular NF solutions called *extreme NF solutions* having respectively smallest value of total cost and max-min distance. Similar to [10], in this current paper we also introduce the concept of NF solutions in the context of the AP. But, the main contribution of our work in this paper is a stronger result than in [10]: we provide an algorithm for finding all NF solutions in polynomial time.

The paper is organized as follows. In Sect. 2, we introduce a linear programming (LP) formulation for the BAP. The concept of NF solutions is presented in Sect. 3. In particular, we prove the existence of NF solutions for the AP and show that they are optimal solutions of a weighted sum objective problem. In Sect. 4, an algorithm for finding all NF solutions as well as computational results on various instances of the AP is given. Finally, conclusion and future works of this paper are discussed in Sect. 5.

2   LP Formulation for BAP

We consider an AP with a $n \times n$ cost matrix whose elements $c_{i,j}$ are positive and they represent the cost assignments between worker i and job j. We first present the linear programming (LP) formulation for the BAP as follows

$$\min Q = u - l \tag{1a}$$

$$\text{s.t.} \sum_{j \in [n]} x_{j,i} = \sum_{j \in [n]} x_{i,j} = 1 \qquad \forall i \in [n] \tag{1b}$$

$$u \geq \sum_{j \in [n]} c_{i,j} x_{i,j} \qquad \forall i \in [n] \tag{1c}$$

$$l \leq \sum_{j \in [n]} c_{i,j} x_{i,j} \qquad \forall i \in [n] \tag{1d}$$

$$x_{i,j} \geq 0 \qquad \forall i, j \in [n]. \tag{1e}$$

where $[n] = \{1, ..., n\}$ and $x_{i,j}$ represents the assignment between worker i and job j corresponding to the cost assignment $c_{i,j}$. In order to calculate the max-min distance Q, we need to determine the maximum and the minimum assignment costs u and l in the assignment solution. Constraints (1c) obviously allow to bound u from below by the maximum assignment cost in the assignment solution. Similarly, constraints (1d) allow to bound l from above by the minimum assignment cost in the assignment solution. As $Q = u - l$ is minimized, u and l will respectively take the values of the maximum and the minimum assignment costs. We will show that this LP formulation has integral optimal solution corresponding to an assignment solution (i.e., bipartite perfect matching).

Theorem 1. *This LP formulation always has an optimal solution where the variables take integer values.*

Proof. The objective function assures that u and l will be equal respectively to the maximum and the minimum assignment costs in the optimal solution. Consequently, the optimal solution of this LP is always integral because the constraints matrix of (1b) is totally unimodular (e.g see [6]) and the constraints (1c) and (1d) are simply bound constraints.   □

In the following we solve the classic AP as well as the BAP for several instances of the AP where we generate random uniform $c_{i,j}$ in $[1, 10^2]$. Optimal solutions of there instances are shown in Table 1 where *assignx* represents an instance of the AP with a cost matrix of dimension $x \times x$ and P, Q represent respectively the total cost and the max-min distance in the optimal solution. We use CPLEX 12.10 on a PC Intel Core i5-9500 3.00 GHz with 6 cores and 6 threads for solving the classic AP and the BAP. We can see in each instance of the AP that the optimal solutions for the classic AP may be undesirable with respect to those for the BAP and vice versa: inefficient values of Q in the optimal solutions for the classic AP comparing with those in the optimal solutions for the BAP and inefficient values of P in the optimal solutions for the BAP comparing with those in the optimal solutions for the classic AP.

Hence, the purpose of this paper is using a fair way to inject the total cost into the objective function for finding assignment solutions having a better trade-off between the two objectives.

Table 1. Optimal solutions for the calssic AP and the BAP

Instance	Classic AP			BAP		
	P	Q	Time	P	Q	Time
assign3	100	6	0.01	200	3	0.01
assign6	114	15	0.01	173	10	0.01
assign17	68	10	0.01	242	3	2.43
assign25	189	27	0.03	2004	11	2.44
assign30	157	18	0.04	643	7	32.4

3 Nash Fairness Solutions for the AP

We have introduced the concept of Nash fairness (NF) solution for the Balanced Traveling Salesman Problem (BTSP) [10]. In this section, we restate the concept of NF solutions in the context of the AP and we put the proofs of theorems in Appendix.

3.1 Proportional Fairness

The concept of NF solutions for the AP is closely related to the concept of proportional fairness for multiple players problem [1]. In the context of multiple players problem, let U be a set of possible *states of the world* or *alternatives* and let I be a finite set, representing a collection of individuals. For each $i \in I$, $u_i : U \longrightarrow \mathbb{R}_+$ be a utility function, describing the amount of happiness an individual i derives from each possible state such that we prefer the alternative x to the alternative y if and only if $u_i(x) \geq u_i(y), \forall i \in I$.

NF solutions for two-player problem [3] are defined by using the Nash standard of comparison: a transfer of utilities between the two players is considered to be fair if the percentage increase in the utility of one player is larger than the percentage decrease in utility of the other player [1].

Proportional fairness introduced by Bertsimas et al. [1] is a generalized NF solution for multiple players. In that setting, the fair allocation should be such that, if compared to any other feasible allocation of utilities, the aggregate proportional change is less than or equal to 0 [1,3,4].

Definition 1 [1]. *$x^{NF} \in U$ is a NF solution for n players problem if and only if*

$$\sum_{j=1}^{n} \frac{u_j(x) - u_j(x^{NF})}{u_j(x^{NF})} \leq 0, \ \forall x \in U, \quad (2)$$

where n is the number of players and $u_j(x) > 0, \ \forall j \in I, \forall x \in U$.

3.2   Characterization of NF Solutions for the AP

Let P, Q represent the total cost and the max-min distance in a feasible assignment solution for the AP. From the definitions of P and Q we have $P \geq Q \geq 0$. We first suppose that $Q > 0$. As P, Q now are two strictly positive functions, we have a two-player problem. In the usual definition of NF solutions [1,3], an alternative assigned a greater value is preferred. However, in the context of the AP we prefer the alternative assigned a smaller value for P and Q (i.e., P and Q now can be considered as two cost functions instead of two utility functions). Thus, the aggregate proportional change should be greater than or equal to 0 in the definition of NF solutions for the AP. That is to say, the sum of relative gains when switching from NF solutions to another feasible solution is not negative in the context of the AP.

We denote the value solution for the total cost and the max-min distance corresponding to a feasible assignment solution by (P, Q). Let (P^*, Q^*) be a NF solution for the AP, condition (2) can be translated into the context of the AP as follows

$$\frac{P - P^*}{P^*} + \frac{Q - Q^*}{Q^*} \geq 0, \ \forall (P, Q) \in \mathcal{S}, \tag{3}$$

which is equivalent to

$$PQ^* + QP^* \geq 2P^*Q^*, \ \forall (P, Q) \in \mathcal{S}, \tag{4}$$

where $\mathcal{S}$ is the set of solutions (P, Q) corresponding to all feasible assignment solutions for the AP.

Note that in case $Q^* = 0$, the condition (4) is also satisfied. Hence, NF solution for the AP can be generally stated as follows

Lemma 1 [10]. $(P^*, Q^*) \in \mathcal{S}$ *is a NF solution for the AP if and only if* $PQ^* + QP^* \geq 2P^*Q^*, \ \forall (P, Q) \in \mathcal{S}.$

Remark 1. An assignment solution having all equal assignment costs (i.e., $Q = 0$) is a NF solution.

3.3   Existence of NF Solutions

In this section, we first show the existence of NF solutions for the AP. Let us recall that in the multiple players problem mentioned in Sect. 3.1, NF solution does not always exist [3]. If exists, it can be obtained equivalently as the optimal solution of the problem

$$\max \sum_{j=1}^{n} \log u_j,$$

provided that U is convex. Notice that the above NF solution (if exists) is equivalently the one maximizing the product of the utilities over U [1].

On the contrary in the AP, there exists NF solutions which can be obtained by minimizing instead of maximizing the product of the utilities.

Theorem 2 [10]. $(P^*, Q^*) = \arg\min_{(P,Q) \in \mathcal{S}} PQ$ *is a NF solution.*

Proof. See Appendix.   □

Theorem 2 proves the existence of NF solutions for the AP that minimize PQ, or equivalently minimize $(\log P + \log Q)$. We call such solutions *Product Nash Fairness* (PNF) solutions. However, finding PNF solutions may be difficult as it requires to minimize a concave function. In the following, we show that all NF solutions can be obtained by solving the following optimization problem

$$\mathcal{P}(\alpha) = \min \ \alpha P + Q \text{ s.t } (P, Q) \in \mathcal{S},$$

where $\alpha \in [0, 1]$ is a coefficient to be determined. For solving $\mathcal{P}(\alpha)$, we solve the LP formulation in Sect. 2 with $\alpha P + Q$ as the objective function instead of Q.

Let $\alpha \in \mathbb{R}_+$ and (P_α, Q_α) be an optimal solution of $\mathcal{P}(\alpha)$. Denote $T_\alpha := \alpha P_\alpha - Q_\alpha$ and $\mathcal{C}_0 := \{\alpha \in \mathbb{R}_+ | T_\alpha = 0\}$. Hence, if $\alpha \in \mathcal{C}_0$ (i.e., $T_\alpha = 0$) then $\alpha \in [0, 1]$, otherwise $T_\alpha > P_\alpha - Q_\alpha \geq 0$.

Theorem 3 [10]. $(P^*, Q^*) \in \mathcal{S}$ *is a NF solution if and only if* (P^*, Q^*) *is an optimal solution of* $\mathcal{P}(\alpha^*)$ *with* $\alpha^* = \frac{Q^*}{P^*}$.

Proof. See Appendix.   □

Consequently, we state the following corollary.

Corollary 1. *If* (P^*, Q^*) *is an optimal solution of* $\mathcal{P}(\alpha^*)$ *and* $T_{\alpha^*} := \alpha^* P^* - Q^* = 0$ *then* (P^*, Q^*) *is a NF solution.*

We call (P, Q) a Pareto-optimal solution for the AP if (P, Q) is an optimal solution of $\mathcal{P}(\alpha)$ where $\alpha \in [0, 1]$. By Theorem 3, a NF solution is necessarily a Pareto-optimal solution but not vice versa.

Proposition 1. *There may be more than one NF solution for the AP.*

Proof. See Appendix.   □

The main question now is how to determine the coefficients of $\mathcal{C}_0$ corresponding one-to-one to all the NF solutions according to Theorem 3. In the next section, we present an algorithm for finding all NF solutions in polynomial time.

4   Finding All NF Solutions for the AP

In [10], we proposed an algorithm converging to *extreme NF solutions* having respectively the smallest value of P and Q. In this section, we introduce another one to find all NF solutions in polynomial time. Obviously, they include the PNF solutions minimizing PQ.

4.1   Upper Bound for the Number of NF Solutions

We call $(P, Q), (P', Q')$ two distinct solutions if $(P, Q) \not\equiv (P', Q')$. We will show that the number of NF solutions for the AP is at most $C_{n^2}^2 + n$ where $C_{n^2}^2 = \frac{n^2(n^2-1)}{2}$ by the following lemma and theorem.

Lemma 2. *If $(P, Q) \not\equiv (P', Q')$ are two distinct NF solutions having $Q, Q' > 0$ then $P \neq P$ and $Q \neq Q'$.*

Proof. Suppose that $Q = Q' > 0$. Using the definition of NF solution we have

$$P'Q + Q'P \geq 2PQ \quad \text{and} \quad P'Q + Q'P \geq 2P'Q'.$$

which is equivalent to

$$P' + P \geq 2P \quad \text{and} \quad P' + P \geq 2P'.$$

Hence we obtain $P = P'$ which leads to a contradiction. By repeating the same argument for $P = P'$, we also have a contradiction. $\square$

Theorem 4. *The number of NF solutions for the AP is at most $C_{n^2}^2 + n$.*

Proof. If (P, Q) is a NF solution and $Q = 0$, the corresponding assignment solution has n equal assignment costs. For the AP with $n \times n$ cost matrix, we have n^2 assignments and consequently there are at most n distinct NF solutions having the same value $Q = 0$.

We consider now the NF solutions with $Q > 0$. We will show that the number of NF solutions having $Q > 0$ is at most $C_{n^2}^2$.

Let c_i^{max} and c_i^{min} be the maximum and the minimum assignment cost in the assignment solution corresponding to (P_i, Q_i) then $Q_i = c_i^{max} - c_i^{min}$. As shown in Lemma 2, for two distinct NF solutions $(P_i, Q_i), (P_j, Q_j)$ with both Q_i and Q_j strictly positive we obtain $Q_i \neq Q_j$ which is equivalent to $c_i^{max} - c_i^{min} \neq c_j^{max} - c_j^{min}$. We have then $(c_i^{max}, c_i^{min}) \not\equiv (c_j^{max}, c_j^{min})$. Thus, the assignment solutions corresponding to $(P_i, Q_i), (P_j, Q_j)$ have distinct pair of assignments representing the maximum and the minimum assignment cost. As we have at most n^2 distinct assignments, the number of distinct pairs of assignments is at most $C_{n^2}^2$. Consequently, the number of NF solutions having $Q > 0$ is at most $C_{n^2}^2$. Hence, the total number of NF solutions for the AP is at most $C_{n^2}^2 + n$. $\square$

By Theorem 4, the number of Pareto-optimal solutions having distinct values of Q is at most $C_{n^2}^2$.

4.2   Algorithm for Finding All NF Solutions

As shown in Theorem 3, each element $\alpha^* \in \mathcal{C}_0$ corresponds to a NF solution and vice versa. For finding all NF solutions, we aim at finding all elements of $\mathcal{C}_0$. Our main idea is that from each $\alpha_0 \in [0, 1]$, we first use a procedure called *Find()* to find $\alpha_k \in \mathcal{C}_0$ satisfying α_k is the unique element $\in \mathcal{C}_0$ between α_0 and α_k.

Thus, let $\mathcal{I}$ be the set containing the intervals $[\alpha_i, \alpha_j]$ corresponding to distinct Pareto-optimal solutions $(P_i, Q_i), (P_j, Q_j)$. We use another procedure called *Test()* for verifying whether there exists a NF solution corresponding to $c_k \in [\alpha_i, \alpha_j]$ or not. Using there procedures, the algorithm for finding all NF solutions can be stated as follows.

Algorithm 1. Finding all NF solutions

Input: An instance of the AP with positive values in a $n \times n$ cost matrix.
Output: Set $\mathcal{C}_0$ whose elements correspond to all NF solutions for this instance.

1: $c_0 \leftarrow Find(0)$
2: $c_1 \leftarrow Find(1)$
3: **if** $c_1 = c_0$ **then**
4:      $\mathcal{C}_0 = \{c_0\}$
5: **else**
6:      $\mathcal{I} = \{[c_0, c_1]\}, \mathcal{C}_0 = \{c_0, c_1\}$
7:      **for** $[c_i, c_j] \in \mathcal{I}$ **do**
8:          $Test([c_i, c_j])$
9:      **end for**
10: **end if**
11: **procedure** FIND(α)
12:     **repeat**
13:         solve $\mathcal{P}(\alpha)$ to obtain (P, Q)
14:         $T \leftarrow \alpha P - Q$
15:         $\alpha \leftarrow Q/P$
16:     **until** $T = 0$
17:     Return α.
18: **end procedure**
19: **procedure** TEST($[c_i, c_j]$)
20:     $\alpha_k = \frac{Q_i - Q_j}{P_j - P_i}$
21:     **if** $\mathcal{P}(\alpha_k)$ has an optimal solution different to (P_i, Q_i) and (P_j, Q_j) **then**
22:         $c_k \leftarrow Find(\alpha_k)$
23:         **if** $c_i = c_k$ **then**
24:             $[c_i, c_j] \leftarrow [\alpha_k, c_j]$ ▷ Update the elements of $\mathcal{I}$
25:             $Test([\alpha_k, c_j])$
26:         **else if** $c_k = c_j$ **then**
27:             $[c_i, c_j] \leftarrow [c_i, \alpha_k]$ ▷ Update the elements of $\mathcal{I}$
28:             $Test([c_i, \alpha_k])$
29:         **else**   ▷ c_k is a new element of $\mathcal{C}_0$
30:             $[c_i, c_j] \leftarrow [c_i, c_k], [c_k, c_j]$ ▷ Update the elements of $\mathcal{I}$
31:             $\mathcal{C}_0 \leftarrow \mathcal{C}_0 \cup c_k$
32:             $Test([c_i, c_k]), Test([c_k, c_j])$
33:         **end if**
34:     **end if**
35: **end procedure**

Let α_0 be the initial point, $T_i := \alpha_i P_i - Q_i$ and $\{\alpha_i\}$ denote the sequence constructed by Procedure $Find(\alpha_0)$. We show that Algorithm 1 explores all NF solutions in polynomial time by the following lemmas and theorem. Due to lack of space, we put some proofs in Appendix.

Lemma 3 [10]. *Let $\alpha, \alpha' \in \mathbb{R}_+$ and (P_α, Q_α), $(P_{\alpha'}, Q_{\alpha'})$ be the optimal solutions of $\mathcal{P}(\alpha)$ and $\mathcal{P}(\alpha')$ respectively, if $\alpha \leq \alpha'$ then $P_\alpha \geq P_{\alpha'}$ and $Q_\alpha \leq Q_{\alpha'}$.*

Proof. See Appendix. □

As a consequence of Lemma 3, if (P_α, Q_α), $(P_{\alpha'}, Q_{\alpha'})$ are optimal solutions of $\mathcal{P}(\alpha)$ and $\mathcal{P}(\alpha')$ and $P_\alpha < P_{\alpha'}$ (or $Q_\alpha > Q_{\alpha'}$) then we obtain $\alpha \geq \alpha'$.

Lemma 4 [10]. *During the execution of Procedure Find(α_0) in Algorithm 1, $\alpha_i \in [0,1]$, $\forall i \geq 1$. Moreover, if $T_0 \geq 0$ then the sequence $\{\alpha_i\}$ is non-increasing and $T_i \geq 0$, $\forall i \geq 0$. Otherwise, if $T_0 \leq 0$ then the sequence $\{\alpha_i\}$ is non-decreasing and $T_i \leq 0$, $\forall i \geq 0$.*

Proof. See Appendix. □

As a consequence of Lemma 4, if $T_k = 0$ then $\alpha_k = \alpha_{k+1}$, if $T_k > 0$ then $T_i > 0$, $\alpha_i > \alpha_{i+1}$, $\forall 0 \leq i \leq k$ and if $T_k < 0$ then $T_i < 0$, $\alpha_i < \alpha_{i+1}$, $\forall 0 \leq i \leq k$.

Lemma 5 [10]. *From each $\alpha_0 \in [0,1]$, Procedure Find(α_0) converges to a coefficient $\alpha_k \in \mathcal{C}_0$ satisfying α_k is the unique element $\in \mathcal{C}_0$ between α_0 and α_k.*

Proof. See Appendix. □

Lemma 6. *Procedure Find(α_0) converges in polynomial time.*

Proof. Suppose that Procedure Find(α_0) converges to $\alpha_k \in \mathcal{C}_0$ in $k+1$ iterations. We only consider the nontrivial case where $k > 0$ (i.e., $T_0 \neq 0$).

If $T_0 > 0$ we have $T_i > 0$, $\forall 0 \leq i \leq k-1$.

By contradiction, we will show that if $T_{i+1} > 0$ then $P_i < P_{i+1}$ and $Q_i > Q_{i+1}$.

Let assume that $P_i \geq P_{i+1}$. According to Lemma 4, $\alpha_i \geq \alpha_{i+1}$ that implies $P_i \leq P_{i+1}$ as the result of Lemma 3. Thus, $P_i = P_{i+1}$. Moreover, $T_{i+1} = \alpha_{i+1}P_{i+1} - Q_{i+1} = \frac{Q_iP_{i+1}-Q_{i+1}P_i}{P_i} > 0$ leads to $Q_iP_{i+1} > Q_{i+1}P_i$. Since $P_i = P_{i+1} > 0$, we get $Q_i > Q_{i+1}$.

On the other hand, as (P_i, Q_i) is the optimal solution $\mathcal{P}(\alpha_i)$, it shows that $\alpha_iP_i + Q_i \leq \alpha_iP_{i+1} + Q_{i+1}$. Using $P_i = P_{i+1}$, we obtain $Q_i \leq Q_{i+1}$ which leads to a contradiction.

By repeating the same argument for $Q_i \leq Q_{i+1}$, we also have a contradiction.

Similarly, if $T_0 < 0$ we obtain the same conclusion.

Consequently, the execution of Procedure Find(α_0) explores k Pareto-optimal solutions having distinct values of Q. As the number of Pareto-optimal solutions having distinct value of Q is at most $C_{n^2}^2$, Procedure Find(α_0) converges after a polynomial number of iterations. Hence, Procedure Find(α_0) converges in polynomial time cause the LP formulation in Sect. 2 for $\mathcal{P}(\alpha)$ can be solved in polynomial time. □

Now by using the following lemma, we show that Procedure *Test()* can be used for verifying the existence of a Pareto-optimal solution (as well as NF solution) in each interval $[\alpha_i, \alpha_j]$.

Lemma 7. *Given an interval $[\alpha_i, \alpha_j]$ defined by $0 \leq \alpha_i < \alpha_j \leq 1$ corresponding to two distinct Pareto-optimal solutions (P_i, Q_i) and (P_j, Q_j). Let $\alpha^* = \frac{Q_j - Q_i}{P_i - P_j}$, if $\mathcal{P}(\alpha^*)$ has no Pareto-optimal solution which is different to (P_i, Q_i) and (P_j, Q_j), there does not have another one in $[\alpha_i, \alpha_j]$.*

Proof. Using Lemma 3 with $\alpha_i < \alpha_j$ and $(P_i, Q_i), (P_j, Q_j)$ be the two distinct Pareto-optimal solutions, we have $Q_i < Q_j, P_i > P_j$.

We first show that $\alpha^* \in [\alpha_i, \alpha_j]$.

Due to the optimality of (P_i, Q_i) and (P_j, Q_j) we obtain

$$\alpha_i P_i + Q_i \leq \alpha_i P_j + Q_j,$$
$$\alpha_j P_j + Q_j \leq \alpha_j P_i + Q_i,$$

Hence, $\alpha_i \leq \frac{Q_j - Q_i}{P_i - P_j} \leq \alpha_j$ which leads to $\alpha_i \leq \alpha^* \leq \alpha_j$.

Now suppose that we do not obtain any Pareto-optimal solution by solving $\mathcal{P}(\alpha^*)$ which is different to (P_i, Q_i) and (P_j, Q_j), we will show that there does not have another one in $[\alpha_i, \alpha_j]$.

Since $\alpha^* = \frac{Q_j - Q_i}{P_i - P_j}$, we have $\alpha^* P_i + Q_i = \alpha^* P_j + Q_j$. That means in this case (P_i, Q_i) and (P_j, Q_j) are two optimal solutions of $\mathcal{P}(\alpha^*)$. If there exists another Pareto-optimal solution (P, Q) of $\mathcal{P}(\alpha)$ where $\alpha \in [\alpha_i, \alpha_j]$, we have then $Q_i < Q < Q_j$ and $P_i > P > P_j$. Applying the consequence of Lemma 3 with $Q < Q_j$ we obtain $\alpha \leq \alpha^*$. Similarly, from $Q > Q_i$ we obtain $\alpha \geq \alpha^*$.

Hence, $\alpha = \alpha^*$ and then $\mathcal{P}(\alpha^*)$ has the Pareto-optimal solution (P, Q) which is different to (P_i, Q_i) and (P_j, Q_j). It leads to a contradiction.   □

Theorem 5. *Algorithm 1 explores all NF solutions in polynomial time.*

Proof. As a consequence of Lemma 5, the interval $[c_0, c_1]$ contains all elements of $\mathcal{C}_0$.

We known that the number of Pareto-optimal solutions having distinct values of Q is at most $C_{n^2}^2$. Consequently, $[c_0, c_1]$ can be separated by at most $C_{n^2}^2 - 1$ intervals $[c_i, c_j]$ such that $c_i < c_j$ correspond to two Pareto-optimal solutions having distinct values of Q and there intervals have no common points except the endpoints. By using Procedure *Test()*, each recursive call give us a Pareto-optimal solution or show that we have explored an interval having no Pareto-optimal solution and consequently no NF solution inside. As we use Procedure *Find(α)* in each recursive call, Procedure *Test()* also terminates in polynomial time. Moreover, we obtain a NF solution from each Pareto-solution found with a corresponding coefficient $\in [c_0, c_1]$. Since Algorithm 1 terminated as the interval $[c_0, c_1]$ is totally explored, it found all NF solutions in polynomial time. Obviously, the PNF solutions minimizing PQ can be easily determined by comparing the products of all NF solutions.   □

4.3   Numerical Results

Let us denote NFAP as the problems of finding all NF solutions for the AP. In this section, we conduct several experiments aiming at solving the NFAP with CPLEX 12.10 on some instances from the data sets of the AP [9] as well as on some instances that we generate random uniform their cost matrix. We also solve the classic AP and the BAP on the same instances. All the experiments are conducted on a PC Intel Core i5-9500 3.00 GHz with 6 cores and 6 threads.

Table 2. Numerical results for the classic AP, BAP and NFBAP

Instance	Classic AP			BAP			PNF			NFAP	
	P	Q	Time	P	Q	Time	P	Q	Time	α	All NF solutions
assign3	100	6	0.01	200	3	0.05	140	4	0.11	0.028	(100, 6), (140, 4), (200, 3)
assign4	70	9	0.01	196	3	0.25	196	3	1.20	0.015	(70, 9), (80, 8) (120, 5), (196, 3)
assign6	114	15	0.01	173	10	0.34	118	12	2.54	0.101	(118, 12)
assign17	68	10	0.01	262	3	2.43	71	8	14.8	0.112	(71, 8), (80, 7), (130, 4)
assign25	189	27	0.14	2004	11	7.12	189	27	41.2	0.142	(189, 27), (452, 14)
assign30	157	18	0.04	643	7	32.4	158	16	74.1	0.101	(158, 16), (473, 8)
assign45	6212	200	0.15	40937	54	85.0	6240	185	574	0.296	(6240, 185), (7133, 160) (9394, 112), (12766, 75)
assign75	8828	65	0.28	63860	36	122	9741	49	336	0.005	(9741, 49)
assign100	305	6	0.58	661	3	34.3	310	3	85.6	0.009	(310, 3)

Table 3. Aggregate proportional change when switching from PNF solutions to the optimal solutions of the classic AP and the BAP

Instance	Aggregate proportional change	
	PNF vs classic AP	PNF vs BAP
assign3	0.214	0.178
assign4	1.357	0.000
assign6	0.216	0.299
assign17	0.207	2.065
assign25	0.000	9.010
assign30	0.118	2.507
assign45	0.076	4.852
assign75	0.232	5.290
assign100	0.983	1.132

Table 2 presents the numerical results in several instances with a range of dimension of cost matrix from 3 × 3 to 100 × 100. We also provide the PNF solution minimizing PQ, its corresponding value of α and the running time for finding the PNF solution. We can see by the values of P and Q in this table that the PNF solution strikes a better trade-off between the total cost and the max-min distance comparing with those for the classic AP and the BAP. In particular, when the solutions for the classical AP and for the BAP are quite different: inefficient values of Q in the optimal solutions for the classic AP comparing with those in the optimal solutions for the BAP and inefficient values of P in the optimal solutions for the BAP comparing with those in the optimal solutions for the classic AP, the PNF solution offers almost a better alternative than the solution for the classic AP (respectively for the BAP) with a significant drop on the values of Q (respectively P) and a slight growth on the values of P (respectively Q). More precisely, Table 3 presents the sums of relative gains when switching from PNF solutions to the optimal solutions for the classic AP and the BAP. Note that their values are not negative as we mentioned in Sect. 3.2 and values further from 0 are preferable for the PNF solutions because they have then a much better trade-off between P and Q. Table 2 also indicates that we only have several NF solutions and in most cases, the PNF solution is one of the extreme NF solutions having smallest value of P or Q. One important issue is that the CPU time for solving the NFAP (almost equivalent to the running time for finding the PNF solution) is quite huge comparing with the CPU time spent for solving the classic AP and the BAP. A deeper analysis on the iterations of Procedure $Find(\alpha)$ tells us that the CPU time spent for solving $P(\alpha)$ with small value of α occupies a very big part of the overall CPU time. Hence, a special-purpose algorithm for solving $P(\alpha)$ may be more interesting than simply optimizing a linear function over the LP given in Sect. 2.

5   Conclusion

In this paper, we have made use of Nash fairness equilibrium to achieve a trade-off between the efficiency estimated by the total cost and the balancedness estimated by the max-min distance in solutions for the *Assignment Problem* (AP). We have proven first the existence of *Nash Fairness* (NF) solutions for the AP. Second, we have designed an algorithm to find all NF solutions including the PNF solutions minimizing the product of total cost and max-min distance. Numerical results conducted on instances of the AP have shown that comparing with the optimal solutions for the BAP, the NF solutions found by our algorithm have almost much smaller total cost with a reasonable augmentation of the max-min distance and vice versa comparing with the optimal solutions for the classic AP. An important notice is that the results in this paper can be also applied to various balanced optimization problems with two cost functions to be minimized such as the balanced traveling salesman problem [10], the balanced spanning tree problem [8], etc. The future developments of our work are improving time complexity for Algorithm 1 by developing a special-purpose algorithm for solving $\mathcal{P}(\alpha)$. Moreover, we are also interested in finding a better upper bound for the number of NF solutions as well as generating the concept of NF solutions for multi-objective optimization problems having positive objective functions.

Appendix

Proposition 1. *There may be more than one NF solution for the AP.*

Proof. Let us illustrate this by an instance of the AP having the following cost matrix

$$A = \begin{bmatrix} 71 & 91 & 132 \\ 128 & 102 & 123 \\ 106 & 104 & 107 \end{bmatrix}$$

By verifying all feasible assignment solutions in this instance, we obtain easily three assignment solutions $(1 - 1, 2 - 2, 3 - 3), (1 - 2, 2 - 3, 3 - 1), (1 - 3, 2 - 2, 3 - 1)$ and $(1 - 3, 2 - 1, 3 - 2)$ corresponding to 4 NF solutions $(280, 36)$, $(320, 32), (340, 30)$ and $(364, 28)$. Note that $i - j$ where $1 \le i, j \le 3$ represents the assignment between worker i and job j in the solution of this instance. □

We recall below the proofs of some recent results that we have published in [10]. They are needed to prove the new results presented in this paper.

Theorem 2 [10]. $(P^*, Q^*) = \arg\min_{(P,Q)\in S} PQ$ *is a NF solution.*

Proof. Obviously, there always exists a solution $(P^*, Q^*) \in S$ such that

$$(P^*, Q^*) = \arg\min_{(P,Q)\in S} PQ.$$

Now $\forall (P', Q') \in S$ we have $P'Q' \ge P^*Q^*$. Then

$$P'Q^* + Q'P^* \ge 2\sqrt{P'Q'P^*Q^*} \ge 2P^*Q^*,$$

The first inequality holds by the Cauchy-Schwarz inequality.
Hence, (P^*, Q^*) is a NF solution. □

Theorem 3 [10]. $(P^*, Q^*) \in S$ *is a NF solution if and only if* (P^*, Q^*) *is an optimal solution of* $\mathcal{P}(\alpha^*)$ *where* $\alpha^* = \frac{Q^*}{P^*}$.

Proof. Firstly, let (P^*, Q^*) be a NF solution and $\alpha^* = \frac{Q^*}{P^*}$. We will show that (P^*, Q^*) is an optimal solution of $\mathcal{P}(\alpha^*)$.
Since (P^*, Q^*) is a NF solution, we have

$$P'Q^* + Q'P^* \ge 2P^*Q^*, \ \forall (P', Q') \in S, \tag{6}$$

Since $\alpha^* = \frac{Q^*}{P^*}$, we have $\alpha^* P^* + Q^* = 2Q^*$.
Dividing two sides of (6) by $P^* > 0$ we obtain

$$2Q^* \le \frac{Q^*}{P^*} P' + Q', \ \forall (P', Q') \in S, \tag{7}$$

So we deduce from (7)

$$\alpha^* P^* + Q^* \leq \alpha^* P' + Q', \ \forall (P', Q') \in \mathcal{S},$$

Hence, (P^*, Q^*) is an optimal solution of $\mathcal{P}(\alpha^*)$.

Now suppose $\alpha^* = \frac{Q^*}{P^*}$ and (P^*, Q^*) is an optimal solution of $\mathcal{P}(\alpha^*)$, we show that (P^*, Q^*) is a NF solution.

If (P^*, Q^*) is not a NF solution, there exists a solution $(P', Q') \in \mathcal{S}$ such that

$$P'Q^* + Q'P^* < 2P^*Q^*,$$

We have then

$$\alpha P' + Q' = \frac{P'Q^* + Q'P^*}{P^*} < \frac{2P^*Q^*}{P^*} = \alpha^* P^* + Q^*,$$

which contradicts the optimality of (P^*, Q^*). $\qquad\square$

Lemma 3 [10]. *Let $\alpha, \alpha' \in \mathbb{R}_+$ and (P_α, Q_α), $(P_{\alpha'}, Q_{\alpha'})$ be the optimal solutions of $\mathcal{P}(\alpha)$ and $\mathcal{P}(\alpha')$ respectively, if $\alpha \leq \alpha'$ then $P_\alpha \geq P_{\alpha'}$ and $Q_\alpha \leq Q_{\alpha'}$.*

Proof. The optimality of (P_α, Q_α) and $(P_{\alpha'}, Q_{\alpha'})$ gives

$$\alpha P_\alpha + Q_\alpha \leq \alpha P_{\alpha'} + Q_{\alpha'}, \text{ and} \tag{8a}$$
$$\alpha' P_{\alpha'} + Q_{\alpha'} \leq \alpha' P_\alpha + Q_\alpha \tag{8b}$$

By adding both sides of (8a) and (8b), we obtain $(\alpha - \alpha')(P_\alpha - P_{\alpha'}) \leq 0$. Since $\alpha \leq \alpha'$, it follows that $P_\alpha \geq P_{\alpha'}$.

On the other hand, inequality (8a) implies $Q_{\alpha'} - Q_\alpha \geq \alpha(P_\alpha - P_{\alpha'}) \geq 0$ that leads to $Q_\alpha \leq Q_{\alpha'}$. $\qquad\square$

Lemma 4 [10]. *During the execution of Procedure Find(α_0) in Algorithm 1, $\alpha_i \in [0,1], \forall i \geq 1$. Moreover, if $T_0 \geq 0$ then the sequence $\{\alpha_i\}$ is non-increasing and $T_i \geq 0, \forall i \geq 0$. Otherwise, if $T_0 \leq 0$ then the sequence $\{\alpha_i\}$ is non-decreasing and $T_i \leq 0, \forall i \geq 0$.*

Proof. Since $P \geq Q \geq 0, \forall (P,Q) \in \mathcal{S}$, it follows that $\alpha_{i+1} = \frac{Q_i}{P_i} \in [0,1], \forall i \geq 0$.

We first consider $T_0 \geq 0$. We proof $\alpha_i \geq \alpha_{i+1}, \forall i \geq 0$ by induction on i. For $i = 0$, we have $T_0 = \alpha_0 P_0 - Q_0 = P_0(\alpha_0 - \alpha_1) \geq 0$, it follows that $\alpha_0 \geq \alpha_1$. Suppose that our hypothesis is true until $i = k \geq 0$, we will prove that it is also true with $i = k + 1$.

Indeed, we have

$$\alpha_{k+1} - \alpha_{k+2} = \frac{Q_k}{P_k} - \frac{Q_{k+1}}{P_{k+1}} = \frac{Q_k P_{k+1} - P_k Q_{k+1}}{P_k P_{k+1}},$$

The inductive hypothesis gives $\alpha_k \geq \alpha_{k+1}$ that implies $P_{k+1} \geq P_k > 0$ and $Q_k \geq Q_{k+1} \geq 0$ according to Lemma 3. It leads to $Q_k P_{k+1} - P_k Q_{k+1} \geq 0$ and then $\alpha_{k+1} - \alpha_{k+2} \geq 0$.

Hence, we have $\alpha_i \geq \alpha_{i+1}$, $\forall i \geq 0$.

Consequently, $T_i = \alpha_i P_i - Q_i = P_i(\alpha_i - \alpha_{i+1}) \geq 0$, $\forall i \geq 0$.

Similarly, if $T_0 \leq 0$ we obtain that the sequence $\{\alpha_i\}$ is non-decreasing and $T_i \leq 0$, $\forall i \geq 0$. That concludes the proof. □

Lemma 5 [10]. *From each $\alpha_0 \in [0,1]$, Procedure Find(α_0) converges to a coefficient $\alpha_k \in \mathcal{C}_0$ satisfying α_k is the unique element $\in \mathcal{C}_0$ between α_0 and α_k.*

Proof. As a consequence of Lemma 4, Procedure *Find*(α_0) converges to a coefficient $\alpha_k \in [0,1]$, $\forall \alpha_0 \in [0,1]$.

By the stopping criteria of Procedure *Find*(α_0), when $T_k = \alpha_k P_k - Q_k = 0$ we obtain $\alpha_k \in \mathcal{C}_0$ and (P_k, Q_k) is a NF solution. (Theorem 3)

If $T_0 = 0$ then obviously $\alpha_k = \alpha_0$. We consider $T_0 > 0$ and the sequence $\{\alpha_i\}$ is now non-negative, non-increasing. We will show that $[\alpha_k, \alpha_0] \cap \mathcal{C}_0 = \alpha_k$.

Suppose that we have $\alpha \in (\alpha_k, \alpha_0]$ and $\alpha \in \mathcal{C}_0$ corresponding to a NF solution (P,Q). Then there exists $1 \leq i \leq k$ such that $\alpha \in (\alpha_i, \alpha_{i-1}]$. Since $\alpha \leq \alpha_{i-1}$, $P \geq P_{i-1}$ and $Q \leq Q_{i-1}$ due to Lemma 3. Thus, we get

$$\frac{Q}{P} \leq \frac{Q_{i-1}}{P_{i-1}} \qquad (9)$$

By the definitions of α and α_i, inequality (9) is equivalent to $\alpha \leq \alpha_i$ which leads to a contradiction.

By repeating the same argument for $T_0 < 0$, we also have a contradiction. □

References

1. Bertsimas, D., Farias, V.F., Trichakis, N.: The price of fairness. Oper. Res. January–February **59**(1), 17–31 (2011)
2. Martello, S., Pulleyblank, W.R., Toth, P., De Werra, D.: Balanced optimization problems. Oper. Res. Lett. **3**(5), 275–278 (1984)
3. Kelly, F.P., Maullo, A.K., Tan, D.K.H.: Rate control for communication networks: shadow prices, proportional fairness and stability. J. Oper. Res. Soc. **49**(3), 237–252 (1997). https://doi.org/10.1057/palgrave.jors.2600523
4. Ogryczak, W., Luss, H., Pioro, M., Nace, D., Tomaszewski, A.: Fair optimization and networks: a survey. J. Appl. Math. **2014**, 1–26 (2014)
5. Marler, R.T., Arora, J.S.: The weighted sum method for multi-objective optimization: new insights. Struct. Multi. Optim. **41**(6), 853–862 (2010)
6. Heller, I., Tompkins, C.B.: An extension of a theorem of Dantzig's. Ann. Math. Stud. (38), 247–254 (1956)
7. Kuhn, H.W.: The Hungarian method for assignment problem. Naval Res. Logist. Q. **2**(1–2), 83–97 (1955)
8. Martello, S.: Most and least uniform spanning trees. Discrete Appl. Math. **15**(2), 181–197 (1986)
9. Beasley, J.E.: Linear programming on Clay supercomputer. J. Oper. Res. Soc. **41**, 133–139 (1990)
10. Nguyen, M.H, Baiou, M., Nguyen, V.H., Vo, T.Q.T.: Nash fairness solutions for balanced TSP. In: International Network Optimization Conference (INOC2022) (2022)

Graphs and Trees

On the Thinness of Trees

Flavia Bonomo-Braberman[1,2], Eric Brandwein[1(✉)],
Carolina Lucía Gonzalez[1,2], and Agustín Sansone[1]

[1] Facultad de Ciencias Exactas y Naturales, Departamento de Computación,
Universidad de Buenos Aires, Buenos Aires, Argentina
{fbonomo,ebrandwein,cgonzalez,asansone}@dc.uba.ar
[2] Instituto de Investigación en Ciencias
de la Computación (ICC), CONICET-Universidad de Buenos Aires, Buenos Aires,
Argentina

Abstract. The study of structural graph width parameters like tree-width, clique-width and rank-width has been ongoing during the last five decades, and their algorithmic use has also been increasing [Cygan et al., 2015]. New width parameters continue to be defined, for example, mim-width in 2012, twin-width in 2020, and mixed-thinness, a generalization of thinness, in 2022.

The concept of *thinness* of a graph was introduced in 2007 by Mannino, Oriolo, Ricci and Chandran, and it can be seen as a generalization of interval graphs, which are exactly the graphs with thinness equal to one. This concept is interesting because if a representation of a graph as a k-thin graph is given for a constant value k, then several known NP-complete problems can be solved in polynomial time. Some examples are the maximum weighted independent set problem, solved in the seminal paper by Mannino et al., and the capacitated coloring with fixed number of colors [Bonomo, Mattia and Oriolo, 2011].

In this work we present a constructive $\mathcal{O}(n \log(n))$-time algorithm to compute the thinness for any given n-vertex tree, along with a corresponding thin representation. We use intermediate results of this construction to improve known bounds of the thinness of some special families of trees.

Keywords: Trees · Thinness · Polynomial time algorithm

1 Introduction

A graph G is *k-thin* if there exists an ordering $\sigma = v_1, \ldots, v_n$ of $V(G)$ and a partition S of $V(G)$ into k classes such that, for each triple (r, s, t) with $r < s < t$, if v_r, v_s belong to the same class and $(v_t, v_r) \in E(G)$, then $(v_t, v_s) \in E(G)$. An order and a partition satisfying those properties are said to be *consistent*. We call the tuple (σ, S) a *consistent solution* or *consistent layout*. The minimum k such that G is k-thin is called the *thinness* of G, and denoted by $\mathrm{thin}(G)$.

Partially supported by CONICET (PIP 11220200100084CO) and UBACyT (20020170100495BA and 20020160100095BA).

I. Ljubić et al. (Eds.): ISCO 2022, LNCS 13526, pp. 189–200, 2022.
https://doi.org/10.1007/978-3-031-18530-4_14

A wide family of problems can be solved in XP parameterized by thinness, given a consistent representation. This family includes weighted variations of list matrix assignment problems with matrices of bounded size, and the possibility of adding bounds on the weight of the sets and their unions and intersections [3].

For a given order of the vertices of G, there exists an algorithm to compute a consistent partition of the vertices of G in the lowest amount of classes with time complexity $\mathcal{O}(n^3)$, where n is the number of vertices of G [3], since the problem can be reduced in linear time to the optimal coloring of an auxiliary co-comparability graph of n vertices, and the latter can be solved in $\mathcal{O}(n^3)$ time [8]. On the other hand, computing a consistent ordering of the vertices for a given partition, or detect it does not exist, is NP-complete [3]. Very recently, by a reduction from that problem, it was proven that deciding whether the thinness of a graph is at most k, without any given order or partition, is NP-complete [14]. In this work we solve this problem in polynomial time for trees. This is the first non-trivial efficient algorithm to compute the thinness (and consistent order and partition of the vertices within a graph class).[1] Some efforts were made before to study the thinness of trees in [13].

The design of this algorithm was heavily inspired by the proof and the algorithm by Høgemo, Telle, and Vågset for another graph invariant, the *linear maximum induced matching width* (linear mim-width) [10]. The linear mim-width is a known lower bound for the thinness [3], and there are families with bounded linear mim-width and unbounded thinness [5]. However, we prove here that, for trees, the two parameters behave alike and the difference between thinness and linear mim-width is at most 1.

2   Definitions and Preliminary Results

All graphs in this work are finite, undirected and have no loops or multiple edges.

Let G be a graph, we denote by $V(G)$ its vertex set and by $E(G)$ its edge set. We denote by $N(v)$ and $N[v]$, respectively, the neighborhood and closed neighborhood of a vertex $v \in V(G)$. Let $X \subseteq V(G)$. We denote by $N(X)$ the set of vertices not in X having at least one neighbor in X, and by $N[X]$ the closed neighborhood $N(X) \cup X$.

We denote by $G[X]$ the subgraph of G induced by X, and by $G - W$ or $G \backslash W$ the graph $G[V(G) \backslash W]$. We use $G \backslash (u, v)$ to denote the graph with vertices $V(G)$ and edges $E(G) \backslash \{(u, v)\}$. A subgraph H of G is a *spanning subgraph* if $V(H) = V(G)$.

A *tree* is a connected graph with no cycles. A *leaf* of a tree T is a vertex with degree one in T. The *diameter* of a tree is the maximum number of edges in a simple path joining two vertices.

A *rooted tree on vertex* r is a tree in which vertex r is labeled as the root, and we will usually denote it by T_r. The *ancestors* of a vertex v in a rooted tree

---

[1] A linear-time algorithm and forbidden induced subgraphs characterization are known for thinness of cographs [3], but the algorithm and proofs follow from their decomposition theorem without much more complication.

T_r are all vertices in *the* simple path between v and r which are not v. Note that r has no ancestors in T_r. The *descendants* of a vertex v in T_r are all vertices for which v is an ancestor in T_r. The *children* of a vertex v are those neighbors of v which are also descendants. Conversely, the *parent* of a vertex $v \neq r$ is the only neighbor of v which is also an ancestor of v, if any. In a rooted tree T_r, the vertex r has no parent. The *grandchildren* of a vertex v are the children's children, and the *grandparent* is the parent's parent, if any. The *height* of a rooted tree is the maximum number of edges in a simple path from the root to a leaf.

Let T be a tree containing the adjacent vertices v and u. The *dangling tree* from v in u, $T\langle v, u \rangle$, is the component of $T\backslash(u, v)$ containing u.

The *pathwidth* of a graph G, denoted by $\mathrm{pw}(G)$, is the minimum clique number of an interval supergraph of G minus one [11]. The *linear mim-width* of a graph G, denoted by $\mathrm{lmimw}(G)$, is the smallest integer k such that $V(G)$ can be arranged in a linear layout $v_1, \ldots, v_n$ in such a way that for every $1 \leq i \leq n - 1$, the size of a maximum induced matching in the bipartite graph formed by the edges of G with an endpoint in $\{v_1, \ldots, v_i\}$ and the other one in $\{v_{i+1}, \ldots, v_n\}$ is at most k [15].

The following relations are known.

Theorem 1 *[3, 12]. For every graph G, $\mathrm{lmimw}(G) \leq \mathrm{thin}(G) \leq \mathrm{pw}(G) + 1$.*

3 Structural Characterization and Polynomial Time Algorithm for Thinness of Trees

A considerable part of the ideas related to the construction of the algorithm to compute the thinness (and a consistent ordering and partition of the vertices) we are presenting in this section was inspired by an algorithm to compute the linear mim-width of a tree and an optimal layout [10], which was at the same time inspired by the framework behind the pathwidth algorithm presented in [7].

Even if the structure results for trees are very similar to those for linear mim-width, the arguments in the proofs are different, because the definitions of the concepts are different.

Lemma 1 (Path Layout Lemma). *Let T be a tree. If there exists a path $P = (x_1, \ldots, x_p)$ in T such that every connected component of $T\backslash N[P]$ has thinness less than or equal to k then $\mathrm{thin}(T) \leq k + 1$. Moreover, given the consistent orderings and partitions for the components in at most k classes we can in linear time compute a consistent ordering and partition for T in at most $k + 1$ classes.*

The proof for this Lemma describes an algorithm that constructs a consistent solution for T given the consistent solutions of the connected components of $T\backslash N[P]$. The algorithm merges a consistent solution Ω that uses just one class for $N[P]$ with the consistent solutions for each of the connected components. It takes the classes used in the connected components and renames them so that only k different classes are used in all of the partitions for the connected components, merges them into one partition, and adds the class used in Ω to the partition. It also maintains the relative order of the vertices inside each of the connected components, and interleaves it with the order used in Ω, maintaining consistency. Thus, the result is a consistent solution for T using $k + 1$ classes.

Definition 1 (k-neighbor). *Let x be a vertex of a tree T, and v a neighbor of x in T. If there exists a vertex $u \neq x$ neighbor of v such that $\mathrm{thin}(T\langle v, u \rangle) \geq k$, then v is a k-neighbor of x.*

Definition 2 (k-component index, k-saturation). *The k-component index of x is the number of k-neighbors of x, and we note it as $D(x, k)$. If $D(x, k) \geq 3$ for some vertex x in T, we say that x is k-saturated in T.*

The thinness of a tree can be characterized in terms of the component index as follows.

Theorem 2 (Classification of Thinness of Trees). *Let T be a tree, then $\mathrm{thin}(T) \geq k + 1$ if and only if $D(x, k) \geq 3$ for some vertex x in T.*

Corollary 1 (Bound on the Number of Vertices). *The thinness of an n-vertex tree T is $\mathcal{O}(\log(n))$. In fact $\mathrm{thin}(T) \leq \log_3(n + 2)$.*

Corollary 2 (Bound on the Number of Leaves). *A nontrivial tree of thinness k has at least $\frac{3^{k-1}+3}{2}$ leaves. In particular, the thinness of a tree with ℓ leaves is at most $\log_3(6\ell - 9)$.*

It was proven in [3] that for a fixed value m, the thinness of a complete m-ary tree on n vertices is $\Theta(\log(n))$, and it was also proven in [13] that the thinness of a non-trivial tree is less than or equal to its height; but, until now, it was an open problem to compute the exact thinness of a complete m-ary tree. As a consequence of Theorem 2, we have the following results.

Theorem 3 (Thinness of Complete m-ary Trees). *Let be $m \geq 3$ and T a complete m-ary tree with height h, then $\mathrm{thin}(T) = \left\lceil \frac{h+1}{2} \right\rceil$.*

Theorem 4 (Thinness of Complete Binary Trees). *Let T be a complete binary tree with height h, then $\mathrm{thin}(T) = \left\lceil \frac{h+1}{3} \right\rceil$.*

From [13] it can be shown by construction that it is always possible to have a consistent solution for a given tree with approximately $\frac{diameter}{2}$ classes. Using the theorems above, we can adjust this bound.

Theorem 5 (Bound on the Diameter). *Let T be a tree and d its diameter, then $\mathrm{thin}(T) \leq \left\lceil \frac{d+1}{4} \right\rceil$. Moreover, if the maximum degree of a vertex in T is at most 3, then $\mathrm{thin}(T) \leq \left\lceil \frac{d+3}{6} \right\rceil$.*

By comparing Theorem 2 with the results in [10], we can inductively prove the following.

Corollary 3. *For any given tree T,* $\mathrm{thin}(T) - \mathrm{lmimw}(T) \leq 1$.

The difference arises from the fact that every graph has thinness at least one, while edgeless graphs have linear mim-width zero. Indeed, Theorem 2 also suggests how to build the minimum trees for each thinness value:

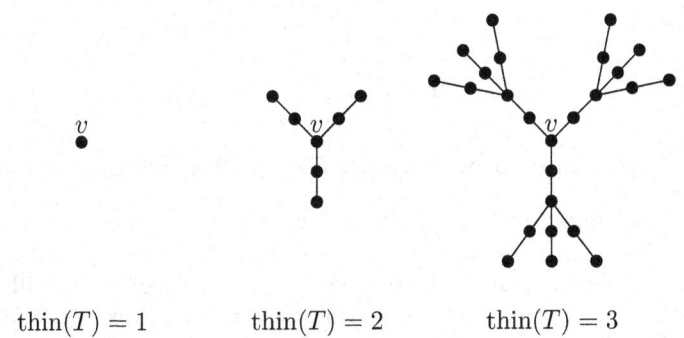

thin(T) = 1   thin(T) = 2   thin(T) = 3

For each thinness value k, the vertex v in the figure is such that $D(v, k-1) = 3$. The minimum tree with thinness k can be constructed by replacing each leaf in the minimum tree with thinness 2 into the minimum tree with thinness $k-1$, thus achieving $D(v, k) = 3$ with the minimum amount of vertices.

Compare this with the minimum trees with linear mim-width 1, 2 and 3 [9]:

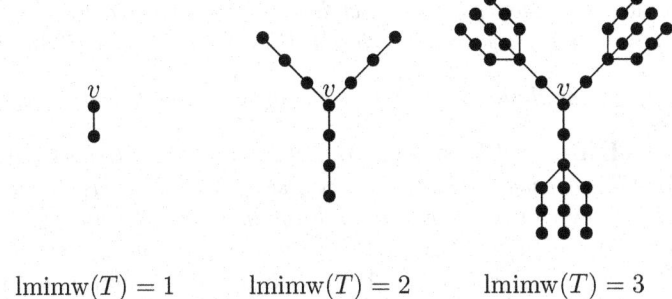

lmimw(T) = 1   lmimw(T) = 2   lmimw(T) = 3

These are pretty similar, except that the leaves in the trees with thinness k are replaced by two vertices. This is because a theorem very similar to Theorem 2 is also true for linear mim-width [10], and the smallest tree with linear mim-width 1 is the path of two vertices, while for thinness 1 it is a single vertex. This produces slightly bigger trees than for the thinness, which corresponds with the fact that the linear mim-width is a lower bound for the thinness.

Regarding the pathwidth, instead, the minimum trees are smaller, which also corresponds with the fact that the pathwidth plus one is an upper bound for the thinness. Again, a theorem similar to Theorem 2 holds for pathwidth [7], but with subtrees instead of dangling trees.

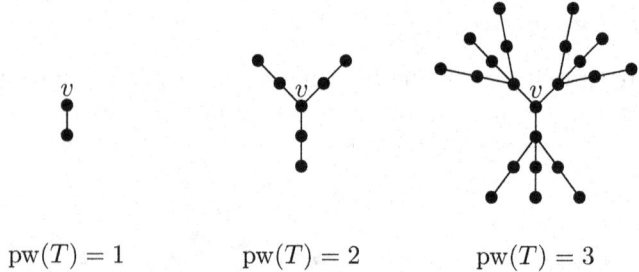

$$\mathrm{pw}(T) = 1 \qquad\qquad \mathrm{pw}(T) = 2 \qquad\qquad \mathrm{pw}(T) = 3$$

3.1   The Algorithm: Rooted Trees, k-critical Vertices and Labels

Our algorithm computing thinness will work on a rooted tree, processing it bottom-up. We will choose an arbitrary vertex r of the tree T and denote by T_r the tree rooted at r. During the bottom-up processing of T_r we will compute a label for various subtrees. The notion of a k-critical vertex is crucial for the definition of labels.

Definition 3 (Rooted Complete Subtree). *We define the* rooted complete subtree $T_r[x]$ *of* T_r *as the subtree of* T_r *rooted at* x *induced by* x *and the descendants of* x.

Definition 4 (k -critical Vertex). *Let* T_r *be a rooted tree. We call a vertex* x *in* T_r k-critical *if it has exactly two children* v_1 *and* v_2 *that have at least one child each,* u_1 *and* u_2 *respectively, such that* $\mathrm{thin}(T_r[u_1]) = \mathrm{thin}(T_r[u_2]) = k$.

Lemma 2. *If* T_r *has thinness* k, *then it has at most one* k-critical vertex.

Definition 5 (Label and Last Type). *Let* T_r *be a rooted tree with* $\mathrm{thin}(T_r) = k$. *Then* $\mathrm{label}(T_r)$ *consists of a list of decreasing numbers,* $(a_1, \dots, a_p)$, *where* $a_1 = k$, *and* $\mathrm{lastType}(T_r)$ *is an integer between 0 and 3 which will have the information of where in the tree an* a_p-critical vertex lies, if it exists at all, *according to the following list. If* $p = 1$ *then we define the label as being* simple, *otherwise it is* complex. *The* $\mathrm{label}(T_r)$ *and* $\mathrm{lastType}(T_r)$ *are defined recursively, with type 0 being a base case for singletons and for stars, and with type 4 being the only one defining a complex label. Note that each tree falls into one and only one of these types, and that for trees of type 4, the* $\mathrm{lastType}(T_r)$ *does not equal the type.*

- **Type 0:** *In this type of trees,* T_r *is a singleton, or all children of* r *are leaves.* $\mathrm{label}(T_r) = (1)$ *and* $\mathrm{lastType}(T_r) = 0$.
- **Type 1:** *Trees of this type are not Type 0 trees, and have no* k-critical vertex. $\mathrm{label}(T_r) = (k)$ *and* $\mathrm{lastType}(T_r) = 1$.
- **Type 2:** r *is the* k-critical vertex of trees of this type. $\mathrm{label}(T_r) = (k)$ *and* $\mathrm{lastType}(T_r) = 2$.
- **Type 3:** *In these trees a child of* r *is* k-critical. $\mathrm{label}(T_r) = (k)$ *and* $\mathrm{lastType}(T_r) = 3$.

- **Type 4:** *There is a k-critical vertex u_k in T_r that is neither r nor a child of r. Let w be the parent of u_k. Then $label(T_r) = k \oplus label(T_r \backslash T_r[w])$, and $lastType(T_r) = lastType(T_r \backslash T_r[w])$.*

In type 4 we note that $\text{thin}(T_r \backslash T_r[w]) < k$ since otherwise u_k would have three k-neighbors (two children in the tree and also its parent) and by Theorem 2 we would then have $\text{thin}(T_r) = k + 1$. Therefore, all numbers in $label(T_r \backslash T_r[w])$ are smaller than k and a complex label is a list of decreasing numbers. We also note that each element of a complex label corresponds to the thinness of some subtree of T_r, with the first element being the thinness of T_r. We now give a proposition that for any vertex x in T_r will be used to compute $label(T_r[x])$ and $lastType(T_r[x])$ based on the labels of the subtrees rooted at the children and grandchildren of x. The subroutine underlying this proposition will be used when reaching vertex x in the bottom-up processing of T_r.

Proposition 1. *Let x be a vertex of T_r with children $Child(x)$, and assume we are given $label(T_r[v])$ and $lastType(T_r[v])$ for all $v \in Child(x)$. Define $k = \max_{v \in Child(x)}\{\text{thin}(T_r[v])\}$, meaning, the maximum thinness of a subtree rooted on a child of x. Also define $N_k = \{v \in Child(x) \mid \text{thin}(T_r[v]) = k\}$, meaning, the set of children for whom the subtrees rooted at them have thinness k. Denote $N_k = \{v_1, \ldots, v_q\}$, $l_i = label(T_r[v_i])$, and $t_i = lastType(T_r[v_i])$. Define $d_k = D_{T_r[x]}(x, k)$ by noting that $d_k = |\{v_i \in N_k \mid v_i \text{ has child } u_j \text{ with } \text{thin}(T_r[u_j]) = k\}|$. Given this information, we can find $label(T_r[x])$ and $lastType(T_r[x])$ as follows.*

- **Case 0:** *x is a leaf or all children of x are leaves, and then $label(T_r[x]) = (1)$ and $lastType(T_r[x]) = 0$.*
- **Case 1:** *x is not a leaf and not all children of x are leaves, and for every $v_i \in N_k$, l_i is simple and t_i is equal to 1 or 0, and $d_k \leq 1$. Then, $label(T_r[x]) = (k)$, and $lastType(T_r[x]) = 1$.*
- **Case 2:** *For every $v_i \in N_k$, l_i is simple and t_i is equal to 1 or 0, but $d_k = 2$. Then, $label(T_r[x]) = (k)$ and $lastType(T_r[x]) = 2$.*
- **Case 3:** *For every $v_i \in N_k$, l_i is simple and t_i is equal to 1 or 0, but $d_k \geq 3$. Then, $label(T_r[x]) = (k + 1)$, and $lastType(T_r[x]) = 1$.*
- **Case 4:** *$|N_k| \geq 2$ and for some $v_i \in N_k$, either l_i is a complex label, or t_i is equal to either 2 or 3. Then, $label(T_r[x]) = (k + 1)$, and $label(T_r[x]) = 1$.*
- **Case 5:** *$|N_k| = 1$, l_1 is a simple label and t_1 is equal to 2. Then, $label(T_r[x]) = (k)$, and $lastType(T_r[x]) = 3$.*
- **Case 6:** *$|N_k| = 1$, l_1 is either complex or t_1 is equal to 3, and $k \notin label(T_r[x] \backslash T_r[w])$, where w is the parent of the k-critical vertex in $T_r[v_1]$. Then, $label(T_r[x]) = k \oplus label(T_r[x] \backslash T_r[w])$, and $lastType(T_r[x]) = lastType(T_r[x] \backslash T_r[w])$.*
- **Case 7:** *$|N_k| = 1$, l_1 is either complex or t_1 is equal to 3, and $k \in label(T_r[x] \backslash T_r[w])$, where w is the parent of the k-critical vertex in $T_r[v_1]$. Then, $label(T_r[x]) = k + 1$, and $lastType(T_r[x]) = 1$.*

3.2   Computing Thinness of Trees and Finding a Consistent Solution

The subroutine underlying Proposition 1 will be used in a bottom-up algorithm that starts out at the leaves and works its way up to the root, computing labels and lastTypes of subtrees $T_r[x]$. However, in cases 6 and 7 we need the label and lastType of $T_r[x]\backslash T_r[w]$, which is not a complete subtree rooted at any vertex of T_r. Note that the label and lastType of $T_r[x]\backslash T_r[w]$ are again given by a recursive call to Proposition 1, and then the label is stored as a suffix of the complex label of $T_r[x]$, and the lastType is the same. We will compute these labels and lastTypes by iteratively calling Proposition 1, substituting the recursion by iteration. We first need to carefully define the subtrees involved when dealing with complex labels.

From the definition of labels it is clear that only type 4 trees lead to a complex label. In that case we have a tree $T_r[x]$ of thinness k and a k-critical vertex u_k that is neither x nor a child of x, and the recursive definition gives $label(T_r[x]) = k \oplus label(T_r[x]\backslash T_r[w])$ for w the parent of u_k. Unravelling this recursive definition, we have the following:

Definition 6. *Let x be a vertex in T_r, and let $h = |label(T_r[x])|$. Denote $label(T_r[x]) = (a_1, \ldots, a_h)$. Then $\omega(T_r[x])$ is a list $(\omega_1, \ldots, \omega_h)$ of vertices in $T_r[x]$ in which $\omega_h = x$, and every other ω_i with $1 \le i < h$ is the parent of the a_i-critical vertex in $T_r[x]\backslash(T_r[\omega_1] \cup \cdots \cup T_r[\omega_{i-1}])$. We will use $\omega(T_r[x])_i$ to denote the element number i of the list, or simply use ω_i when it is clear which tree we are referring to.*

Now, in the first level of a recursive call to Proposition 1 the role of $T_r[x]$ is taken by $T_r[x]\backslash T_r[\omega_1]$, and in the next level it is taken by $(T_r[x]\backslash T_r[\omega_1])\backslash T_r[\omega_2]$ etc. The following definition gives a shorthand for denoting these trees.

Definition 7. *Let x be a vertex in T_r, and $label(T_r[x]) = (a_1, a_2, \ldots, a_p)$. For any non-negative integer s, the tree $T_r[x, s]$ is the subtree of $T_r[x]$ obtained by removing all trees $T_r[\omega_i]$ from $T_r[x]$, where $a_i \ge s$. In other words, if q is such that $a_q \ge s > a_{q+1}$, then $T_r[x, s] = T_r[x]\backslash(T_r[\omega_1] \cup T_r[\omega_2] \cup \cdots \cup T_r[\omega_q])$.*

Lemma 3. *Some important properties of $T_r[x, s]$ are the following. Let $T_r[x, s]$, $label(T_r[x, s])$, and q be as in the definition. Then*

1. *if $s > a_1$, then $T_r[x, s] = T_r[x]$*
2. *$label(T_r[x, s]) = (a_{q+1}, \ldots, a_p)$*
3. *$thin(T_r[x, s]) = a_{q+1} < s$*
4. *$thin(T_r[x, s + 1]) = s$ if and only if $s \in label(T_r[x])$*
5. *$T_r[x, s + 1] \ne T_r[x, s]$ if and only if $s \in label(T_r[x])$*

Lemma 4. *Let $x \in V(T_r)$, and let u be a child of x in T_r. Let $s \in \mathbb{N}$ such that $T_r[x, s]$ and $T_r[u, s]$ are not empty, that is to say, s is greater both than the last element of $label(T_r[x])$ and than the last element of $label(T_r[u])$. Let $T_s^* = T_r[x, s] \cap T_r[u]$. Then $T_s^* = T_r[u, s]$, meaning, the child subtree of $T_r[x, s]$ rooted at u is equal to $T_r[u, s]$.*

Corollary 4. *If $s \in label(T_r[x])$ and $s \in label(T_r[u])$, $T_{s+1}^* = T_r[u, s+1]$.*

Proof. That s be both in $label(T_r[x])$ and in $label(T_r[u])$ means that $s + 1$ is bigger than the last element of both labels, and so the conditions for the Lemma are satisfied for $s + 1$.   □

Note that, for any s, the tree $T_r[x, s]$ is defined only after we know $label(T_r[x])$. In the algorithm, we compute $label(T_r[x])$ by iterating over increasing values of s until $s > \text{thin}(T_r[x])$ since by Lemma 3.1 we then have $T_r[x, s] = T_r[x]$. This poses a problem: we cannot know which subtrees to calculate the labels for until we have finished with all subtrees. To solve this, each iteration of the loop will correctly compute the label of another subtree called $T_{union}[x, s]$, which is not always equal to $T_r[x]$. Nonetheless, for $s > \text{thin}(T_r[x])$, the equality $T_{union}[x, s] = T_r[x, s] = T_r[x]$ will hold, and so calculating labels for these subtrees will aid in calculating labels for bigger subtrees.

Definition 8. *Let x be a vertex in T_r with children $v_1, \ldots, v_d$. $T_{union}[x, s]$ is then equal to the tree induced by x and the union of all $T_r[v_i, s]$ for $1 \le i \le d$. More technically, $T_{union}[x, s] = T_r[V']$ where $V' = x \cup V(T_r[v_1, s]) \cup \cdots \cup V(T_r[v_d, s])$.*

Given a tree T, we find its thinness by rooting it at an arbitrary vertex r, and computing labels by processing T_r bottom-up. The answer is given by the first element of $label(T_r[r])$, which by definition is equal to $\text{thin}(T)$. At a vertex x of T_r which is a leaf or all their children are leaves we initialize by $label(T_r[x]) \leftarrow (1)$, according to Definition 5. When reaching a higher vertex x we compute the label of $T_r[x]$ by calling function MAKELABEL(T_r, x).

Lemma 5. *Given labels at descendants of vertex x in T_r, MAKELABEL(T_r, x) computes $label(T_r[x])$ as the value of cur_label and $lastType(T_r[x])$ as the value of cur_type.*

Theorem 6. *Given any tree T, $\text{thin}(T)$ can be computed in $\mathcal{O}(n \log(n))$-time.*

Proof. We find $\text{thin}(T)$ by bottom-up processing of T_r and returning the first element of $label(T_r)$. After correctly initializing at leaves and vertices whose children are all leaves, we make a call to MAKELABEL for each of the remaining vertices. Correctness follows by Lemma 5 and induction on the structure of the rooted tree. We will now show that each call runs in $\mathcal{O}(\log(n))$ time to prove the time complexity.

Let m be the biggest number in any label of children of x, which is $\mathcal{O}(\log(n))$ by Corollary 1. For every integer s from 1 to m, the algorithm checks how many labels of children of x contain s to compute N_s, and how many labels of grandchildren of x contain s to compute t_s. The labels are sorted in descending order; therefore the whole loop goes only once through each of these labels, each of length $\mathcal{O}(\log(n))$. Other than this, MAKELABEL only does a constant amount of work. Therefore, MAKELABEL(T_r, x), if x has a children and b grandchildren, takes time proportional to $\mathcal{O}(\log(n)(a+b))$. As the sum of the number of children and grandchildren over all vertices of T_r is $\mathcal{O}(n)$ we conclude that the total runtime to compute $\text{thin}(T)$ is $\mathcal{O}(n \log(n))$.   □

Algorithm 1. Compute $cur\_label = label(T_r[x])$ and $cur\_type = lastType$ $(T_r[x])$

---

function MAKELABEL(T_r: tree, x : vertex)
  $cur\_label \leftarrow (1)$
  $cur\_type \leftarrow 1$
  $\{v_1, \ldots, v_d\} \leftarrow$ children of x in T_r
  for $s \leftarrow 1, \max_{i=1}^{d}\{$ first element of $label(T_r[v_i])$ $\}$ **do**
    $\{l'_1, \ldots, l'_d\} \leftarrow \{label(T_r[v_i, s+1]) \mid 1 \le i \le d\}$
    $\{t'_1, \ldots, t'_d\} \leftarrow \{lastType(T_r[v_i, s+1]) \mid 1 \le i \le d\}$
    $N_s \leftarrow \{v_i \mid 1 \le i \le d, s \in l'_i\}$
    $d_s \leftarrow |\{v_i \mid v_i \in N_s, v_i \text{ has a child } u_j \text{ s.t. } s \in label(T_r[u_j, s+1])\}|$
    if $N_s \ne \emptyset$ **then**
      $case$ $\leftarrow$ the   case   from   Prop.   1   applying   to
$s, \{l'_1, \ldots, l'_d\}, \{t'_1, \ldots, t'_d\}, N_s$ and d_s
        $cur\_label \leftarrow$ as given by $case$ in Prop. 1 ($s \oplus cur\_label$ if Case 6)
        $cur\_type \leftarrow$ as given by $case$ in Prop. 1
    end if
  end for
end function

Theorem 7. *An optimal consistent solution can be found in $\mathcal{O}(n\log(n))$-time.*

Proof. Given T we first run the algorithm computing thin(T) finding the label and lastType of every full rooted subtree in T_r. We give a recursive layout algorithm that uses these labels in tandem with CONSISTENTLAYOUT presented in the Path Layout Lemma. We call it on a rooted tree where labels of all subtrees are known. For simplicity we call this rooted tree T_r even though in recursive calls this is not the original root r and tree T. The layout algorithm goes as follows:

1. Let thin$(T_r) = k$ and find a path P in T_r such that all trees in $T_r \backslash N[P]$ have thinness lower than k. The path depends on the type of T_r as explained in detail below.
2. Call this layout algorithm recursively on every rooted tree in $T_r \backslash N[P]$ to obtain linear layouts; to this end, we need the correct label for every vertex in these trees.
3. Call CONSISTENTLAYOUT on T_r, P and the layouts provided in step 2.

Every tree in the forest $T \backslash N[P]$ is equal to a dangling tree $T\langle v, u \rangle$ where v is a neighbor of some $x \in P$.

We observe that if thin$(T) = k$, then by definition thin$(T\langle v, u \rangle) = k$ if and only if v is a k-neighbor of x. It follows that every tree in $T \backslash N[P]$ has thinness at most $k-1$ if and only if no vertex in P has a k-neighbor that is not in P. We use this fact to show that for every type of tree we can find a satisfying path in the following way:

- *Type 0 trees*: Choose $P = (r)$. Since $|T \backslash N[r]| = 0$ in these trees, this must be a satisfying path.
- *Type 1 trees*: These trees contain no k-critical vertices, which by definition means that for any vertex x in T_r, at most one of its children is a k-neighbor of x. Choose P to start at the root r, and as long as the last vertex in P has a k-neighbor v, v is appended to P. This set of vertices is obviously a path in T_r. No vertex in P can possibly have a k-neighbor outside of P, therefore all connected components of $T \backslash N[P]$ have thinness lower or equal to $k - 1$. Furthermore, all components of $T \backslash N[P]$ are full rooted subtrees of T_r and so the labels are already known.
- *Type 2 trees*: In these trees the root r is k-critical. We look at the trees rooted in the two k-neighbors of r, $T_r[v_1]$ and $T_r[v_2]$. By Remark 2 these must both be Type 1 trees, and so we find paths P_1, P_2 in $T_r[v_1]$ and $T_r[v_2]$ respectively, as described above. Gluing these paths together at r we get a satisfying path for T_r, and we still have correct labels for the components $T \backslash N[P]$.
- *Type 3 trees*: In these trees, r has exactly one child v such that $T_r[v]$ is of type 2 and none of its other children have thinness k. We choose P as we did above for $T_r[v]$. Vertex r is clearly not a k-neighbor of v, or else $D_T(v, k) = 3$. Every other vertex in P has all their neighbors in $T_r[v]$. Again, every tree in $T \backslash N[P]$ is a full rooted subtree, and every label is known.
- *Type 4 trees*: In these trees, T_r contains precisely one vertex $w \neq r$ such that w is the parent of a k-critical vertex, x. This w is easy to find using the labels and lastTypes by annotating each subtree $T_r[v]$ with one new piece of information: its only k'-critical vertex, with $k' = \text{thin}(T_r[v])$, if it has one. To calculate it for some subtree $T_r[v]$, given that we have already calculated it for all its child subtrees, a simple check of every child subtree will give the answer for each type:
  - *Type 0 and 1 trees*: These do not contain any k'-critical vertex, so there is nothing to annotate.
  - *Type 2 trees*: v is the k'-critical vertex, so we annotate this tree with v.
  - *Type 3 trees*: We know that some child u of v is the root of a type 2 tree. We check every child of v to find the only child subtree with thinness k' that is a type 2 tree, and we copy its annotation into $T_r[v]$.
  - *Type 4 trees*: Some child subtree will have thinness k' and will also have a k'-critical vertex, by definition of a type 4 tree. We can then do the same as with the type 3 trees, checking the lastTypes of the child subtrees with thinness k' to see which one has a k'-critical vertex.

This procedure can be done in $\mathcal{O}(n)$ by traversing the whole tree in a bottom-up fashion, so the time complexity is not affected by it. After finding the k-critical vertex of T_r, w is simply its only parent, also easy to find.

Clearly, the tree $T_r[w]$ is a type 3 tree with thinness k. We find a path P that is satisfying in $T_r[w]$ as described above. w is still not a k-neighbor of x, therefore P is a satisfying path. In this case, we have one connected component of $T \backslash N[P]$ that is not a full rooted subtree of T_r, that is $T_r \backslash T_r[w]$. Thus, for every ancestor y of w, $T_r[y] \backslash T_r[w]$ is not a full rooted subtree either, and we need to update the labels of these trees.

As each $T_r[y]$ contains the k-critical vertex x, it has thinness greater or equal to k. Also, as $T_r[y]$ is a subtree of T_r, it has thinness lower or equal to k. These two facts tell us that $\text{thin}(T_r[y]) = k$, which means that k is the first element of its label. Also, they are all type 4 trees, because they each have the parent w of x as a descendant. This means that $w = \omega(T_r[y])_1$ for each y. With this, we see that $T_r[y]\backslash T_r[w]$ is by definition equal to $T_r[y, k]$, whose label is equal to $label(T_r[y])$ without its first number. Thus we quickly find the correct labels to do the recursive call.   □

References

1. Balabán, J., Hlinený, P., Jedelský, J.: Twin-width and transductions of proper k-mixed-thin graphs. arXiv: 2202.12536 [math.CO]
2. Bonnet, É., Kim, E.J., Thomassé, S., Watrigant, R.: Twin-width I: tractable FO model checking. In: 2020 Proceedings of the 61st IEEE Annual Symposium on Foundations of Computer Science - FOCS, pp. 601–612 (2020)
3. Bonomo, F., De Estrada, D.: On the thinness and proper thinness of a graph. Discret. Appl. Math. **261**, 78–92 (2019)
4. Bonomo, F., Mattia, S., Oriolo, G.: Bounded coloring of co-comparability graphs and the pickup and delivery tour combination problem. Theoret. Comput. Sci. **412**(45), 6261–6268 (2011)
5. Bonomo-Braberman, F., Brettell, N., Munaro, A., Paulusma, D.: Solving problems on generalized convex graphs via MIM-width. In: Lubiw, A., Salavatipour, M. (eds.) WADS 2021. LNCS, vol. 12808, pp. 200–214. Springer, Cham (2021). https://doi.org/10.1007/978-3-030-83508-8_15
6. Cygan, M., et al.: Parameterized Algorithms. Springer, Cham (2015). https://doi.org/10.1007/978-3-319-21275-3
7. Ellis, J., Sudborough, I., Turner, J.: The vertex separation and search number of a graph. Inf. Comput. **113**(1), 50–79 (1994)
8. Golumbic, M.: The complexity of comparability graph recognition and coloring. Computing **18**, 199–208 (1977). https://doi.org/10.1007/BF02253207
9. Høgemo, S.: On the linear MIM-width of trees. Master's thesis, The University of Bergen, Bergen, Norway (2019)
10. Høgemo, S., Telle, J.A., Vågset, E.R.: Linear MIM-width of trees. In: Sau, I., Thilikos, D.M. (eds.) WG 2019. LNCS, vol. 11789, pp. 218–231. Springer, Cham (2019). https://doi.org/10.1007/978-3-030-30786-8_17
11. Kaplan, H., Shamir, R.: Pathwidth, bandwidth, and completion problems to proper interval graphs with small cliques. SIAM J. Comput. **25**(3), 540–561 (1996)
12. Mannino, C., Oriolo, G., Ricci, F., Chandran, S.: The stable set problem and the thinness of a graph. Oper. Res. Lett. **35**, 1–9 (2007)
13. Rabinowicz, L.: Sobre la thinness de árboles. Master's thesis, Departamento de Computación, FCEyN, Universidad de Buenos Aires, Buenos Aires (2019)
14. Shitov, Y.: Graph thinness is NP-complete (2021). Manuscript
15. Vatshelle, M.: New width parameters of graphs. Ph.D. thesis, Department of Informatics, University of Bergen, Bergen (2012)

Generating Spanning-Tree Sequences of a Fan Graph in Lexicographic Order and Ranking/Unranking Algorithms

Ro-Yu Wu[1], Cheng-Chia Tseng[2], Ling-Ju Hung[3], and Jou-Ming Chang[2(✉)]

[1] Department of Industrial Management,
Lunghwa University of Science and Technology, Taoyuan 333326, Taiwan
`eric@mail.lhu.edu.tw`
[2] Institute of Information and Decision Sciences,
National Taipei University of Business, Taipei 100025, Taiwan
`{10966010,spade}@ntub.edu.tw`
[3] Department of Product Innovation and Entrepreneurship,
National Taipei University of Business, Taoyuan 324022, Taiwan
`ljhung@ntub.edu.tw`

Abstract. Cameron et al. [27th Int. Conf. Computing and Combinatorics (COCOON 2021), LNCS 13025, pp. 49–60] recently presented an algorithm for generating all spanning trees of a fan graph in $\mathcal{O}(1)$-amortized time. The listing of spanning trees fulfills the so-called pivot Gray code property so that successive trees differ by pivoting a single edge around a vertex. They also presented algorithms for ranking and unranking a spanning tree in the listing in $\mathcal{O}(n)$ time using $\mathcal{O}(n)$ space. In this paper, we first observe that all spanning trees of a fan graph can be naturally represented by integer sequences so that their coding tree has properties with regularity. Then, we propose a simple algorithm for generating spanning-tree sequences in lexicographic order in $\mathcal{O}(1)$-amortized time according to these properties. Additionally, based on the lexicographic order, we develop ranking and unranking algorithms in $\mathcal{O}(n)$-time using $n + \mathcal{O}(1)$ space (i.e., the size of the space is just slightly larger than n).

Keywords: Spanning trees · Fan graphs · Generation algorithms ranking/unranking algorithms · Constant amortized time algorithms

1 Introduction

Exhaustively generating a set of combinatorial objects is an important work in the field of combinatorics and computing, which has many applications such as searching for the desired entity with a particular feature among all generated objects, looking for a counterexample to a certain conjecture, or analyzing the average performance of an algorithm over all possible inputs. According to the list of generated objects, two practical functions, called *ranking* and *unranking*,

© The Author(s), under exclusive license to Springer Nature Switzerland AG 2022
I. Ljubić et al. (Eds.): ISCO 2022, LNCS 13526, pp. 201–211, 2022.
https://doi.org/10.1007/978-3-031-18530-4_15

are usually developed. The former can determine the rank of a given object in this listing order, and the latter can produce the object corresponding to a given rank. Since efficient ranking and unranking are essential for storing and retrieving elements in the large class of combinatorial objects, these techniques have been extended to the uses of database indexing and data compression [10,13].

Remarkably, there is a wide range of demands for generating all spanning trees of a graph since Minty [15] discovered its application in electrical circuits and network systems in his early work. Afterward, many generation algorithms have been proposed for undirected graphs [3,9,11,14,19] and directed graphs [9,12], respectively. Especially, efficiently generating spanning trees for special classes of graphs has also received much attention, such as complete graphs [7], Halin graphs [16], and fan graphs [2] etc. For ranking and unranking spanning trees of an arbitrary undirected graph with n vertices, the first algorithm requiring $\mathcal{O}(n^3)$ time was developed by Colbourn et al. [5]. See also [6] for two algorithms of a directed graph version with n vertices. For more results related to the ranking and unranking of spanning trees of other special graphs, please refer to [8,17,18].

Recently, Cameron et al. [2] proposed an algorithm for generating all spanning trees of a fan graph so that the list of generated trees with the so-called pivot Gray code property, where the succeeding tree is obtained from the previous one by pivoting an edge around a vertex. Then, they showed that by applying a greedy approach with the priority that first chooses vertex to be pivoted on and follows by considering the adjusting endpoints for addition/removal, the generation algorithm can be run in $\mathcal{O}(1)$-amortized time. In addition, they also showed that ranking and unranking a spanning tree in the greedy listing can be done in $\mathcal{O}(n)$ time using $\mathcal{O}(n)$ space.

From the structure of fan graphs with labels on the vertices (formally defined later in Sect. 2), we can naturally find that each spanning tree possesses a representation of an integer sequence. This paper takes the unit-cost RAM model as the computational model and assumes that every integer fits in one unit. Then, we adopt a systematic way to describe all spanning-tree sequences using a coding tree. According to the regularity in the coding tree, we can easily design an algorithm for generating all spanning-tree sequences in lexicographical order to run in $\mathcal{O}(1)$-amortized time with no more than two data change operations per generation. Furthermore, we propose ranking and unranking algorithms based on the lexicographical order in $\mathcal{O}(n)$ time. Each of the above three algorithms takes $n + \mathcal{O}(1)$ space (i.e., the required space is slightly more than n).

2   Preliminary

For a graph $G = (V, E)$, a *spanning tree* T is a connected acyclic subgraph of G such that $V(T) = V(G)$. A *rooted tree* is a directed tree so that one of its vertex, called the *root*, is distinguished from the others. A rooted tree is *ordered* if all vertices are presented according to a designed order. Two integer sequences $x(1..n)$ and $y(1..m)$ are said to be in *lexicographic order* if there is an integer

$i \in \{1, 2, \ldots, \min\{n, m\}\}$ such that $x(j) = y(j)$ for all $j \in \{1, 2, \ldots, i - 1\}$ and $x(i) < y(i)$.

Let P_n be a simple path on n vertices with labels $1, 2, \ldots, n$. The *fan graph* F_n (or called an *n-fan*) is obtained from P_n by adding an additional vertex with label 0 such that this vertex is adjacent to every vertex of P_n. Let t_n denote the number of spanning trees of F_n. By the celebrated Kirchhoff's Matrix-tree Theorem, Bogdanowicz [1] showed that

$$t_n = f_{2n} = 2\frac{((3 - \sqrt{5})/2)^{n+1} - ((3 + \sqrt{5})/2)^{n-1}}{5 - 3\sqrt{5}}, \tag{1}$$

where f_n is the nth number of the Fibonacci sequence with $f_1 = f_2 = 1$.

For convenience, a spanning tree of a fan graph F_n is called a *fan tree* and let $\mathcal{T}_n$ be the set consisting of all fan trees of F_n. Intuitively, a fan tree can be perceived as an orderly rooted tree by taking the vertex labeled 0 as the root. Thus, by describing the parent's label of each vertex (except the root) in order, this ordered exhibition naturally results in each fan tree having a specific representation called a *fan-tree sequence*. For ease of generating all sequences of $\mathcal{T}_n$, we use $Seq(1..n)$ to represent a fan-tree sequence, and thus the edge set of the fan tree is $\{(i, Seq(i)) \mid 1 \le i \le n\}$. For example, Fig. 1 shows all the fan trees in the set $\mathcal{T}_4$ and their corresponding tree sequences in lexicographic order.

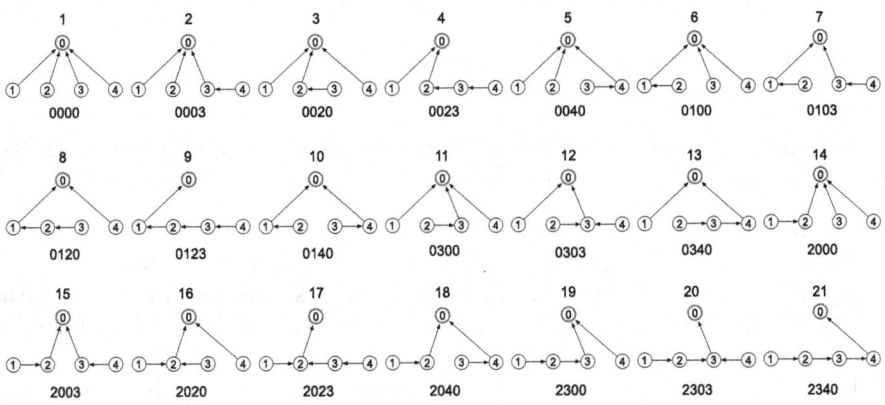

Fig. 1. The set $\mathcal{T}_4$ contains 21 fan trees rooted at vertex 0 and each tree has a specific tree sequence.

From experience, we know that to systematically represent a massive number of objects (such as trees), it is necessary to resort to the so-called *coding tree* [20] (or called the *recursive tree* [4]). In a coding tree, each node on level i (except for the root on level 0) is associated with a label $Seq(i)$ so that the labels collected along each path from a node on level 1 to a leaf (i.e., a node on level n) represents a fan-tree sequence $Seq(1..n)$ of $\mathcal{T}_n$. Two nodes on the same level with a common parent are called *siblings*, and a *series* is a set consisting of all siblings. Notably, a series is called a *direct series* on level i if the common parent of nodes in the series has the label i, and is called a *collateral series* otherwise. Figure 2 exhibits the coding tree of $\mathcal{T}_4$ in which the labels from left to right are in increasing order for each series. According to this arrangement, fan-tree sequences in the coding tree result in lexicographic order. For example, a series with two nodes labeled by 0 and 4 having the parent with label 3 is a direct series (represented by a solid line ellipse). A series with three nodes labeled by 0, 2, and 4 having the parent with label 0 or 1 is a collateral series (represented by a dashed line ellipse).

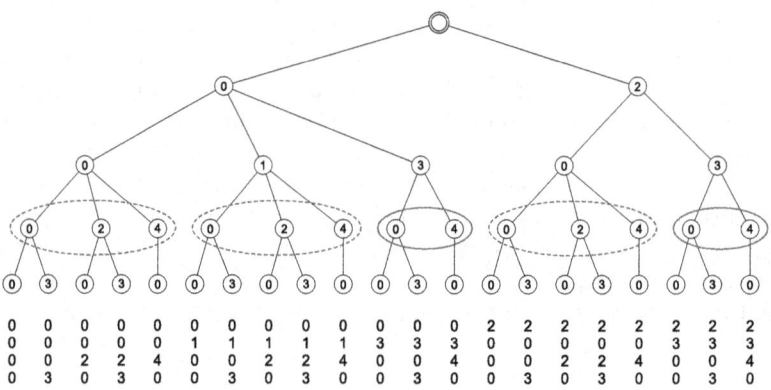

Fig. 2. The layout of the coding tree of $\mathcal{T}_4$ in lexicographic order.

For the coding tree, the following observations are easy to obtain from its layout and the structure of fan trees:

1. The root is a dummy node.
2. The level 1 has only one series with two nodes labeled by 0 and 2.
3. For level $i \in \{2, 3, \ldots, n-1\}$, each collateral series contains three nodes with labels 0, $i-1$, and $i+1$, and each direct series contains two node with labels 0 and $i+1$.
4. For level n, each collateral series contains two nodes with labels 0 and $n-1$, and each direct series has only one node with label 0.

For $i \in \{1, 2, \ldots, n\}$, let ℓ_i be the number of nodes on the ith level in the coding tree of $\mathcal{T}_n$. More precisely, for $i \in \{1, 2, \ldots, n-1\}$, let ℓ_i^c (resp. ℓ_i^d) be the number of nodes on the ith level of $\mathcal{T}_n$ in which each node expands its sons to form a collateral series (resp. direct series) on the $(i+1)$-th level of $\mathcal{T}_n$. Clearly, $\ell_i = \ell_i^c + \ell_i^d$ for $i \in \{1, 2, \ldots, n-1\}$.

Lemma 1. *The following results hold in the coding tree of* $\mathcal{T}_n$.

(a) *For* $i \in \{1, 2, \ldots, n-1\}$, $\ell_i^d = f_{2i-1}$, $\ell_i^c = f_{2i}$, *and* $\ell_i = f_{2i-1} + f_{2i} = f_{2i+1}$.
(b) $\ell_n = t_n = f_{2n}$.

Proof. We prove the first assertion by induction on i. For $i = 1$, it is evident that $\ell_1^{\mathbf{d}} = f_1 = 1$, $\ell_1^{\mathbf{c}} = f_2 = 1$, and thus $\ell_1 = f_1 + f_2 = 1 + 1 = 2$. Suppose that $i \in \{2, 3, \ldots, n-1\}$ and the assertion holds on the jth level of $\mathcal{T}_n$ for $j \leq i - 1$. By induction hypothesis, we have $\ell_{i-1}^{\mathbf{d}} = f_{2i-3}$ and $\ell_{i-1}^{\mathbf{c}} = f_{2i-2}$. From the coding tree structure, we observe that each direct series (resp. each collateral series) on the $(i-1)$th level of $\mathcal{T}_n$ contains two nodes (resp. three nodes), one expands its sons to form a direct series and the other expands its sons to form a collateral series (resp. the other two expand their sons to form two collateral series) simultaneously on the ith level of $\mathcal{T}_n$. Thus, we have

$$\ell_i^{\mathbf{d}} = \ell_{i-1}^{\mathbf{d}} + \ell_{i-1}^{\mathbf{c}} = f_{2i-3} + f_{2i-2} = f_{2i-1}$$

and

$$\ell_i^{\mathbf{c}} = \ell_{i-1}^{\mathbf{d}} + 2 \cdot \ell_{i-1}^{\mathbf{c}} = f_{2i-3} + 2f_{2i-2} = f_{2i}$$

Hence, $\ell_i = \ell_i^{\mathbf{d}} + \ell_i^{\mathbf{c}} = f_{2i-1} + f_{2i} = f_{2i+1}$ and the first assertion holds.

Particularly, for level n, each collateral series contains two nodes and each direct series has a singleton. Thus, we have $\ell_n = \ell_{n-1}^{\mathbf{d}} + 2\ell_{n-1}^{\mathbf{c}} = f_{2n-3} + 2f_{2n-2} = f_{2n}$, and the second assertion directly follows from Eq. (1). $\qquad\square$

3   Generating Fan-Tree Sequences

The algorithm for generating fan-tree sequences relies on a procedure called NextSeq(), which is designed to generate the next fan-tree sequence based on the previous one. Except for the space $Seq(0..n)$, the algorithm requires an additionally global variable i indicating where the position will be dealt with in the current sequence. Initially, set $Seq(1..n) = 00 \cdots 0$ and $i = n$. Particularly, let $Seq(0) = 1$, which corresponds to the root in the coding tree, and it acts as a sentinel to detect whether the task of NextSeq() is still going on. Accordingly, we regard the series on level 1 as a direct series. Figure 3 demonstrates the generation algorithm and NextSeq() procedure.

We now illustrate the details of NextSeq() as follows. From the observations mentioned above, we first note that a series on level i ($\neq n$) starts at a node with label 0 and ends at a node with label $i + 1$. In contrast, unless facing the end of a direct series, the variation of $Seq(n)$ on level n is carried out frequently and will keep changing. Indeed, $Seq(n)$ has only two values, 0 or $n - 1$, depending on whether the change occurs at the end of a collateral series. If it is not (i.e., $i = n$ and the condition $Seq(n-1) \neq n$ holds in line 4), it performs the setting $Seq(n) = n - 1$, and the indicator i will be changed to $n - 1$ (see lines 3–5); otherwise, we have $i = n - 1$ in the current situation. Then, it performs the setting $Seq(n) = 0$ (see line 9) and another setting (see lines 10–12 and illustrate later), and the indicator i will be n (see line 13). Along with $Seq(n)$ change, the status of indicator i is either $n - 1$ or n, depending on the process that enters or returns from NextSeq(). Since indicator i is a global variable, its value will be kept into the next call of NextSeq().

Algorithm 1: Generating Fan-Tree Sequences	Procedure NextSeq( )

Algorithm 1 (left box):

```
Seq(0) = 1;
for i ← 1 to n do Seq(i) ← 0;
i ← n;
while i > 0 do
    Print-Sequence(Seq(1..n));
    NextSeq();
end
```

Procedure NextSeq( ) (right box):

```
1 begin
2     if i = n then
3         i ← n − 1;
4         if Seq(n − 1) ≠ n then
5             Seq(n) ← n − 1;  return;
6         while Seq(i) = i + 1 do
7             Seq(i) ← 0;  i ← i − 1;
8         if i < 0 then return;
9     else Seq(n) ← 0;
10    if Seq(i − 1) = i then Seq(i) ← i + 1;
11    else if Seq(i) = 0 then Seq(i) ← i − 1;
12    else Seq(i) ← i + 1
13    i ← n;
14    return;
```

Fig. 3. An algorithm for generating fan-tree sequences and the NextSeq() procedure.

Next, let's look in detail into the end of a direct series. In this case, there exists an integer $k \in \{0, 1, \ldots, n-1\}$ such that the previous sequence has $Seq(i) = i + 1$ for $i \in \{k, k + 1, \ldots, n - 1\}$. Then, the process will make the setting $Seq(i) = 0$ for all i from $n - 1$ down to k to produce the next sequence, which we call the state of *returning to zero* (see lines 6–7). Note that if $k = 0$, the presetting of sentinel will be changed to $Seq(0) = 0$, and thus $i = -1$. In this situation, the algorithm will be terminated (see line 8). Normally, an additional setting needs to be performed after the above processing or in the situation of $i = n - 1$. If the current change occurs at a direct series on a certain level (i.e., the condition $Seq(i-1) = i$ holds in line 10), the setting $Seq(i) = i+1$ is required. On the other hand, if the change is complete at the first collateral series on a certain level (i.e., the condition $Seq(i) = 0$ holds in line 11, where the value of $Seq(i)$ is the initial setting or set by the previous state of returning to zero), the setting $Seq(i) = i-1$ is required. Finally, if the change is complete at the second collateral series on a certain level, then the setting $Seq(i) = i + 1$ is required (see line 12).

To present all the detailed work in the algorithm, we provide an example for generating all fan-tree sequences of $\mathcal{T}_4$ as shown in Fig. 4 that illustrates the variation of $Seq(1..n)$ and the indicator i. The algorithm generates tree sequences from left to the right in lexicographic order. Each column represents a tree sequence, and a value (either n or $n - 1$) under a line between two columns indicates the current status of the indicator i when a call NextSeq() is newly invoked. Then, we write changes above a line from bottom to up in which each cell with a label showing the line number where the NextSeq() executes.

Obviously, the generation algorithm requires $\mathcal{O}(n)$ space. In what follows, we give an amortized analysis of the time complexity of the algorithm. Let $T(n)$ denote the time required for generating all fan-tree sequences of $\mathcal{T}_n$. For $i \in \{1, 2, \ldots, n\}$, let c_i be the number of changes of $Seq(i)$ in the algorithm. Clearly, $c_i = \ell_i$ (i.e., c_i is equal to the number of nodes on the ith level of the coding tree) for $i \in \{1, 2, \ldots, n-1\}$ and $c_n = \ell_n - \ell_{n-2}$. By Lemma 1, we have

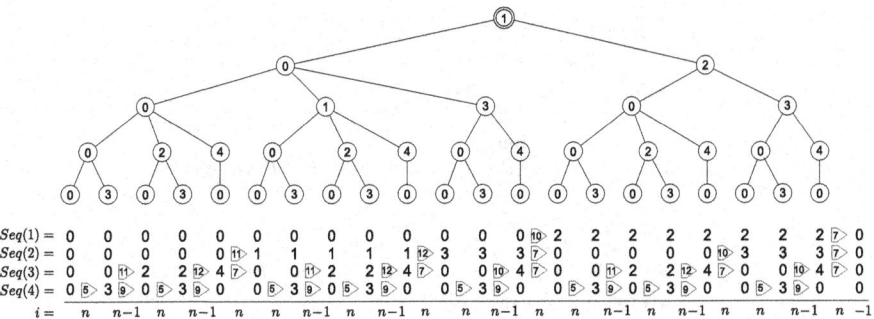

```
Seq(1) = 0    0    0    0    0    0    0    0    0    0    0    0    0 [10> 2    2    2    2    2    2    2    2 [7> 0
Seq(2) = 0    0    0    0    0 [11> 1    1    1    1    1 [12> 3    3    3 [7> 0    0    0    0    0 [10> 3    3    3 [7> 0
Seq(3) = 0    0 [11> 2    2 [12> 4 [7> 0    0 [11> 2    2 [12> 4 [7> 0    0 [10> 4 [7> 0    0 [11> 2    2 [12> 4 [7> 0    0 [10> 4 [7> 0
Seq(4) = 0 [5> 3 [9> 0 [5> 3 [9> 0    0 [5> 3 [9> 0 [5> 3 [9> 0    0 [5> 3 [9> 0    0 [5> 3 [9> 0 [5> 3 [9> 0 [5> 3 [9> 0    0
   i =    n   n-1   n   n-1   n    n   n-1   n   n-1   n    n   n-1   n    n   n-1   n    n   n-1   n  -1
```

Fig. 4. Illustration of the variation of $Seq(1..n)$ and the indicator i in the algorithm for generating fan-tree sequences of T_4.

$$c_i = \begin{cases} f_{2i+1} & \text{if } i \in \{1, 2, \ldots, n-1\}; \\ f_{2n} - f_{2n-3} & \text{if } i = n. \end{cases}$$

Since $f_1 + f_3 + \cdots + f_{2n-1} = f_{2n}$, the time complexity can be calculated as follows:

$$T(n) = \sum_{i=1}^{n} c_i = \left(\sum_{i=1}^{n-1} f_{2i+1} \right) + \left(f_{2n} - f_{2n-3} \right) = 2f_{2n} - f_{2n-3} - 1 < 2|T_n|.$$

Thus, the amortized cost of time for each generation is $T(n)/f_{2n} < 2$. We summarize all results described in this section in the following theorem.

Theorem 1. *All fan-tree sequences of T_n can be generated in lexicographic order in $\mathcal{O}(|T_n|)$ time using $n + \mathcal{O}(1)$ memory space. In particular, each generation requires only constant amortized time with no more than two data change operations.*

4   Ranking and Unranking Algorithms

The ranking algorithm inputs a fan-tree sequence $Seq(1..n)$ and outputs an integer R to indicate the rank of $Seq(1..n)$ in lexicographic order of T_n. The algorithm calculates this ranking through the distribution of series at each level in the coding tree. It requires three additional variables described below. The indicator i points out where the current level is being dealt with in the coding tree. The variable $j \in \{2, 3\}$ is the length (i.e., the number of nodes) of the current series. And, k corresponds to the term of the Fibonacci sequence and is determined by the indicator i. The process that carries out the calculation in the coding tree adopts the top-down approach, and thus initially, we have $i = 1$, $j = 2$, and $R = 0$. Figure 5 shows the ranking algorithm.

Algorithm 2: Ranking($Seq(1..n)$)

1 $j \leftarrow 2$; $R \leftarrow 0$;
2 **for** $i \leftarrow 1$ **to** $n-1$ **do**
3 $\quad k \leftarrow 2n - (2i - 1)$;
4 $\quad$ **if** $j = 2$ **then**
5 $\quad\quad$ **if** $Seq(i) = i + 1$ **then** $R \leftarrow R + F(k)$;
6 $\quad\quad$ **else** $j \leftarrow 3$;
7 $\quad$ **else if** $Seq(i) = i + 1$ **then**
8 $\quad\quad$ $R \leftarrow R + 2 \cdot F(k)$; $j \leftarrow 2$;
9 $\quad$ **else if** $Seq(i) = i - 1$ **then** $R \leftarrow R + F(k)$;
10 **if** $Seq(n) = 0$ **then** $R \leftarrow R + 1$;
11 **else** $R \leftarrow R + 2$;
12 **return** R.

Fig. 5. A ranking algorithm of fan tree sequences in lexicographic order.

For each level i from 1 to $n - 1$, we first calculate $k = 2n - (2i - 1)$, where $F(k)$ denotes the number of leaves in the subtree rooted at the first node in the current series. Then, based on where the position of $Seq(i)$ occurred in the current series, one of the following three cases is carried out:

- The current series is a direct series (i.e., the condition $j = 2$ holds in line 4): if $Seq(i)$ presents at the end of the direct series (i.e., the condition $Seq(i) = i+1$ holds in line 5), we update R by increasing $F(k)$; otherwise, set $j = 3$ for the next iteration (see line 6).
- The value of $Seq(i)$ presents at the end of a collateral series (i.e., the condition $Seq(i) = i+1$ holds in line 7): we update R by increasing two times of $F(k)$, and then set $j = 2$ for the next iteration (see line 8).
- The value of $Seq(i)$ presents at the middle position of a collateral series (i.e., the condition $Seq(i) = i-1$ holds in line 9): we update R by increasing $F(k)$.

Finally, if $Seq(n) = 0$, we increase R by one. Otherwise, we increase R by two (see lines 10–11).

Example 1. We perform Ranking(2023) in $\mathcal{T}_4$. Initially, $j = 2$ and $R = 0$.
When $i = 1$, we have $k = 7$. Since $j = 2$ and $Seq(1) = 2$, we update $R = R + F(7) = 13$ (lines 4–5).
When $i = 2$, we have $k = 5$. Since $j = 2$ and $Seq(2) \neq 3$, we set $j = 3$ (line 6).
When $i = 3$, we have $k = 3$. Since $j \neq 2$ and $Seq(3) = 2$, we update $R = R + F(3) = 13 + 2 = 15$ (line 9).
Finally, since $Seq(4) \neq 0$, we update $R = R + 2 = 15 + 2 = 7$ (line 11). Thus, the ranking algorithm outputs $R = 17$.

We conclude this section via the following theorem.

Theorem 2. *Given a fan-tree sequence of $\mathcal{T}_n$, Algorithm 2 can determine the rank of the sequence in lexicographic order in $\mathcal{O}(n)$ time and $n + \mathcal{O}(1)$ space.*

In the following, we present the unranking algorithm, which takes an integer R as the input, and outputs the corresponding fan-tree sequence $Seq(1..n)$ for which the sequence has the rank R in lexicographic order of $\mathcal{T}_n$. Three variables i, j, and k have the same meanings as those in the ranking algorithm. Similarly, the process determines the values of $Seq(1..n)$ in sequence and initially sets up $j = 2$. Figure 6 shows the unranking algorithm.

Algorithm 3: Unranking(R)

1   $j \leftarrow 2$;
2   **for** $i \leftarrow 1$ to n **do**
3   $\quad$ $k \leftarrow 2n - (2i - 1)$;
4   $\quad$ **if** $j = 2$ **then**
5   $\quad\quad$ **if** $R > F(k)$ **then**
6   $\quad\quad\quad$ $R \leftarrow R - F(k)$; $Seq(i) \leftarrow i + 1$;
7   $\quad\quad$ **else**
8   $\quad\quad\quad$ $Seq(i) \leftarrow 0$; $j \leftarrow 3$;
9   $\quad$ **else if** $R > 2 \cdot F(k)$ **then**
10 $\quad\quad$ $R \leftarrow R - 2 \cdot F(k)$; $Seq(i) \leftarrow i + 1$; $j \leftarrow 2$;
11 $\quad$ **else if** $R > F(k)$ **then**
12 $\quad\quad$ $R \leftarrow R - F(k)$; $Seq(i) \leftarrow i - 1$;
13 $\quad$ **else** $Seq(i) \leftarrow 0$;;
14 **return** $Seq(1..n)$.

Fig. 6. An unranking algorithm of fan tree sequences in lexicographic order.

For each level i from 1 to $n - 1$, we first calculate $k = 2n - (2i - 1)$. Then, based on the current values of j and R, one of the following cases is carried out:

- The current series is a direct series (i.e., the condition $j = 2$ holds in line 4): if the current rank R is sufficient to accommodate the arrangement of $F(k)$ nodes before it (i.e., the condition of line 5 holds), we update R by decreasing $F(k)$, and then set $Seq(i) = i + 1$ to indicate that $Seq(i)$ is at the end of the series (see line 6); otherwise, set $Seq(i) = 0$ to mean that $Seq(i)$ is at the front of the series, and then set $j = 3$ for the next iteration (see line 8).
- The current series is a collateral series:
  - If the current rank of R is sufficient to accommodate the arrangement of $2F(k)$ nodes before it (i.e., the condition $R > 2F(k)$ holds in line 9), we update R by decreasing $2F(k)$, set $Seq(i) = i + 1$ to indicate that $Seq(i)$ is at the end of the series, and then set $j = 2$ for the next iteration (see line 10).
  - If $F(k) < R \leqslant 2F(k)$ (i.e., the condition of line 11 holds), we update R by decreasing $F(k)$, and then set $Seq(i) = i - 1$ to mean that $Seq(i)$ is at the middle position of the series (see line 12).
  - If $R \leqslant F(k)$, then set $Seq(i) = 0$ to mean that $Seq(i)$ is at the front of the series (see line 13).

Finally, the unranking algorithm outputs $Seq(1..n)$.

Example 2. We perform Unranking (18) in $\mathcal{T}_4$. Initially, $j = 2$.

When $i = 1$, we have $k = 7$. Since $j = 2$ and $R = 18 > F(7) = 13$, we update $R = R - F(7) = 5$, set $Seq(1) = 2$ (line 6).

When $i = 2$, we have $k = 5$. Since $j = 2$ and $R = 5 = F(5)$, we set $Seq(2) = 0$ and $j = 3$ (line 8).

When $i = 3$, we have $k = 3$. Since $j = 3$ and $R = 5 > 2F(3) = 4$, we update $R = R - 2F(3) = 1$, set $Seq(3) = 4$ and $j = 2$ (line 10).

When $i = 4$, we have $k = 1$. Since $j = 2$ and $R = 1 = F(1)$, we set $Seq(4) = 0$ (line 8).

Thus, the unranking algorithm outputs $Seq(1..4) = 2040$.

We conclude this section by the following theorem.

Theorem 3. *Given a positive integer R, Algorithm 3 can determine the fan-tree sequence corresponding to the rank R in lexicographic order of $\mathcal{T}_n$ in $\mathcal{O}(n)$ time and $n + \mathcal{O}(1)$ space.*

5   Concluding Remarks

This paper proposes a constant amortized-time algorithm for generating all fan-tree sequences in lexicographic order. Moreover, based on such an ordering, we offer efficient ranking and unranking algorithms with time complexity and space requirement $\mathcal{O}(n)$. Finally, it remains an interesting open question about generating all spanning-tree sequences for other classes of graphs, including wheels, prisms, Möbius ladders, and square of cycles.

Acknowledgments. This research was supported by the Ministry of Science and Technology of Taiwan under Grants MOST110-2221-E-262–002 (R.-Y. Wu) and MOST110-2221-E-141–004 (J.-M. Chang).

References

1. Bogdanowicz, Z.R.: Formulas for the number of spanning trees in a fan. Appl. Math. Sci. **2**(16), 781–786 (2008)
2. Cameron, B., Grubb, A., Sawada, J.: A pivot gray code listing for the spanning trees of the fan graph. In: Chen, C.-Y., Hon, W.-K., Hung, L.-J., Lee, C.-W. (eds.) COCOON 2021. LNCS, vol. 13025, pp. 49–60. Springer, Cham (2021). https://doi.org/10.1007/978-3-030-89543-3_5
3. Chakraborty, M., Chowdhury, S., Chakraborty, J., Mehera, R., Pal, R.K.: Algorithms for generating all possible spanning trees of a simple undirected connected graph: an extensive review. Complex Intell. Syst. **5**(3), 265–281 (2019). https://doi.org/10.1007/s40747-018-0079-7
4. Chang, Y.-H., Wu, R.-Y., Lin, C.-K., Chang, J.-M.: A loopless algorithm for generating (k, m)-ary trees in gray code order. Optim. Lett. **15**(4), 1133–1154 (2021). https://doi.org/10.1007/s11590-020-01613-z
5. Colbourn, C.J., Day, R.P., Nel, L.D.: Unranking and ranking spanning trees of a graph. J. Algorithms **10**(2), 271–286 (1989)

6. Colbourn, C.J., Myrvold, W.J., Neufeld, E.: Two algorithms for unranking arborescences. J. Algorithms **20**(2), 268–281 (1996)
7. Datta, S., Chakraborty, S., Chakraborty, M., Pal, R.K.: Algorithm to generate all spanning tree structures of a complete graph. In: Balas, V.E., Hassanien, A.E., Chakrabarti, S., Mandal, L. (eds.) Proceedings of International Conference on Computational Intelligence, Data Science and Cloud Computing. LNDECT, vol. 62, pp. 169–184. Springer, Singapore (2021). https://doi.org/10.1007/978-981-33-4968-1_14
8. Eğecioğlu, Ö., Remmel, J.B., Williamson, S.G.: A class of graphs which has efficient ranking and unranking algorithms for spanning trees and forests. Int. J. Found. Comput. Sci. **15**(4), 619–648 (2004)
9. Gabow, H.N., Myers, E.W.: Finding all spanning trees of directed and undirected graphs. SIAM J. Comput. **7**(3), 280–287 (1978)
10. Golynski, A.: Optimal lower bounds for rank and select indexes. Theor. Comput. Sci. **387**(3), 348–359 (2007)
11. Kapoor, S., Ramesh, H.: Algorithms for enumerating all spanning trees of undirected and weighted graphs. SIAM J. Comput. **24**(2), 247–265 (1995)
12. Kapoor, S., Ramesh, H.: An algorithm for enumerating all spanning trees of a directed graph. Algorithmica **27**(2), 120–130 (2000). https://doi.org/10.1007/s004530010008
13. Mäkinen, V., Navarro, G.: Rank and select revisited and extended. Theor. Comput. Sci. **387**(3), 332–347 (2007)
14. Matsui, T.: A flexible algorithm for generating all the spanning trees in undirected graphs. Algorithmica **18**(4), 530–543 (1997). https://doi.org/10.1007/PL00009171
15. Minty, G.J.: A simple algorithm for listing all the trees of a graph. IEEE Trans. Circuit Theory **12**(1), 120 (1965)
16. Reddy, K.K.M., Renjith, P., Sadagopan, N.: Listing all spanning trees in Halin graphs - sequential and parallel view. Discrete Math. Algorithms Appl. **10**(1), 1850005 (2018)
17. Remmel, J.B., Williamson, S.G.: Spanning trees and function classes. Electron. J. Comb. **9**(1), R34 (2002)
18. Remmel, J.B., Williamson, S.G.: Ranking and unranking trees with a given number or a given set of leaves. arXiv:1009.2060v1
19. Shioura, A., Tamura, A., Uno, T.: An optimal algorithm for scanning all spanning trees of undirected graphs. SIAM J. Comput. **26**(3), 678–692 (1997)
20. Wu, R.-Y., Chang, J.-M., Chang, C.-H.: Ranking and unranking of non-regular trees with a prescribed branching sequence. Math. Comput. Model. **53**(5–6), 1331–1335 (2011)

Cutting and Packing

High Multiplicity Strip Packing
with Three Rectangle Types

Andrew Bloch-Hansen$^{(\boxtimes)}$, Roberto Solis-Oba, and Andy Yu

Western University, London, ON N6A 3K7, Canada
{ablochha,solis,ayu}@uwo.ca, andy.yu@uwaterloo.ca

Abstract. The two-dimensional strip packing problem consists of packing in a rectangular strip of width 1 and minimum height a set of n rectangles, where each rectangle has width $0 < w \leq 1$ and height $0 < h \leq h_{max}$. We consider the high-multiplicity version of the problem in which there are only K different types of rectangles. For the case when $K = 3$, we give an algorithm providing a solution requiring at most height $\frac{3}{2}h_{max} + \epsilon$ plus the height of an optimal solution, where ϵ is any positive constant.

Keywords: LP-relaxation · Two-dimensional strip packing · High multiplicity · Approximation algorithm

1 Introduction

The *two-dimensional strip packing problem* (2DSPP) is defined as follows.

Definition 1. *Given n rectangles with widths w_1, w_2, ..., w_n and heights h_1, h_2, ..., h_n, where $0 < w_i \leq 1$ and $0 < h_i \leq h_{max}$ for $i = 1$, 2, ..., n, the goal is to pack all the rectangles without rotations or overlaps in a rectangular strip of width 1 and minimum height.*

This is a well-studied problem with applications in areas as diverse as resource allocation, scheduling, manufacturing, and transportation, among others. 2DSPP is equivalent to the classical bin packing problem if all rectangles have the same height, and since the bin packing problem is NP-hard [4] then 2DSPP is also NP-hard; therefore, the best possible approximation ratio achievable in polynomial time for 2DSPP is $\frac{3}{2}$ unless P = NP.

Baker et al. (1980) [1] designed the first approximation algorithm for 2DSPP which has approximation ratio 3. Coffman et al. (1980) [3] presented an algorithm with approximation ratio 2.7, Sleator (1980) [12] improved the approximation ratio to 2.5, and Schiermeyer (1994) [11] and Steinberg (1997) [13] further

R. Solis-Oba—The work of this author was partially supported by a Discovery Grant (RGPIN-2020-06423) from the Natural Sciences and Engineering Research Council of Canada.

I. Ljubić et al. (Eds.): ISCO 2022, LNCS 13526, pp. 215–227, 2022.
https://doi.org/10.1007/978-3-031-18530-4_16

reduced the approximation ratio to 2. Harren and Van Stee (2009) [5] later presented an algorithm with approximation ratio 1.9396. The best known approximation algorithm for 2DSPP is from Harren et al. (2014) [6] with approximation ratio $\frac{5}{3} + \epsilon$. Several Asymptotic Polynomial Time Approximation Schemes (APTAS) have been presented as well: Kenyon and Rémila (2000) [10] gave an APTAS with an additive constant of $O(\frac{1}{\epsilon^2})$, and Jansen and Solis-Oba (2009) [8] improved Kenyon and Rémila's additive constant to 1.

In this paper we study the *two-dimensional high multiplicity strip packing problem* (2DHMSPP), in which there is only a fixed number K of different rectangle types. Note that the input to 2DHMSPP can be described using a list of only $3K$ numbers: the width w_i, height h_i, and number n_i of rectangles of each type T_i. Therefore, a challenging issue faced when designing an approximation algorithm for the problem is to ensure that its running time is a polynomial function of the size of the input. Observe that even describing a feasible solution for the problem using a polylogarithmic number of bits is not trivial as this requires specifying the positions of n rectangles in the packing; therefore, it is unknown whether 2DHMSPP belongs to the class NP.

We present an algorithm for 2DHMSPP for the case when $K = 3$ that computes solutions of value at most $OPT + \frac{3}{2}h_{max} + \epsilon$, where OPT is the value of an optimum solution, h_{max} is the height of the tallest rectangle, and ϵ is a positive constant. This is an improvement over the works of Yu and Solis-Oba (2019) [14] and Bloch-Hansen and Solis-Oba (2019) [2] whose algorithms computed solutions of value at most $OPT + \frac{5}{3}h_{max} + \epsilon$. Our approach uses a formulation of 2DHMSPP that allows fractional rectangles in the solution called the *two-dimensional fractional strip packing problem* (2DFSPP). We show that a solution for 2DFSPP can be converted into a solution for 2DHMSPP by a careful shifting, re-shaping, and combining of the fractional rectangles to form whole rectangles while increasing the height of the solution by at most $\frac{3}{2}h_{max} + \epsilon$. Our analysis is nearly tight as it is not hard to see that there are instances for which the corresponding fractional and integral solutions differ by h_{max}.

The rest of the paper is organized in the following way. In Sect. 2 we describe how to compute a near optimum solution for 2DFSPP. In Sect. 3 we present our algorithm for the case when $K = 3$. Finally, in Sect. 4 we describe a polynomial time implementation of the algorithm.

2 Solving 2DFSPP in Polynomial Time

2DHMSPP can be relaxed to the *two-dimensional fractional strip packing problem* (2DFSPP) by allowing horizontal cuts on the rectangles. A solution to 2DFSPP consists of a set of *configurations*. A *base configuration* C_j consists of a multiset of rectangle types whose total width is at most 1 (see Fig. 1). A base configuration can be specified by indicating the number of rectangles of each type T_i in it. For example, the base configuration shown in Fig. 1 consists of 4 rectangles of type T_1, 2 rectangles of type T_2, and 3 rectangles of type T_3, so that base configuration can be represented with the tuple (4, 2, 3).

A group of rectangles following a base configuration can be stacked on top of each other as shown in Fig. 1, so that any horizontal line parallel to the base of the strip drawn across any part of the group will intersect the same multiset of rectangle types. This group of rectangles is called a *configuration*. A vertical line drawn across any part of a configuration will intersect either only rectangles of the same type, or empty space. The height of a vertical line intersecting rectangles of a configuration is called the height of the configuration. The configurations are stacked one on top of the other to form a fractional packing (see Fig. 2b). Note that the number of possible configurations is $O(n^K)$.

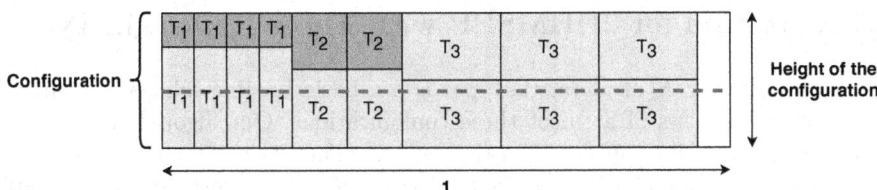

Fig. 1. A configuration with *base configuration* (4, 2, 3). The fractional rectangles are shaded in a darker color.

For a configuration C_j let $n_{i,j}$ be the number of rectangles of type T_i in its base configuration, for $i = 1, 2, ..., k$. Let x_j be a variable denoting the height of C_j. Let J be the set of all possible configurations. 2DFSPP can be expressed as the following linear program, hereafter referred to as linear program (1):

$$\text{Minimize: } \sum_{C_j \in J} x_j$$

$$\text{Subject to: } \sum_{C_j \in J} x_j n_{i,j} \geq n_i h_i, \text{ for each rectangle type } T_i \qquad (1)$$

$$x_j \geq 0, \text{ for each } j \in J$$

where n_i is the number of rectangles of type T_i and h_i is the height of each rectangle of type T_i. The objective function is to minimize the total height of the packing.

We denote with $OPT(I)$ the height of an optimal packing for instance I of 2DHMSPP and denote with $LIN(I)$ an optimal solution to the corresponding instance of 2DFSPP. It is not hard to see that $LIN(I) \leq OPT(I)$.

Note that 2DFSPP is identical to the fractional bin packing problem; in the latter problem a base configuration is a set of items that fit within a single bin and a solution to linear program (1) gives the fractional number of bins needed to pack all the items. Therefore, we can use an algorithm of Karmarkar and Karp [9] to compute a basic feasible solution for linear program (1) in time $O(K^9 \log K \log^2 \frac{K}{\epsilon})$ of value at most $LIN(I) + \epsilon$ for any fixed $\epsilon > 0$.

In any basic feasible solution, the number of nonzero variables is at most the number of constraints [7]. Thus, the number of nonzero variables, and therefore, the number of configurations used in a basic feasible solution for linear program (1) is at most the number of rectangle types, K.

A simple algorithm for 2DHMSPP is to compute a basic feasible solution for linear program (1) and replace each fractional rectangle with a whole one of the corresponding type, shifting surrounding rectangles upwards as needed. Since a basic feasible solution for (1) uses at most K configurations and replacing the fractional rectangles with whole ones increases the height of a configuration by at most h_{max}, this algorithm computes a solution to 2DHMSPP of height at most $OPT(I) + Kh_{max} + \epsilon$.

3   Algorithm for 2DHMSPP with Three Rectangle Types

When the number K of rectangle types is 3 a basic feasible solution for linear program (1) consists of at most three configurations. Our algorithm performs several steps described in detail in the next sections: 1) the fractional solution of the linear program is divided in two parts: S_{Common} and $S_{Uncommon}$, and the fractional rectangles in S_{Common} are rounded up; 2) in $S_{Uncommon}$ the rectangles in each configuration are sorted by non-decreasing fractional value and $S_{Uncommon}$ is further partitioned into vertical sections; 3) the vertical sections are grouped according to the heights of the fractional rectangles in them; and 4) the fractional rectangles in each group are combined and/or rounded into whole ones depending on their heights.

In this paper we only consider the case when the fractional solution computed by solving linear program (1) consists of exactly three configurations and within $S_{Uncommon}$ each configuration contains exactly two different rectangle types. The other cases, when the solution of linear program (1) has one or two configurations or when a configuration has three rectangle types, are discussed in the full paper[1].

3.1   Partitioning the Packing

For notational simplicity, in the sequel we assume $h_{max} = 1$. The three configurations of the solution for linear program (1) are stacked one on top of the other as shown in Fig. 2a. Let the configuration packed at the top be C_1, the one in the middle be C_2, and the one at the bottom be C_3. Rectangles are rearranged horizontally within the configurations so that rectangles of the same type appearing in all three configurations are placed together in a section on the left side of the packing called S_{Common}. In the remaining portion of the packing, called $S_{Uncommon}$, each rectangle type is packed in at most 2 configurations (see Fig. 2b).

The fractional rectangles in S_{Common} are rounded up to form whole rectangles, increasing the height of the packing by at most 1 (see Fig. 2a). In the sequel, we discuss only how to round the fractional rectangles in $S_{Uncommon}$.

---

[1] A full version of this paper is available at www.csd.uwo.ca/~ablochha/2DHMSPP. pdf.

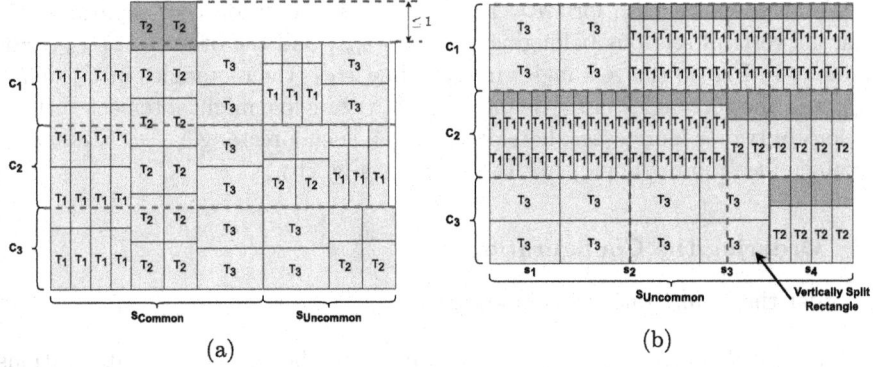

Fig. 2. (a) Rounding the fractional rectangles in S_{Common} increases the height of the packing by at most 1. (b) Within each configuration the rectangles in $S_{Uncommon}$ are sorted according to their fractional values.

Within each configuration, we place the fractional rectangles in $S_{Uncommon}$ at the top of the configuration. Let r be a fractional rectangle. The ratio between the height of r and the height of a rectangle of the same type as r is called the *fractional value* of r. We sort the rectangles so that fractional rectangles are sorted in non-decreasing order of their fractional values (see Fig. 2b).

We draw a vertical line at each point where two rectangles of different types are packed side-by-side within a configuration. These vertical lines partition $S_{Uncommon}$ into *vertical sections* (see Fig. 2b). Vertical sections are indexed from left to right starting at index 1 for the leftmost section. Within some vertical section s_i, let $C_{1(i)}$, $C_{2(i)}$, and $C_{3(i)}$ refer to the part of C_1, C_2, and C_3 that is located within s_i, respectively.

3.2   Grouping Vertical Sections

Within a vertical section s_i, each configuration has a single rectangle type. Let $f_{1(i)}$, $f_{2(i)}$, and $f_{3(i)}$ represent the fractional values of the fractional rectangles packed in s_i at the top of C_1, C_2, and C_3, respectively.

We classify the vertical sections $s_i \in S_{Uncommon}$ into 3 cases, depending on the three fractional values $f_{1(i)}$, $f_{2(i)}$, and $f_{3(i)}$ as follows:

- S_{Case1} includes all sections s_i such that $f_{1(i)} + f_{2(i)} + f_{3(i)} \leq 1$.
- S_{Case2} includes all sections s_i such that $f_{1(i)} + f_{2(i)} + f_{3(i)} > 1$ and either $f_{1(i)} + f_{2(i)} \leq 1$, $f_{1(i)} + f_{3(i)} \leq 1$, or $f_{2(i)} + f_{3(i)} \leq 1$.
- S_{Case3} includes all sections s_i such that $f_{1(i)} + f_{2(i)} > 1$, $f_{1(i)} + f_{3(i)} > 1$, and $f_{2(i)} + f_{3(i)} > 1$. Note that for each $s_i \in S_{Case3}$

$$f_{1(i)} + f_{2(i)} + f_{3(i)} > \frac{3}{2} \tag{2}$$

We denote with $B_{i,j}$, for $i,j = 1$, 2, 3, a vertical division that separates two adjacent vertical sections belonging one to S_{Casei} and the other to S_{Casej}. For example, in Fig. 3a the rectangles in C_1 define $B_{1,1}$, the rectangles in C_2 define $B_{2,3}$, and the rectangles in C_3 define $B_{1,2}$. A rectangle r might intersect vertical sections of two or more cases; hereafter, we call such a rectangle a *vertically split rectangle* (see the rectangle with the arrow in Fig. 2b).

3.3   Ordering the Configurations

We order the configurations as follows:

- If none of S_{Case1}, S_{Case2}, and S_{Case3} are empty, then order the configurations so that the rectangles in C_2 define $B_{2,3}$ and those in C_3 define $B_{1,2}$. Note that the rectangles in C_2 cannot define $B_{2,3}$ and $B_{1,2}$ because C_2 has only rectangles of two different types.
- If S_{Case3} is empty or S_{Case2} is empty then order the configurations so that the rectangles in C_3 define $B_{1,2}$ or $B_{1,3}$, respectively.
- If S_{Case1} is empty then order the configurations so that the rectangles in C_2 define $B_{2,3}$.
- If none of the above criteria are met, the configurations are ordered randomly.

Having the rectangles in C_2 define $B_{2,3}$, if possible, allows flexibility for shifting the rectangles in $C_2 \cap S_{Case3}$ as we show; for some of our algorithm's cases we shift these rectangles downwards into empty space if the rectangles in $C_3 \cap S_{Case3}$ take up less height than the rectangles in $C_3 \cap S_{Case2}$. Therefore, ordering the configurations in the manner described above is important to our algorithm.

Let a and b be the fractional values of the leftmost and rightmost fractions in C_1, respectively. Let c and d be the fractional values of the leftmost and rightmost fractions in C_2, and let e and f be the fractional values of the leftmost and rightmost fractions in C_3, respectively (see Fig. 3).

We use a variable called *count* to track how many wide rectangle types appear in a packing, where a rectangle type is considered to be wide if it is the leftmost type in its configuration and it is packed, at least partially, within S_{Case2} or S_{Case3}. The presence (or absence) of these wide rectangle types is important in deciding whether we can re-use the empty space left behind when fractional rectangles are shifted around in S_{Case1} and S_{Case2}. We initialize variable *count* to 0. If any fractional rectangles with fractional value a are packed within any vertical section of S_{Case2} or S_{Case3}, we increase the value of *count* by one. If any fractional rectangles with fractional value c are packed within any section of S_{Case2} or S_{Case3}, we increase the value of *count* by one.

3.4   Rounding Fractional Rectangles

We provide different algorithms for rounding fractional rectangles into whole ones based on which of S_{Case1}, S_{Case2}, and S_{Case3} are not empty and what the value of *count* is.

Lemma 1. *If none of S_{Case1}, S_{Case2}, and S_{Case3} are empty, then count > 0.*

Proof. Assume that $count = 0$ and none of S_{Case1}, S_{Case2}, and S_{Case3} are empty. Because of how we ordered the configurations, the rectangles in C_2 define the boundary $B_{2,3}$ and therefore at least one of the fractional rectangles in C_2 with fractional value c must be packed in S_{Case2}, contradicting the assumption that $count = 0$. □

By Lemma 1, we do not need consider the case when none of S_{Case1}, S_{Case2}, and S_{Case3} are empty and $count = 0$. The cases we must consider are described below.

3.5 None of S_{Case1}, S_{Case2}, and S_{Case3} are Empty, $count = 1$, and $f_{1(i)} + f_{2(i)} \leq 1$ for all Vertical Sections $s_i \in S_{Case2}$

Our algorithm will produce two solutions and choose the one with shorter height. For the first solution, C_1 and C_2 are *paired* (see Fig. 3a). When pairing C_1 with C_2, we flip C_1 upside down. Let F_1 be the set of fractional rectangles in each vertical section $s_i \in S_{Case1}$, and let F_2 be the set of fractional rectangles from C_1 and C_2 in each vertical section $s_i \in S_{Case2}$. We remove the sets F_1 and F_2 from their original positions in the packing. If $F_1 \cup F_2$ is not empty we shift up the remaining rectangles in C_1 so that the tops of the topmost rectangles in C_1 lie on a common line and the distance between C_1 and C_2 in vertical section s_1 is 1. This creates a region in S_{Case1} and S_{Case2} of height at most 1 between C_1 and C_2 where we will pack F_1 and F_2; we call this region C_{A1} (see Fig. 3a). If $F_1 \cup F_2$ is empty, then region C_{A1} has initial height zero, but its height might be increased later as explained below.

We re-shape the fractional rectangles in $F_1 \cup F_2$ so that their area does not change but they have the full height of a rectangle of the same type.

Lemma 2. *Let C_1 and C_2 be paired as described above. The re-shaped fractional rectangles in $F_1 \cup F_2$ can be packed in region C_{A1}.*

Proof. Let vertical section $s_i \in S_{Case1}$ have width W_i and let $C_{A1(i)}$ be the part of C_{A1} within s_i. The total empty area A_i in $C_{A1(i)}$ is $A_i = W_i * 1 = W_i$. Since each of $C_{1(i)}$, $C_{2(i)}$, and $C_{3(i)}$ has only one fractional rectangle type, the total area a_i of the fractional rectangles in $C_{1(i)}$, $C_{2(i)}$, and $C_{3(i)}$ is

$$a_i \leq (W_i * f_{1(i)}) + (W_i * f_{2(i)}) + (W_i * f_{3(i)}) < W_i = A_i,$$

as the height of a rectangle is at most 1 and $f_{1(i)} + f_{2(i)} + f_{3(i)} \leq 1$ for $s_i \in S_{Case1}$.

A similar argument can be made for the vertical sections $s_i \in S_{Case2}$ for which $f_{1(i)} + f_{2(i)} \leq 1$. □

Corollary 1. *After re-shaping the fractional rectangles in $F_1 \cup F_2$ we can pack them in C_{A1} so that there is at most one fractional rectangle of each type in C_{A1}.*

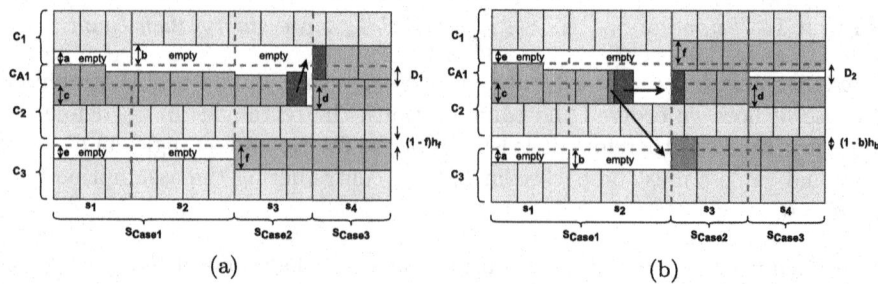

Fig. 3. $count = 1$ and $f_{1(i)} + f_{2(i)} \leq 1$ for all vertical sections $s_i \in S_{Case2}$.

Proof. We combine the fractional rectangles in $F_1 \cup F_2$ such that a whole rectangle is formed whenever a sufficient number of pieces of the same type have been packed. When fractional rectangles of the same type do not form a whole rectangle, they merge to become one larger fractional rectangle. Therefore, at most one fractional rectangle of each type may remain. By Lemma 2 the rectangles can be packed in C_{A1}. □

We *round up* the fractional rectangles for each vertical section $s_i \in S_{Case3}$. Rounding up a fractional rectangle r means replacing it with a whole rectangle of the same type as r and shifting rectangles up as needed to make room for the whole rectangle. When shifting rectangles from C_1 we need ensure that the tops of the topmost rectangles in C_1 lie on a common line. Finally, we round up the fractional rectangles in $C_3 \cap S_{Case2}$ (see Fig. 3a).

For the second solution, re-order the configurations so that fractional rectangles with fractional value a appear in C_3, and fractional rectangles with fractional value e appear in C_1, then pair C_1 and C_2 (see Fig. 3b) using the same process as described above in Sect. 3.5. Note that we can no longer guarantee that $f_{1(i)} + f_{2(i)} \leq 1$ for all $s_i \in S_{Case2}$; therefore, we round up the fractional rectangles in $C_{1(i)}$ and $C_{2(i)}$ for each vertical section $s_i \in S_{Case2}$ for which $f_{1(i)} + f_{2(i)} > 1$.

Note that after pairing two configurations and re-shaping rectangles some whole rectangles might be vertically split by the boundaries $B_{1,2}$ and $B_{2,3}$. Because of the way in which regions C_{A1} and C_{A2} were defined, the two pieces of a whole rectangle that is vertically split by any of those boundaries are placed side-by-side forming a whole rectangle. However, pieces of fractional rectangles that are vertically split might be placed in different parts of the packing. Later we show how to shift these fractional pieces to form whole rectangles.

Let h_i be the height of the rectangles corresponding to fractional value i, for $i = a, b, c, d, e,$ and f.

Lemma 3. *If none of S_{Case1}, S_{Case2}, and S_{Case3} are empty, count $= 1$, and $f_{1(i)} + f_{2(i)} \leq 1$ for all vertical sections $s_i \in S_{Case2}$, then there is an algorithm that produces an integer packing of height at most $\frac{3}{2}$ plus the value of the solution for linear program (1).*

Proof. In the first solution, described above (see Fig. 3a), the height increase in S_{Case1} and S_{Case2} caused by creating C_{A1} is at most $h_1 - ah_a - ch_c \leq h_1 - ch_c$, where $h_1 = max\{h_a, h_b, h_c\}$ (note that C_{A1} re-uses the space that was occupied by the fractional rectangles of fractional values a and c). In S_{Case3}, the height increase caused by rounding up the fractional rectangles with fractional values b and d is at most $(1 - b)h_b + (1 - d)h_d$; hence the height increase from C_1 and C_2 is at most $D_1 = max\{h_1 - ch_c, (1 - b)h_b + (1 - d)h_d\}$. The height increase caused by rounding up the fractional rectangles in C_3 with fractional value f is at most $(1 - f)h_f$ (see Fig. 3a). Therefore, the total height increase is at most $max\{\Delta_A, \Delta_B\}$, where $\Delta_A = h_1 - ch_c + (1 - f)h_f \leq 2 - f - ch_c$, as $h_1 \leq 1$ and $h_f \leq 1$ and $\Delta_B = (1 - b)h_b + (1 - d)h_d + (1 - f)h_f \leq 3 - b - d - f < \frac{3}{2}$ as $h_b \leq 1, h_d \leq 1$, and $b + d + f > \frac{3}{2}$ by (2).

For the second solution we only consider the case when $f + c > 1$ (see Fig. 3b); the case when $f + c \leq 1$ is similar. The height increase caused by creating C_{A1} is at most $h_2 - ch_c - eh_e$, where $h_2 = max\{h_c, h_e\}$. In S_{Case2} and S_{Case3}, the height increase caused by rounding up the fractional rectangles with fractional values c, d, and f is at most $max\{(1 - c)h_c, (1 - d)h_d\} + (1 - f)h_f$; hence the height increase from C_1 and C_2 is at most $D_2 = max\{h_2 - ch_c - eh_e, max\{(1 - c)h_c, (1-d)h_d\}+(1-f)h_f\}$. The height increase caused by rounding up fractional rectangles in C_3 with fractional value b is at most $(1 - b)h_b$. Therefore, the total height increase is at most $max\{\Delta_C, \Delta_D\}$, where $\Delta_C = (1 - b)h_b + h_2 - ch_c - eh_e$ and $\Delta_D = (1 - b)h_b + max\{(1 - c)h_c, (1 - d)h_d\} + (1 - f)h_f$.

Selecting the better of the solutions produces an increase in the height of the solution by $max\{min\{\Delta_A, \Delta_C\}, min\{\Delta_A, \Delta_D\}, min\{\Delta_B, \Delta_C\}, min\{\Delta_B, \Delta_D\}\}$.

- $min\{\Delta_A, \Delta_C\}$: $\Delta_A = 2 - f - ch_c$ and $\Delta_C = (1-b)h_b + max\{h_c, h_e\} - ch_c - eh_e$. Note that $min\{\Delta_A, \Delta_C\}$ will achieve its maximum value when $h_c \approx 0$, $h_b = 1$, and $h_e = 1$, therefore $\Delta_A \leq 2 - f$ and $\Delta_C \leq (1 - b) + (1 - e) = 2 - b - e$. Note that $f + b > 1$ as fractional rectangles with fractional values f and b appear in S_{Case3}; therefore, either $f > \frac{1}{2}$ or $b > \frac{1}{2}$ so $min\{\Delta_A, \Delta_C\} \leq \frac{3}{2}$.
- $min\{\Delta_A, \Delta_D\}$: $\Delta_A = 2 - f - ch_c$ and $\Delta_D = (1 - b)h_b + max\{(1 - c)h_c, (1 - d)h_d\}+(1-f)h_f$. Recall our assumption that $f+c > 1$ (the case when $f+c \leq 1$ is similar), therefore either $f > \frac{1}{2}$ or $c > \frac{1}{2}$. If $f > \frac{1}{2}$ then $\Delta_A < \frac{3}{2} - ch_c < \frac{3}{2}$. If $c > \frac{1}{2}$ then $\Delta_D \leq (1 - b) + (1 - f) + max\{1 - c, 1 - d\} = 2 - b - f + max\{1 - c, 1 - d\}$:
  - If $1 - c > 1 - d$ then $\Delta_D \leq 3 - b - f - c < 3 - \frac{1}{2} - b - f < \frac{3}{2}$ as $b + f > 1$.
  - If $1 - d > 1 - c$ then $\Delta_D \leq 3 - b - d - f \leq \frac{3}{2}$ by (2).
  Therefore, $min\{\Delta_A, \Delta_D\} \leq \frac{3}{2}$.
- $min\{\Delta_B, \Delta_C\} \leq \frac{3}{2}$ and $min\{\Delta_B, \Delta_D\} \leq \frac{3}{2}$ because $\Delta_B \leq \frac{3}{2}$.

□

Observe that in the first solution, depicted in Fig. 3a, there might be a fractional rectangle r in C_1 that is vertically split by $B_{2,3}$ and such that one piece of r is re-shaped and packed as explained in Lemma 2, while the other piece is rounded up to the height of a rectangle of the same type as r. These pieces are marked in Fig. 3a. Note that the two pieces can be placed beside each other to form a whole rectangle without further increasing the height of the packing. In the sequel we will not explicitly explain how fractional rectangles that are vertically split are combined to form whole rectangles; instead the figures will show how to manipulate these rectangles.

3.6   None of S_{Case1}, S_{Case2}, and S_{Case3} are Empty, $count = 1$, and $f_{1(i)} + f_{2(i)} > 1$ for at Least One Vertical Section $s_i \in S_{Case2}$

Lemma 4. *If none of S_{Case1}, S_{Case2}, and S_{Case3} are empty and $count = 1$, then C_1's rectangles cannot define $B_{2,2}$, $B_{2,3}$, or $B_{3,3}$.*

Proof. Note that if C_1's rectangles defined $B_{2,2}$, then the value of $count$ would be 2 because rectangles in C_1 with fractional value a would appear in S_{Case2} and since the rectangles in C_2 define $B_{2,3}$ then rectangles with fractional value c would also appear in S_{Case2}. Similarly, it is not possible that C_1's rectangles define boundaries $B_{2,3}$ or $B_{3,3}$.   □

Lemma 5. *If none of S_{Case1}, S_{Case2}, and S_{Case3} are empty, $count = 1$, and $f_{1(i)} + f_{2(i)} > 1$ for at least one $s_i \in S_{Case2}$, then there is an algorithm that produces an integer packing of height at most $\frac{3}{2}$ plus the value of the solution for linear program (1).*

Proof. By Lemma 4, C_1's rectangles cannot define $B_{2,2}$, $B_{2,3}$, or $B_{3,3}$. Note that if C_1's rectangles defined $B_{1,1}$, then $f_{1(i)} + f_{2(i)} \leq 1$ for all vertical sections $s_i \in S_{Case2}$ since the rectangles in C_2 define $B_{2,3}$ and thus fractional rectangles with fractional values b and c must appear within S_{Case1} and so $b + c \leq 1$. Therefore, the rectangles in C_1 and C_3 must create a coinciding boundary $B_{1,2}$ so that $b + c$ could be larger than 1, as required by the Lemma.

Since $b + c > 1$, then $b > \frac{1}{2}$ and/or $c > \frac{1}{2}$. If $b > c$, then re-order the configurations so that fractional rectangles with fractional value b appear in C_3. Pair C_1 and C_2 and re-shape, pack, and round rectangles as explained in the second solution of Lemma 3. The height increase from C_1 and C_2 is at most 1: for sections s_i where $f_{1(i)} + f_{2(i)} \leq 1$ the height increase caused by creating C_{A1} is at most 1, and for sections s_i where $f_{1(i)} + f_{2(i)} > 1$ the height increase caused by rounding up $f_{1(i)}$ and $f_{2(i)}$ is also at most 1. The height increase caused by rounding up fractional rectangles with fractional value b is at most $\frac{1}{2}$ (see Fig. 4a) so the total height increase is at most $\frac{3}{2}$.

If $c > b$, then re-order the configurations so that fractional rectangles with fractional value c appear in C_3, pair C_1 and C_2, and re-shape, pack, and round rectangles as explained in the second solution of Lemma 3. The height increase from C_1 and C_2 is at most 1 and the height increase caused by rounding up fractional rectangles with fractional value c is at most $\frac{1}{2}$ (see Fig. 4b).   □

For space limitations we leave the remaining cases, which can be solved using similar algorithms to the ones we have described, to the full paper.

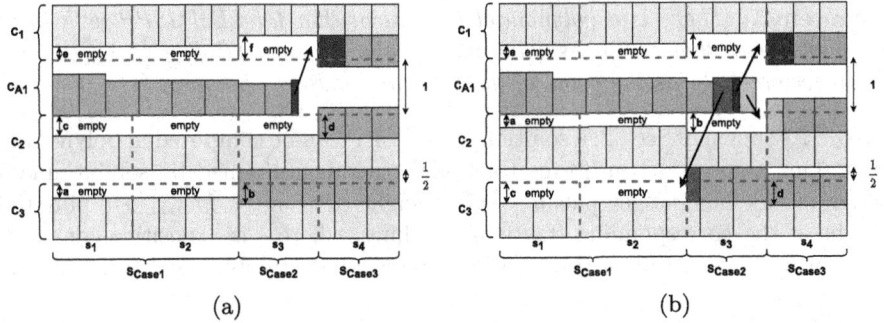

(a)   (b)

Fig. 4. $count = 1$ and $f_{1(i)} + f_{2(i)} > 1$ for at least one vertical section $s_i \in S_{Case2}$.

4   Polynomial Time Implementation

Recall that the input to 2DHMSPP is represented as a list of $3K$ numbers; therefore, any algorithm that specifies individual locations of rectangles in a solution for 2DHMSPP will not run in polynomial time. We overcome this by only specifying locations of rectangle groups.

We represent a configuration as a list of $O(K)$ numbers: for $1 \leq i \leq K$ we specify the rectangle type T_i, the number of rectangles of type T_i packed side-by-side, and the number of rectangles of type T_i packed on top of each other.

Since there are at most K configurations, and we create at most one additional configuration by creating C_{A1} or C_{A2} during the rounding process, then at most $O(K^2)$ numbers are needed to specify the packing in $S_{Uncommon}$. Similarly, the packing in S_{Common} is specified using at most $O(K)$ numbers, for a total of $O(K^2)$ numbers for the entire packing.

The number of rectangles of type T_i that are packed side-by-side in S_{Common} is equal to the minimum of the number of rectangles of type T_i that are packed in each of C_1, C_2, and C_3, and the number of rectangles of that type packed on top of each other in S_{Common} is the sum of the number of rectangles of type T_i packed on top of each other in each of C_1, C_2, and C_3. Therefore, finding the number of rectangles of each type packed in S_{Common} requires $O(K^2)$ operations.

Processing S_{Common} requires $O(K)$ operations as for $1 \leq i \leq K$ our algorithm only needs to round up the fractional values for each rectangle type T_i. Sorting the rectangles in each configuration in $S_{Uncommon}$ by their fractional values requires $O(K^2)$ operations.

Ordering the configurations and computing the value of the *count* variable requires $O(K)$ operations, and checking which cases are present in the fractional packing requires $O(K)$ operations. Re-shaping, packing, and rounding fractional rectangles as described in Sect. 3.4 needs $O(K^2)$ operations. Finally, packing leftover vertically split fractional rectangles as shown in the figures requires $O(K)$ operations.

Theorem 1. *There is a polynomial time algorithm for 2DHMSPP with three rectangle types that computes solutions of value at most $OPT + \frac{3}{2}h_{max} + \epsilon$ for $\epsilon > 0$, where OPT is the value of the solution for linear program (1).*

Proof. As shown in Sect. 2, a solution to 2DFSPP can be computed in polynomial time. Our algorithm transforms fractional packings obtained by solving linear program (1) into integer packings with height of at most $\frac{3}{2}h_{max} + \epsilon$ plus the height of the corresponding fractional packing, where ϵ is a positive constant. As shown above our algorithm can be implemented in polynomial time.   □

5   Conclusion

We presented an algorithm for the *two-dimensional high multiplicity strip packing problem* for the case when $K = 3$ that computes solutions of value at most $OPT + \frac{3}{2}h_{max} + \epsilon$, where OPT is the value of the solution for linear program (1), h_{max} is the height of the tallest rectangle, and ϵ is a positive constant.

Our ideas can be used to design an algorithm for 2DHMSPP for the case when $K = 4$ that computes solutions of value at most $OPT + \frac{7}{3}h_{max} + \epsilon$. Additionally, our ideas can be used to design an algorithm for 2DHMSPP for the case when $K > 3$ is constant that computes solutions of value at most $OPT + \lfloor \frac{3}{4}K \rfloor + 1 + \epsilon$.

References

1. Baker, B., Coffman, E., Rivest, R.: Orthogonal packings in two dimensions. SIAM J. Comput. **9**(4), 846–855 (1980)
2. Bloch-Hansen, A.: High multiplicity strip packing. Electronic Thesis and Dissertation Repository. 6559 (2019)
3. Coffman, E., Garey, M., Johnson, D., Tarjan, R.: Performance bounds for level-oriented two-dimensional packing algorithms. SIAM J. Comput. **9**(4), 808–826 (1980)
4. Garey, M., Johnson, D.: Computers and Intractability. Freeman, San Francisco (1979)
5. Harren, R., van Stee, R.: Improved absolute approximation ratios for two-dimensional packing problems. In: Dinur, I., Jansen, K., Naor, J., Rolim, J. (eds.) APPROX/RANDOM -2009. LNCS, vol. 5687, pp. 177–189. Springer, Heidelberg (2009). https://doi.org/10.1007/978-3-642-03685-9_14
6. Harren, R., Jansen, K., Prädel, L., Van Stee, R.: A ($\frac{5}{3} + \epsilon$)-approximation for strip packing. Comput. Geom. **47**(2), 248–267 (2014)
7. Karloff, H.: Linear Programming. Springer, Heidelberg (2008)
8. Jansen, K., Solis-Oba, R.: Rectangle packing with one-dimensional resource augmentation. Discret. Optim. **6**(3), 310–323 (2009)
9. Karmarkar, N., Karp, R.: An efficient approximation scheme for the one-dimensional bin-packing problem. In: 23rd Annual Symposium on Foundations of Computer Science, pp. 312–320. IEEE (1982)
10. Kenyon, C., Rémila, E.: A near-optimal solution to a two-dimensional cutting stock problem. Math. Oper. Res. **25**(4), 645–656 (2000)

11. Schiermeyer, I.: Reverse-fit: a 2-optimal algorithm for packing rectangles. In: van Leeuwen, J. (ed.) Algorithms – ESA 1994. European Symposium on Algorithms, vol. 855, pp. 290–299. Springer, Berlin, Heidelberg (1994)
12. Sleator, D.: A 2.5 times optimal algorithm for packing in two dimensions. Inform. Process. Lett. **10**(1), 37–40 (1980)
13. Steinberg, A.: A strip-packing algorithm with absolute performance bound 2. SIAM J. Comput. **26**(2), 401–409 (1997)
14. Yu, A.: High multiplicity strip packing problem with three rectangle types. Electronic Thesis and Dissertation Repository. 6684 (2019)

Improved Bounds for Stochastic Extensible Bin Packing Under Distributional Assumptions

Guillaume Sagnol$^{(\boxtimes)}$ and Daniel Schmidt genannt Waldschmidt

Fakultät II, Institut für Mathematik, MA 5-2, Technische Universität Berlin, Straße des 17. Juni 136, 10623 Berlin, Germany
{sagnol,dschmidt}@math.tu-berlin.de

Abstract. In the stochastic extensible bin packing problem, n items of random size must be packed into m bins of unit capacity. The number of bins is fixed, but their capacity can be extended at extra cost. This model plays an important role in surgery scheduling, where the extension of bin capacity represents overtime of operating rooms. It is known that this problem has a $(1 + e^{-1})$-approximation algorithm in expectation for arbitrary probability distributions of the item sizes. Using the theory of second-order stochastic dominance, we obtain improved bounds when the sizes have a bounded Pietra or Gini index, and when they belong to specific parametric families, such as Gamma, Lognormal or Weibull distributions.

Keywords: Approximation algorithm · Stochastic scheduling · Extensible bin packing · Second-order stochastic dominance

1 Introduction

In the *extensible bin packing problem* (EBP), the input consists of n items of size $(p_1, \ldots, p_n)$ and a number m of bins. Each bin has a regular unit capacity, which can be extended at some extra cost. Specifically, a bin holding the items $I \subseteq \{1, \ldots, n\}$ has a cost of $\max(\sum_{i \in I} p_i, 1)$. The goal is to assign each item to one of the m bins so as to minimize the total cost. This model arises in scheduling problems in which the machines are available for some amount of time at a fixed cost, and extra-time can be performed at some additional cost. Throughout we use the scheduling terminology: bins are machines, items are jobs, and item sizes are processing times.

Application to Surgery Scheduling. The EBP model gained attention recently, as it was used to represent the cost of assigning surgery patients to operating rooms [3,5,19]. The model of Denton et al. [5] assumes that the decision maker can choose the number $k \leq m$ of operating rooms (machines) of capacity T to open at a fixed cost c^f, and there is a variable cost c^v for each minute of overtime. Under this model, the total cost of a solution assigning

I. Ljubić et al. (Eds.): ISCO 2022, LNCS 13526, pp. 228–241, 2022.
https://doi.org/10.1007/978-3-031-18530-4_17

the subset of surgeries S_i to the ith operating room ($i = 1, \ldots, k$) becomes $kc^f + c^v \sum_{i=1}^{k} \max(\sum_{j \in S_i} p_j - T, 0)$. The EBP corresponds to the situation in which $T = c^v = c^f = 1$ and all operating rooms must be open, i.e., $k = m$.

When working on approximation algorithms, the primary focus on the EBP is justified by the following observation [3]: If an algorithm has a performance guarantee of $(1+\rho)$ for the EBP, then one can obtain a $(1 + \rho \frac{Tc^v}{c^f})$-approximation algorithm for the aforementioned generalized cost function: To this end, one applies the algorithm on the EBP-instance with rescaled processing times $p'_j = \frac{p_j}{T}$ and k machines, for each candidate number of machines $k \in \{1, \ldots, m\}$, and retains the best solution. This remark applies to the stochastic variant of the problem (stochastic extensible bin packing: SEBP) studied in this paper, too.

In practice, surgical durations are not precisely known in advance, but historical data can be used to fit probability distributions on the processing times. This motivated the study of the stochastic counterpart of the extensible bin packing problem (SEBP), in which the processing time of each job follows some known probability distribution, and the expected costs have to be minimized. This paper builds on [20, 22], where it is shown that there is a simple policy with an approximation ratio of $1 + e^{-1} \approx 1.368$ for the SEBP (this performance guarantee is only proved for the special case of *short jobs* in [22], i.e., for jobs such that $P_j \leq 1$ holds almost surely, and it has been generalized to arbitrary stochastic jobs in [20]). We prove improved bounds under distributional assumptions of the processing times.

Related Work. The (deterministic) EBP was introduced by Dell'Olmo et al. [4], who show that the problem is strongly NP-hard. Moreover, they prove that the *longest processing time first* (LPT) algorithm –which considers the jobs sorted in nonincreasing order of their processing time and assigns them sequentially to the machine with the largest remaining capacity– is a $\frac{13}{12}$-approximation algorithm. Following from a general result [1], there also exists a polynomial time approximation scheme (PTAS) for this problem. For more background on EBP and its variants, we refer to [22] and the references therein.

There is a vast literature on the applications of mathematical optimization techniques for the management of uncertainty in operating rooms. Due to lack of space, we indicate only the recent survey article [23], which presents a comprehensive synthesis of the challenges in this field, based on numerous references.

In stochastic scheduling problems various notions of stochastic dominance have been considered to obtain optimal policies for specific classes of processing time distributions; see e.g. the book by Pinedo [16] and the references therein. In addition, approximative policies have been designed where the performance guarantee is parameterized by a coefficient measuring the dispersion of the processing times, such as their coefficient of variation [25] or their δ-NBUE coefficient [14]. In this work, we use the notion of second-order stochastic dominance and Lorenz dominance as well as the Pietra and Gini index of the processing times. To the best of our knowledge, this is the first work to obtain performance guarantees using these indices or these notions of stochastic dominance in this context.

Organization and Main Results. Our paper is organized as follows. The model is presented in Sect. 2. The necessary background on stochastic dominance, Pietra and Gini indices is given in Sect. 3, where we also introduce the concept of a Lorenz-dominated family (LDF) of random variables and give important examples of such families. The improved performance guarantees are proved in Sect. 4. In particular, we prove that the *longest expected processing time first* (LEPT) policy is 1.2334-approximative when the processing times are lognormally distributed with a squared coefficient of variation $\Delta \leq \frac{1}{4}$, a reasonable assumption in the context of surgery scheduling [19,24]. Other authors suggested to use Gamma or Weibull distributed random variables [8] to approximate surgery durations; our approach can also handle these cases, see Table 2. Finally, we give a distribution-free bound for instances with a bounded Pietra or Gini index in Theorem 2.

2   Stochastic Extensible Bin Packing

Problem Definition. The input of the *stochastic extensible bin packing problem* (SEBP) consists of a vector of non-negative random variables $\boldsymbol{P} = (P_1, \ldots, P_n)$ representing the job processing times and a number m of machines (we assume $m < n$, as the problem is trivial otherwise). The set of jobs and machines is denoted by $\mathcal{J} = \{1, \ldots, n\}$ and $\mathcal{M} = \{1, \ldots, m\}$, respectively. We assume that the P_j's are mutually independent and that their expectation is finite and computable. We do not specify how the processing time distributions should be represented in the input of the problem, as the policy we study only requires the expected value of the processing times.

In stochastic scheduling problems, the appropriate solution concept is a *non-anticipatory policy*; see [15]. Unlike in the deterministic case, a scheduling strategy can take more general forms than an allocation of jobs to machines, as information is gained during the execution of the schedule. More precisely, job durations become known upon completion, and adaptive policies can react to the processing times observed so far. In this work however, we focus on a simple fixed assignment policy Π which only specifies a job-to-machine assignment, thus we write $j \xrightarrow{\Pi} i$ if the policy Π puts job j on machine i. Jobs are started sequentially and without idle time on each machine. The workload X_i^{Π} of machine i, i.e., the latest completion time of a job executed on $i \in \mathcal{M}$, is thus $X_i^{\Pi} = \sum_{j \xrightarrow{\Pi} i} P_j$. The cost incurred on machine i is equal to $\max(X_i^{\Pi}, 1)$, which accounts for the fixed costs, plus the amount by which the regular working time has to be extended. We want to minimize the expected value of the total costs, and thus the objective function of the SEBP can be written as

$$\Phi(\Pi) := \mathbb{E}\left[\sum_{i \in \mathcal{M}} \max(X_i^{\Pi}, 1)\right] = \sum_{i \in \mathcal{M}} \mathbb{E}\left[\max\left(\sum_{j \xrightarrow{\Pi} i} P_j, 1\right)\right],$$

where the second equality only holds for the special case of a fixed assignment policy such as the LEPT-policy studied in this article. For a thorough description of the problem with arbitrary non-anticipatory policies, we refer to [22].

Denote by OPT the optimal value of the SEBP, i.e., the smallest possible value of $\Phi(\Pi)$ over the class of all non-anticipatory policies. We say that a policy Π has a performance guarantee of $\alpha \geq 1$ if $\Phi(\Pi) \leq \alpha \cdot OPT$ holds for all instances $(\boldsymbol{P}, m)$ of the problem. In this paper, we obtain performance guarantees under the assumption that the P_j's belong to certain families of random variables. It is important to keep in mind that, even though the policy we study is a fixed assignment policy, we compare its quality to that of an optimal policy Π^* in which the job-to-machine assignment can depend on the realization.

LEPT Policy. There is no unique way to generalize the LPT rule used in the deterministic case for SEBP. One possibility is to compute the assignment resulting from applying the LPT rule to the deterministic instance in which each random job is replaced by its expected value. This assignment defines a fixed-assignment policy, which we call LEPT. In other words, LEPT precomputes the job to machine assignments offline, as follows: jobs are considered in nonincreasing order of $\mathbb{E}[P_j]$, and sequentially assigned to the least loaded machine (in expectation). LEPT has a performance guarantee of $1 + e^{-1}$ for SEBP, and a matching lower bound is known [22]. The LEPT policy was also used to study semi-adaptive policies for another stochastic load balancing problem in [21].

Notation. We shall now introduce some notation relying on LEPT. Consider an instance $(P_1, \ldots, P_n; m)$ of SEBP. Throughout, we denote by $J_i := \{j \in \mathcal{J} : j \xrightarrow{\text{LEPT}} i\}$ the subset of jobs assigned to a machine $i \in \mathcal{M}$, and we define $n_i := |J_i|$. Note that $n_i \geq 1$ follows from the assumption $n > m$, and there is at least one machine with $n_i \geq 2$. We also denote by x_i the expected workload of machine i under LEPT, i.e., $x_i = \mathbb{E}[X_i^{\text{LEPT}}] = \sum_{j \in J_i} \mathbb{E}[P_j]$.

It will also be useful to split the jobs into a truncated part $P_j' := \min(P_j, 1)$ and an excess part $P_j'' := \max(P_j - 1, 0)$, so that $P_j = P_j' + P_j''$. We define the truncated load of machine i under LEPT by $\alpha_i := \sum_{j \in J_i} \mathbb{E}[P_j']$ and the excess of machine i by $\beta_i := \sum_{j \in J_i} \mathbb{E}[P_j'']$, so that $x_i = \alpha_i + \beta_i$.

We conclude this section on the SEBP with two results taken from [22].

Proposition 1 ([22, Proposition 1 & Lemma 5]; [20, Lemma 4.3]). *For all SEBP instances, it holds*

$$OPT \geq \max \left(\sum_{i \in \mathcal{M}} x_i, \ \ m + \sum_{i \in \mathcal{M}} \beta_i \right).$$

Proposition 2 ([22, Lemma 3]). *Let $\ell := \min\{x_i : i \in \mathcal{M}\}$. Then, for all machines i such that $n_i \geq 2$ it holds $\ell \leq x_i \leq 2\ell$.*

3   Second-Order Stochastic Dominance

The results of this paper heavily rely on the notion of second-order stochastic dominance, which was introduced in the late 60's to model the preferences of decision-makers regarding different gambles. This section gives a short introduction with the necessary background on this topic.

Definition 1. *Let Y and Z be random variables with finite expectation. We say that Y has second-order stochastic dominance over Z, and we write $Y \succeq_{(2)} Z$, if and only if*

$$\int_{-\infty}^{x} (F_Z(t) - F_Y(t))\, dt \geq 0, \quad \forall x \in \mathbb{R},$$

with F_Y and F_Z the cumulative distribution functions of Y and Z, respectively.

Using the well-known fact that $\mathbb{E}[Y] - \mathbb{E}[Z] = \int_{-\infty}^{\infty} (F_Z(t) - F_Y(t))dt$, it is easy to see that a simple sufficient condition for $Y \succeq_{(2)} Z$ is that $\mathbb{E}[Y] \geq \mathbb{E}[Z]$ and that F_Y and F_Z are *single-crossing*, i.e., for some x_0 we have $F_Y(t) - F_Z(t) \leq 0$ on the interval $(-\infty, x_0)$ and $F_Y(t) - F_Z(t) \geq 0$ over $[x_0, \infty)$.

In this work, we mostly use the $\preceq_{(2)}$-ordering to compare random variables with the same mean. In this case, we shall see that $\preceq_{(2)}$ is linked to another dominance relation relying on the concept of Lorenz curves.

Definition 2. *Let Y be a random variable with finite expectation. The Lorenz curve of Y is a function $L_Y : [0,1] \to [0,1]$ with $L_Y(p) := \frac{1}{\mathbb{E}[Y]} \int_0^p F_Y^{-1}(z)dz$, where $F_Y^{-1}(z) = \inf\{y : F_Y(y) \geq z\}$ is the quantile function of Y.*

The Lorenz curve L_Y of a random variable Y is convex and nondecreasing, and it satisfies $L_Y(0) = 0$, $L_Y(1) = 1$. It was introduced by Lorenz [11] to compare the distribution of income across different countries: for a population with continuous distribution of income F_Y, $L_Y(p)$ represents the percentage of the total wealth owned by the bottom $100p\%$ of all individuals. The situation where all individuals own the same wealth corresponds to a deterministic variable ($Y = a$ for some $a \geq 0$) with Lorenz curve $L_Y(p) = p$, called the *line of perfect equality*. Based on the Lorenz curve we can define another dominance relation.

Definition 3. *Let Y and Z be nonnegative random variables with finite expectation. We say that Y Lorenz dominates Z, and we write $Y \succeq_L Z$, if and only if for all $p \in [0,1]$,*

$$L_Y(p) \leq L_Z(p).$$

We next summarize equivalent characterizations of $\preceq_{(2)}$ and $\preceq_L$ under the assumption that Y and Z have equal means:

Proposition 3. *Let Y and Z be nonnegative random variables with $\mathbb{E}[Y] = \mathbb{E}[Z] < \infty$. The following statements are equivalent:*

(i) $Y \succeq_{(2)} Z$
(ii) $\mathbb{E}[u(Y)] \geq \mathbb{E}[u(Z)]$, for all nondecreasing concave utility functions u
(iii) $\mathbb{E}[f(Y)] \leq \mathbb{E}[f(Z)]$, for all nondecreasing convex utility functions f
(iv) $Y \preceq_L Z$.

Note that the assumption that Y and Z are nonnegative and have same mean is not required for each of the equivalences listed above; we refer to [13, Chapter 17] for a more detailed discussion of these results. (i)⇔(ii) is proved in the seminal papers by Hadar and Russel [7] and Rothschild and Stiglitz [18]. (i)⇔(iii) is shown in Li and Wong [10], and (i)⇔(iv) is due to Atkinson [2].

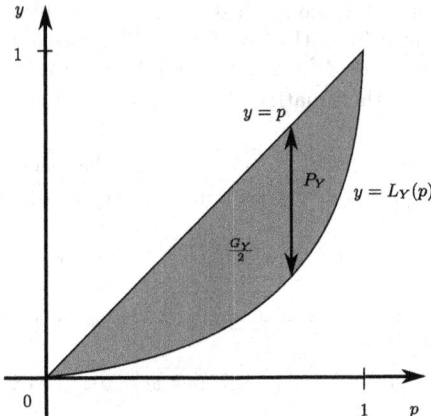

Fig. 1. Lorenz curve of a nonnegative random variable Y. The Gini index $G_Y \in [0, 1]$ corresponds to twice the shaded area, and the Pietra index P_Y corresponds to the maximal vertical distance between the Lorenz curve and the line of perfect equality.

We recall that in economics, risk aversion is commonly modeled by the fact that risk-averse agents seek to maximize a concave increasing utility function of their wealth, while risk-lovers have a convex increasing utility function. The above proposition tells us that $Y \succeq_{(2)} Z$ means that risk-averse expected-utility maximizers prefer gamble Y over gamble Z, while risk-lovers prefer Z over Y, and explains why $Y \preceq_L Z$ can be interpreted as "Y is less dispersed than Z". We warn the reader that a $\preceq$-inequality does not systematically imply a $\leq$-inequality. Instead, the direction of the implied inequality depends on the convexity of the utility function. As a result, some inequalities can seem counter-intuitive but this cannot be avoided because we need both properties (ii) and (iii) in this work.

The Lorenz curve of a nonnegative random variable with finite expectation can be used to define several dispersion indices.

Definition 4. *Let Y be a nonnegative random variable with finite expectation. The Gini index G_Y of Y is defined to be twice the area between the line of perfect equality and the Lorenz curve of Y, and the Pietra index P_Y is defined to be the maximal distance between the line of perfect equality and the Lorenz curve, i.e.,*

$$G_Y := 2 \int_0^1 (p - L_Y(p)) \, dp \qquad and \qquad P_Y := \max_{p \in [0,1]} p - L_Y(p).$$

Both indices are depicted in Fig. 1. Many other equivalent expressions are known for the above indices, see e.g. [6], notably

$$G_Y = \frac{1}{2\mu} \mathbb{E}\big[|Y^{(1)} - Y^{(2)}|\big] \qquad and \qquad P_Y = \frac{1}{2\mu} \mathbb{E}\big[|Y - \mu|\big],$$

where $\mu := \mathbb{E}[Y]$ and $Y^{(1)}, Y^{(2)}$ are independent copies of Y. Thus, it follows from Jensen's inequality that $0 \leq P_Y \leq G_Y \leq 1$.

Table 1. Example of families of nonnegative random variables $\mathcal{D}_\Delta$ with squared coefficient of variation bounded by Δ that are Lorenz-dominated, together with their $\preceq_L$-maximal element of unit mean $Z^{\mathcal{D}_\Delta}$. For Weibull distribution, the coefficient k_Δ is the unique positive solution of the equation $\frac{\Gamma(1+2/k)}{\Gamma(1+1/k)^2} = \Delta + 1$.

$\mathcal{D}_\Delta$	Family	Maximal element $Z^{\mathcal{D}_\Delta}$
Log-normal $\mathcal{L}_\Delta$	$\{\mathcal{LN}(\mu,\sigma) : \mu \in \mathbb{R}, \sigma^2 \leq \log(\Delta+1)\}$	$\mathcal{LN}\left(\log(\frac{1}{\sqrt{\Delta+1}}), \sqrt{\log(\Delta+1)}\right)$
Gamma $\mathcal{G}_\Delta$	$\{\mathrm{Gamma}(k,\theta) : k \geq \frac{1}{\Delta}, \theta > 0\}$	$\mathrm{Gamma}(\frac{1}{\Delta}, \Delta)$
Weibull $\mathcal{W}_\Delta$	$\left\{\mathrm{Weibull}(k,\lambda) : \frac{\Gamma(1+2/k)}{\Gamma(1+1/k)^2} \leq \Delta+1\right\}$	$\mathrm{Weibull}\left(k_\Delta, \frac{1}{\Gamma(1+1/k_\Delta)}\right)$
Uniform $\mathcal{U}_\Delta$	$\left\{\mathcal{U}(a,b) : 0 \leq a \leq b \leq a + \sqrt{3\Delta}(a+b)\right\}$	$\mathcal{U}(1 - \sqrt{3\Delta}, 1 + \sqrt{3\Delta})$
Two-point $\mathcal{B}_\Delta$	$\left\{a \cdot \mathrm{Bernoulli}(p) : a > 0, \frac{1}{\Delta+1} \leq p \leq 1\right\}$	$(\Delta+1) \cdot \mathrm{Bernoulli}(\frac{1}{\Delta+1})$

Returning to SEBP, we will show in the next section that we can obtain improved performance guarantees for LEPT when all processing time distributions come from a certain family. We next introduce the concept of Lorenz-dominated family, and show that most common families of nonnegative two-parameter probability distributions satisfy this property when we bound their coefficient of variation. Recall that the squared coefficient of variation of a random variable X with mean μ and variance σ^2 is $\Delta = \frac{\sigma^2}{\mu^2}$.

Definition 5. *Let $\mathcal{D}$ be a family of nonnegative random variables with finite expectation. We say that $\mathcal{D}$ is a Lorenz dominated family (LDF) if there exists a nonnegative random variable $Z^{\mathcal{D}} \in \mathcal{D}$ such that*

$$\mathbb{E}[Z^{\mathcal{D}}] = 1 \qquad and \qquad X \preceq_L Z^{\mathcal{D}}, \quad \forall X \in \mathcal{D}. \qquad (1)$$

When the above holds, we use the shorthand expression "$\mathcal{D}$ is an LDF with maximal element $Z^{\mathcal{D}}$".

Proposition 4. *Let $\mathcal{D}_\Delta$ be one of the families of nonnegative random variables with squared coefficient of variation bounded by Δ listed in Table 1. Then, $\mathcal{D}_\Delta$ is an LDF with maximal element $Z^{\mathcal{D}_\Delta}$ given in the table.*

Proof. Let $\mathcal{D}$ be any of the families listed in the table. It follows from standard formulas that if $X \in \mathcal{D}$ is a random variable with mean $\mathbb{E}[X] = \mu$ and squared coefficient of variation $\mathbb{V}[X]/\mu^2 = \delta$, then it holds $X \overset{d}{=} \mu Z^{\mathcal{D}_\delta}$, where the equality holds in distribution. Therefore, to show (1), it suffices to establish that $(Z^{\mathcal{D}_\delta})_{\delta \geq 0}$ is monotonically increasing for the relation of Lorenz dominance, i.e., we have $(\delta \leq \delta' \iff Z^{\mathcal{D}}_\delta \preceq_L Z^{\mathcal{D}}_{\delta'})$.

For the cases of lognormal distributions, gamma distributions, and Weibull distributions, this monotonicity property is proved in [9,17] and [12], respectively. For uniform and two-point distributions, it is straightforward to verify that $F_{Z^{\mathcal{D}_\delta}}$ and $F_{Z^{\mathcal{D}_{\delta'}}}$ satisfy the single-crossing property. □

Remark 1. If $X' \overset{d}{=} a + X$ for some $a \geq 0$ and $X \in \mathcal{D}$, where $\mathcal{D}$ is an LDF with maximal element $Z^{\mathcal{D}}$, then it holds $\frac{X'}{\mathbb{E}[X']} \preceq_L \frac{X}{\mathbb{E}[X]} \preceq_L Z^{\mathcal{D}}$, which can be checked with the single-crossing property. As a result, the families of Table 1 can be extended by allowing an additional nonnegative location parameter.

4   Restriction to a Family of Processing Time Distributions

We start with a lemma giving an upper bound on the cost of a machine for a fixed assignment policy, when the processing times come from an LDF. For the case of SEBP, we only need the following result for the function $f : x \mapsto \max(x, 1)$, but we give it in the following form as it holds for a larger class of functions.

Lemma 1.   *For an LDF $\mathcal{D}$ with maximal element $Z^{\mathcal{D}}$, let $P_1, \ldots, P_k \in \mathcal{D}$ with $\sum_{j=1}^{k} \mathbb{E}[P_j] = x$. Then, for every nondecreasing convex function f, it holds*

$$\mathbb{E}[f(P_1 + \ldots + P_k)] \leq \mathbb{E}[f(xZ^{\mathcal{D}})].$$

Proof. Denote by μ_j the expectation of P_j, so that, since the Lorenz curve is scale-invariant by construction, we have $P_j \preceq_L Z^{\mathcal{D}} \iff \frac{P_j}{\mu_j} \preceq_L Z^{\mathcal{D}} \iff \frac{P_j}{\mu_j} \succeq_{(2)} Z^{\mathcal{D}}$, for all $j \in [k]$, where the last equivalence is Proposition 3 (iv). Then, we know from Theorem 10 of the work by Li and Wong [10] that $\sum_{j=1}^{k} P_j \succeq_{(2)} \sum_{j=1}^{k} \mu_j Z^{(j)}$, where $Z^{(1)}, \ldots, Z^{(k)}$ are independent copies of $Z^{\mathcal{D}}$. Now, we can apply Theorem 12 of [10], which states that any convex combination of independent copies of some random variable X has second-order stochastic dominance over X itself. Hence,

$$\sum_{j=1}^{k} P_j \succeq_{(2)} \sum_{j=1}^{k} \mu_j Z^{(j)} = x \sum_{j=1}^{k} \frac{\mu_j}{x} Z^{(j)} \succeq_{(2)} xZ^{\mathcal{D}}.$$

Finally, we observe that $\sum_{j=1}^{k} P_j$ and $xZ^{\mathcal{D}}$ have the same mean $(= x)$, hence we obtain the desired result from Proposition 3 (iii). □

Lemma 2.   *Let $\mathcal{D}$ be an LDF with maximal element $Z^{\mathcal{D}}$, and define $g^{\mathcal{D}}(x) := \mathbb{E}[\max(xZ^{\mathcal{D}}, 1)]$. Then, the function $x \mapsto 1 + x - g^{\mathcal{D}}(x)$ is nondecreasing, and for all $X \in \mathcal{D}$ it holds*

$$1 + \mathbb{E}[X] - g^{\mathcal{D}}(\mathbb{E}[X]) \leq \mathbb{E}[\min(X, 1)].$$

Proof. First, note that $g^{\mathcal{D}}$ is a convex function as the expectation of a convex function, hence, its right derivative $g_{\mathcal{D}}^{\prime+}(x)$ exists for all $x > 0$ and is a nondecreasing function. Since $\max(t,1) \le 1 + t$ holds for all $t \ge 0$, we have $g^{\mathcal{D}}(x) = \mathbb{E}[\max(xZ^{\mathcal{D}}, 1)] \le \mathbb{E}[1 + xZ^{\mathcal{D}}] = 1 + x$, for all $x \ge 0$. This implies that the right derivatives of $g^{\mathcal{D}}$ satisfy $g_{\mathcal{D}}^{\prime+}(x) \le 1, \forall x > 0$, as otherwise the bound $g^{\mathcal{D}}(\tilde{x}) \le 1 + \tilde{x}$ would not hold for some sufficiently large $\tilde{x}$. Hence the function $x \mapsto 1 + x - g^{\mathcal{D}}(x)$ is nondecreasing.

Using the equality $1 + t = \max(t, 1) + \min(t, 1)$, we obtain

$$1 + \mathbb{E}[X] = \mathbb{E}\big[1 + \mathbb{E}[X]Z^{\mathcal{D}}\big] = \underbrace{\mathbb{E}[\max(\mathbb{E}[X]Z^{\mathcal{D}}, 1)]}_{=g^{\mathcal{D}}(\mathbb{E}[X])} + \underbrace{\mathbb{E}[\min(\mathbb{E}[X]Z^{\mathcal{D}}, 1)]}_{\le \mathbb{E}[\min(X,1)]},$$

where the inequality follows from Proposition 3 (ii), using the fact that $\mathcal{D}$ is an LDF, so $\mathbb{E}[X] \cdot Z^{\mathcal{D}} \succeq_L X$, and that $t \mapsto \min(t, 1)$ is concave nondecreasing.   □

Now, we are going to apply these lemmas in order to get improved performance guarantees for SEBP instances with processing times in an LDF.

Theorem 1. *Let $P_1, \dots, P_n \in \mathcal{D}$ for an Lorenz-dominated family $\mathcal{D}$ with maximal element $Z^{\mathcal{D}}$, and let $g^{\mathcal{D}}(x) := \mathbb{E}[\max(xZ^{\mathcal{D}}, 1)]$. Then, we have*

$$\frac{\Phi(LEPT)}{OPT} \le \sup_{t \in [0,1]} \left(2 - \frac{1}{t}\right) \cdot g^{\mathcal{D}}(t) + \left(\frac{1}{t} - 1\right) \cdot g^{\mathcal{D}}(2t).$$

Numerical values of this bound for several distribution families $\mathcal{D}$ are indicated in Table 2.

Proof. Denote by ℓ the minimum expected load of a machine, as in Proposition 2. The main work in this proof will be to show that there exists a subset of machines $\mathcal{M}'$ of cardinality m', and some $x_i' \in [\ell, 2\ell], \forall i \in \mathcal{M}'$, such that

$$\frac{\Phi(\text{LEPT})}{\text{OPT}} \le \frac{\sum_{i \in \mathcal{M}'} g^{\mathcal{D}}(x_i')}{\max(\sum_{i \in \mathcal{M}'} x_i', m')}. \tag{2}$$

Let us first prove the theorem assuming that the above claim is valid. For each $i \in \mathcal{M}'$ we express x_i as a convex combination of ℓ and 2ℓ, writing $x_i' = (2 - \frac{x_i'}{\ell}) \cdot \ell + (\frac{x_i'}{\ell} - 1) \cdot 2\ell$. By convexity of $g^{\mathcal{D}}$, we have $g^{\mathcal{D}}(x_i') \le (2 - \frac{x_i'}{\ell}) \cdot g^{\mathcal{D}}(\ell) + (\frac{x_i'}{\ell} - 1) \cdot g^{\mathcal{D}}(2\ell)$. Now, let $s := \sum_{i \in \mathcal{M}'} x_i'$. Summing over all $i \in \mathcal{M}'$ we have

$$\frac{\Phi(\text{LEPT})}{\text{OPT}} \le \frac{(2m' - \frac{s}{\ell})g^{\mathcal{D}}(\ell) + (\frac{s}{\ell} - m')g^{\mathcal{D}}(2\ell)}{\max(s, m')} = \frac{(2 - \frac{\rho}{\ell})g^{\mathcal{D}}(\ell) + (\frac{\rho}{\ell} - 1)g^{\mathcal{D}}(2\ell)}{\max(\rho, 1)},$$

where we have set $\rho := s/m'$. We will now show that the above bound is maximized for $\rho = 1$. To see this, notice that $f : x \mapsto \max(x, 1)$ satisfies $f(x) \le f(2x) \le 2f(x)$ for any $x \ge 0$. Hence, taking the expectation we obtain $g^{\mathcal{D}}(x) \le g^{\mathcal{D}}(2x) \le 2g^{\mathcal{D}}(x)$ for any $x \ge 0$. Then, our claim follows from the derivative of the above bound with respect to ρ, which is equal to $\frac{1}{\ell}(g^{\mathcal{D}}(2\ell) - g^{\mathcal{D}}(\ell)) \geq 0$ for all $\rho \leq 1$, and is equal to $-\frac{1}{\rho^2}(2g^{\mathcal{D}}(\ell) - g^{\mathcal{D}}(2\ell)) \leq 0$ for all $\rho \geq 1$.

To obtain the statement of the lemma, we show that the supremum of the function $h^{\mathcal{D}} : \mathbb{R}_{\geq 0} \to \mathbb{R}_{\geq 0}$ with $t \mapsto (2 - \frac{1}{t})g^{\mathcal{D}}(t) + (\frac{1}{t} - 1)g^{\mathcal{D}}(2t)$ is attained in the interval $[0,1]$. For $t > 1$, we express t as a convex combination of 1 and $2t$, that is, $t = \frac{t}{2t-1} \cdot 1 + \frac{t-1}{2t-1} \cdot 2t$. We have $g^{\mathcal{D}}(t) \leq \frac{t}{2t-1}g^{\mathcal{D}}(1) + \frac{t-1}{2t-1}g^{\mathcal{D}}(2t)$ by convexity of $g^{\mathcal{D}}$. Multiplying both sides of this inequality by $2 - \frac{1}{t}$, we obtain

$$(2 - \frac{1}{t})g^{\mathcal{D}}(t) \leq g^{\mathcal{D}}(1) + (1 - \frac{1}{t})g^{\mathcal{D}}(2t) \implies h^{\mathcal{D}}(t) \leq g^{\mathcal{D}}(1) = h^{\mathcal{D}}(1).$$

It remains to show that (2) holds for some $\mathcal{M}' \subseteq \mathcal{M}$ of cardinality m' and some $x' \in [\ell, 2\ell]^{m'}$. Applying Lemma 1 to the function $f : x \mapsto \max(x, 1)$, we obtain $\Phi(\text{LEPT}) \leq \sum_{i \in \mathcal{M}} g^{\mathcal{D}}(x_i)$. Then, we readily observe that if $x_i \leq 2\ell$ holds for all machines, we could simply set $\mathcal{M}' = \mathcal{M}$ and $x_i' = x_i$, $\forall i \in \mathcal{M}$, and (2) would follow from Proposition 1 and $\sum_i \beta_i \geq 0$. Moreover, we know from Proposition 2 that $x_i \leq 2\ell$ holds whenever $n_i \geq 2$. Thus, we only need to take special care of those machines where LEPT assigns a single job and $x_i > 2\ell$. To this end, we introduce a partition of the machines relying on the truncated loads α_i's:

$$\mathcal{M}_0 := \{i \in \mathcal{M} : n_i = 1, \ x_i > 2\ell, \ \alpha_i > 2\ell\},$$
$$\mathcal{M}_1 := \{i \in \mathcal{M} : n_i = 1, \ x_i > 2\ell, \ \alpha_i \leq 2\ell\},$$
$$\mathcal{M}_2 := \mathcal{M} \backslash (\mathcal{M}_0 \cup \mathcal{M}_1).$$

Define $x_i' := \max(\ell, \alpha_i)$ for all $i \in \mathcal{M}_0 \cup \mathcal{M}_1$, and $x_i' := x_i$ for all $i \in \mathcal{M}_2$. Furthermore, we define $\mathcal{M}' := \mathcal{M} \setminus \mathcal{M}_0$ with $m' := |\mathcal{M}'|$, and we note that x_i' lies in the interval $[\ell, 2\ell]$ for all $i \in \mathcal{M}'$, as required.

Let us bound the cost induced by LEPT on each machine. Let $\delta_i := x_i - x_i' \geq 0$, $\forall i \in \mathcal{M}$. We claim that $\mathbb{E}[\max(X_i^{\text{LEPT}}, 1)] \leq g^{\mathcal{D}}(x_i') + \delta_i$ holds for each machine. For the machines $i \in \mathcal{M}_2$, this results from Lemma 1, together with $x_i' = x_i$ and $\delta_i = 0$. For the other machines, we have $\mathbb{E}[\max(X_i^{\text{LEPT}}, 1)] = 1 + \beta_i$, because these machines host a single job. Now, we distinguish two cases: (1) If $\alpha_i \geq \ell$, then $1 + \beta_i \leq g^{\mathcal{D}}(x_i') + \delta_i$ follows from $1 \leq g^{\mathcal{D}}(x_i')$ and $\delta_i = x_i - \alpha_i = \beta_i$; (2) If $\alpha_i < \ell$, then we obtain from Lemma 2 and $x_i > \ell$ that

$$1 + \ell - g^{\mathcal{D}}(\ell) \leq 1 + x_i - g^{\mathcal{D}}(x_i) \leq \mathbb{E}[\min(X_i^{\text{LEPT}}, 1)] = \alpha_i,$$

which implies $1 + \beta_i \leq \alpha_i + \beta_i - \ell + g^{\mathcal{D}}(\ell) = g^{\mathcal{D}}(\ell) + x_i - \ell = g^{\mathcal{D}}(x_i') + \delta_i$. Hence, the claim is proved, and summing the bound over all machines yields

$$\Phi(\text{LEPT}) \leq \sum_{i \in \mathcal{M}} g^{\mathcal{D}}(x_i') + \delta_i. \tag{3}$$

On the other hand, we have

$$OPT \geq \max(\sum_{i \in \mathcal{M}} x_i, m + \sum_{i \in \mathcal{M}} \beta_i) \geq \max(\sum_{i \in \mathcal{M}} x_i', m) + \sum_{i \in \mathcal{M}} \delta_i \tag{4}$$

Table 2. Approximation guarantees from Theorem 1 for several families of probability distributions (cf. Table 1 for a description of the symbols used in the first column) and upper bound Δ on the squared coefficient of variation. Boldface entried improve the bound of $(1 + e^{-1})$ from [22]. Note that $\Delta \le \frac{1}{3}$ always holds for a nonnegative uniform random variable.

Δ	0	$1/8$	$1/6$	$1/4$	$1/3$	$1/2$	1
$\mathcal{L}_\Delta$	**1.1716**	**1.1990**	**1.2112**	**1.2334**	**1.2526**	**1.2843**	**1.3485**
$\mathcal{G}_\Delta$	**1.1716**	**1.2012**	**1.2148**	**1.2401**	**1.2629**	**1.3023**	1.3896
$\mathcal{W}_\Delta$	**1.1716**	**1.2044**	**1.2186**	**1.2450**	**1.2685**	**1.3080**	1.3896
$\mathcal{U}_\Delta$	**1.1716**	**1.2041**	**1.2210**	**1.2526**	**1.2812**	–	–
$\mathcal{B}_\Delta$	**1.1716**	**1.2222**	**1.2385**	**1.2702**	**1.3005**	**1.3573**	1.5000

where the first inequality is Proposition 1, and the second one follows from $\beta_i \ge \delta_i, \forall i$. Now, we combine (3) and (4), and use the fact that removing $\sum_{i \in \mathcal{M}} \delta_i \ge 0$ from both the numerator and the denominator can only worsen the ratio, to obtain

$$\frac{\Phi(\text{LEPT})}{OPT} \le \frac{\sum_{i \in \mathcal{M}} g^{\mathcal{D}}(x_i')}{\max(\sum_{i \in \mathcal{M}} x_i', m)}.$$

At this stage, observe that we have already proved that (2) holds in the case where $\mathcal{M}_0$ is empty. So it only remains to handle the case $\mathcal{M}_0 \ne \emptyset$. In this case, we have $2\ell < \alpha_i \le 1$ for all $i \in \mathcal{M}_0$, which implies $x_i' \le 1$ for all machines $i \in \mathcal{M}$. Thus, $\sum_{i \in \mathcal{M}} x_i' \le m$ and $\sum_{i \in \mathcal{M}'} x_i' \le m'$. Altogether, we obtain

$$\frac{\Phi(\text{LEPT})}{OPT} \le \frac{\sum_{i \in \mathcal{M}} g^{\mathcal{D}}(x_i')}{m} \le \frac{\sum_{i \in \mathcal{M}'} g^{\mathcal{D}}(x_i')}{m'} = \frac{\sum_{i \in \mathcal{M}'} g^{\mathcal{D}}(x_i')}{\max(\sum_{i \in \mathcal{M}'} x_i', m')}, \quad (5)$$

where we have used $g^{\mathcal{D}}(x_i') \le 1$ for all $i \in \mathcal{M}_0$ in the second inequality, so we could remove the constant $|\mathcal{M}_0| = m - m'$ from both the numerator and the denominator in order to increase the ratio. This concludes the proof.   □

Remark 2. As Table 2 shows, the bound of Theorem 1 is not tight. For example, the bound can get larger than $1 + e^{-1}$ for large values of Δ. Moreover, if $\Delta = 0$ (corresponding to the deterministic version EBP), the bound equals $4 - 2\sqrt{2} \simeq 1.1716$, but a tight bound of $\frac{13}{12} \simeq 1.0833$ is known in this case [4].

We will now use Theorem 1 to derive a distribution-free bound that depends only on the Pietra index of the processing times.

Theorem 2. *Let $P_1, \ldots, P_n$ be nonnegative random variables with finite expectation and Pietra index at most ϱ. Then,*

$$\frac{\Phi(\text{LEPT})}{OPT} \le 2\left(2 - \sqrt{2} + \varrho(\sqrt{2} - 1)\right).$$

Remark 3. Due to the inequality $P_Y \leq G_Y$, the above result also holds if all processing times have Gini index at most ϱ.

Proof. We first prove the result for the case where all random variables have bounded support, and the result will follow by standard continuity arguments. For a constant θ large enough (we require $\theta \cdot (1 - \varrho) > 1$), define $\mathcal{D}_\theta^\varrho$ as the set of all nonnegative random variables with finite expectation and Pietra index at most ϱ such that $Y \leq \theta \mathbb{E}[Y]$ holds almost surely. Henceforth we assume $P_j \in \mathcal{D}_\theta^\varrho$, for all jobs j.

Our assumption implies that $F_{P_j}^{-1}(1) \leq \theta \mathbb{E}[P_j]$, hence the left derivative of the Lorenz function at $p = 1$ is $L_{P_j}'^{-}(1) = \frac{1}{\mathbb{E}[P_j]} F_{P_j}^{-1}(1) \leq \theta$. Using this and the convexity of L_{P_j}, we obtain $L_{P_j}(p) \geq 1 + \theta(p - 1)$ for all $p \in [0, 1]$. By definition of the Pietra index, we know for all $p \in [0, 1]$ that $L_{P_j}(p) \geq p - \varrho$. So we have

$$L_{P_j}(p) \geq \max(0, p - \varrho, 1 + \theta(p - 1)), \quad \forall p \in [0, 1].$$

It is easy to see that the right-hand side of the above expression coincides with the Lorenz curve of the random variable Z_θ^ϱ such that

$$\mathbb{P}[Z_\theta^\varrho = 0] = \varrho, \qquad \mathbb{P}[Z_\theta^\varrho = 1] = 1 - \varrho\frac{\theta}{\theta - 1}, \quad \text{and} \quad \mathbb{P}[Z_\theta^\varrho = \theta] = \frac{\varrho}{\theta - 1},$$

where these probabilities are nonnegative since we assumed $\theta \cdot (1 - \varrho) > 1$. This shows that $P_j \preceq_L Z_\theta^\varrho$, and since $\mathbb{E}[Z_\theta^\varrho] = 1$, the family $\mathcal{D}_\theta^\varrho$ is Lorenz-dominated with maximal element Z_θ^ϱ. So, by Theorem 1:

$$\frac{\Phi(\text{LEPT})}{OPT} \leq \sup_{t \in [0,1]} (2 - \frac{1}{t})g_\theta(t) + (\frac{1}{t} - 1)g_\theta(2t),$$

where $g_\theta(t) = \mathbb{E}[\max(1, t \cdot Z_\theta^\varrho)] = \max(1, \varrho + 1 + \frac{\varrho\theta}{\theta-1}(t - 1), \varrho + t)$. It is easy to see that for all $t > 0$, $g_\theta(t)$ is nondecreasing with respect to θ. As a consequence, we obtain $g_\theta(t) \leq g_\infty(t) := \max(1 + \varrho t, \varrho + t)$ for all $t \geq 0$. Finally, simple calculus shows that the function $t \mapsto (2 - \frac{1}{t})g_\infty(t) + (\frac{1}{t} - 1)g_\infty(2t)$ reaches its maximum over $t \in [0, 1]$ at $t = \frac{1}{\sqrt{2}}$, and we get the desired result after substitution. □

This theorem improves the bound of $1 + e^{-1}$ from [22] for all instances with a Pietra index bounded by $\varrho \leq \frac{1+\sqrt{2}}{2}e^{-1} + \frac{1-\sqrt{2}}{2} \approx 0.237$.

Conclusion and Outlook. We have shown that the concept of second-order stochastic dominance can be used to obtain improved performance guarantees for a stochastic scheduling problem, when it is known that the processing times belong to an LDF, an assumption which we believe is reasonable for many applications. For future work, it would be interesting to investigate whether this approach can be generalized to other stochastic load balancing problems.

Acknowledgement. The authors wish to thank anonymous reviewers for their careful reading and suggestions to improve the presentation of this manuscript.

References

1. Alon, N., Azar, Y., Woeginger, G., Yadid, T.: Approximation schemes for scheduling on parallel machines. J. Sched. **1**(1), 55–66 (1998)
2. Atkinson, A.: On the measurement of inequality. J. Econ. Theory **2**(3), 244–263 (1970)
3. Berg, B., Denton, B.: Fast approximation methods for online scheduling of outpatient procedure centers. INFORMS J. Comput. **29**(4), 631–644 (2017)
4. Dell'Olmo, P., Kellerer, H., Speranza, M., Tuza, Z.: A 13/12 approximation algorithm for bin packing with extendable bins. Inf. Proc. Lett. **65**(5), 229–233 (1998)
5. Denton, B., Miller, A., Balasubramanian, H., Huschka, T.: Optimal allocation of surgery blocks to operating rooms under uncertainty. Oper. Res. **58**(4–1), 802–816 (2010)
6. Eliazar, I.: A tour of inequality. Ann. Phys. **389**, 306–332 (2018)
7. Hadar, J., Russell, W.: Rules for ordering uncertain prospects. Am. Econ. Rev. **59**(1), 25–34 (1969)
8. Joustra, P., Meester, R., van Ophem, H.: Can statisticians beat surgeons at the planning of operations? Empir. Econ. **44**(3), 1697–1718 (2013). https://doi.org/10.1007/s00181-012-0594-0
9. Levy, H.: Stochastic dominance among log-normal prospects. Int. Econ. Rev. **14**, 601–614 (1973)
10. Li, C.K., Wong, W.K.: Extension of stochastic dominance theory to random variables. RAIRO-Oper. Res. **33**(4), 509–524 (1999)
11. Lorenz, M.: Methods of measuring the concentration of wealth. Publ. Am. Stat. Assoc. **9**(70), 209–219 (1905)
12. Lubrano, M., Protopopescu, C.: Density inference for ranking European research systems in the field of economics. J. Econometr. **123**(2), 345–369 (2004)
13. Marshall, A.W., Olkin, I., Arnold, B.C.: Inequalities: Theory of Majorization and Its Applications, 2nd edn. Springer, New York (2011). https://doi.org/10.1007/978-0-387-68276-1
14. Megow, N., Uetz, M., Vredeveld, T.: Models and algorithms for stochastic online scheduling. Math. Oper. Res. **31**(3), 513–525 (2006)
15. Möhring, R., Radermacher, F., Weiss, G.: Stochastic scheduling problems I-general strategies. Z. Oper. Res. **28**(7), 193–260 (1984). https://doi.org/10.1007/BF01919323
16. Pinedo, M.L.: Scheduling: Theory, Algorithms, and Systems. Springer, Heidelberg (2016). https://doi.org/10.1007/978-3-642-46773-8_5
17. Ramos, H., Ollero, J., Sordo, M.: A sufficient condition for generalized Lorenz order. J. Econ. Theory **90**(2), 286–292 (2000)
18. Rothschild, M., Stiglitz, J.: Increasing risk: I. A definition. J. Econ. Theory **2**(3), 225–243 (1970)
19. sagnol, G., et al.: Robust allocation of operating rooms: a cutting plane approach to handle lognormal case durations. Eur. J. Oper. Res. **271**(2), 420–435 (2018)
20. Sagnol, G., Schmidt genannt Waldschmidt, D.: Stochastic extensible bin packing (2020). arXiv:2002.00060 [cs.DS]
21. Sagnol, G., Schmidt genannt Waldschmidt, D.: Restricted adaptivity in stochastic scheduling. In: European Symposium on Algorithms, ESA 2021. LIPIcs, vol. 204, pp. 1–14 (2021)
22. Sagnol, G., Schmidt genannt Waldschmidt, D., Tesch, A.: The price of fixed assignments in stochastic extensible bin packing. In: Approximation and Online Algorithms WAOA 2018. LNCS, vol. 11312, pp. 327–347 (2018)

23. Shehadeh, K.S., Padman, R.: Stochastic optimization approaches for elective surgery scheduling with downstream capacity constraints: models, challenges, and opportunities. Comput. Oper. Res. **137**, 105523 (2022)
24. Strum, D., May, J., Vargas, L.: Surgical procedure times are well modeled by the lognormal distribution. Anesth. Analg. **86**(2S), 47S (1998)
25. Uetz, M.: Algorithms for Deterministic and Stochastic Scheduling. Cuvillier (2001)

Applications

One Transfer per Patient Suffices: Structural Insights About Patient-to-Room Assignment

Tabea Brandt[1], Christina Büsing[1], and Sigrid Knust[2]([✉])

[1] Combinatorial Optimization, RWTH Aachen University, Pontdriesch 10-12, 52062 Aachen, Germany
{brandt,buesing}@combi.rwth-aachen.de
[2] Institute of Computer Science, Osnabrück University, Wachsbleiche 27, 49090 Osnabrück, Germany
sigrid@informatik.uni-osnabrueck.de

Abstract. Assigning patients to rooms is a fundamental task in hospitals and, especially, within wards. For this so-called patient-to-room assignment problem (PRA) many heuristics have been proposed with a large variety of different practical constraints. However, a thorough investigation of the problem's structure itself has been neglected so far. In this paper, we present insights about the basic, underlying combinatorial problem of PRA forbidding gender-mixed room assignments with a focus on minimizing the number of patient transfers which occur if patients have to change rooms during their stay. Particularly, we prove that in the case of double bedrooms, each patient has to be transferred at most once.

Keywords: Combinatorial optimization · Patient-to-room assignment

1 Introduction

Hospital beds are next to medical trained staff one of the most important resources in hospitals [19]. Whether or not a patient can be treated in an hospital often depends on the existence of a free bed. Assigning patients to rooms as efficiently as possible is therefore very important in a well-functioning health care system [14].

There are two ways patients enter a hospital for hospitalization: as emergency patients or as elective patients. The allocation of emergency patients to beds is usually centrally organized for the complete hospital, depending on the allocation of present patients, planned elective patients and the diagnosis made

---

This work was supported by the Freigeist-Fellowship of the Volkswagen Stiftung and by the German research council (DFG) Research Training Group 2236 UnRAVeL. This work was partially supported by the German Federal Ministry of Education and Research (grant no. 05M16PAA) within the project "HealthFaCT - Health: Facility Location, Covering and Transport".

I. Ljubić et al. (Eds.): ISCO 2022, LNCS 13526, pp. 245–259, 2022.
https://doi.org/10.1007/978-3-031-18530-4_18

in the emergency room. In contrast, the allocation of elective patients is usually organized by each ward independently [15]. As soon as an appointment for a patient is fixed, the patient is directly assigned to a specific ward. Afterwards, the actual assignment of the patient to a room in this ward is performed by experienced nurses or the ward's head.

In this paper, we consider the planning of elective patients to be independent of the accommodation of emergency patients and focus on the problem of assigning elective patients only. This is impractical for specialities with a high rate of emergency patients, however, there do exist specialities with nearly 100% elective patients. Thus, this assumption is not only helpful to gain first structural insights, but also has a direct practical application.

Apart from ensuring that all patients always have a bed during their stay in rooms separated by gender, reassigning patients to different rooms should be avoided. These so-called *patient transfers* mean additional work for the medical staff and inconvenience for the patient.

Formally, in the *patient-to-room assignment problem* (PRA), a set $\mathcal{P} = \{1, \ldots, P\}$ of patients, divided into a set $\mathcal{P}^f$ of female and a set $\mathcal{P}^m$ of male patients, is given. Furthermore, we have a set of rooms $\mathcal{R} = \{1, \ldots, R\}$, where room r has capacity C_r, i.e., r contains C_r beds. We denote by T the length of the planning horizon and by $\mathcal{T} = \{1, \ldots, T\}$ the set of all time periods. For each patient $p \in \mathcal{P}$, an arrival period $a_p \in \mathcal{T}$, and a departure period $b_p \in \mathcal{T}$ are specified. The goal of PRA is to assign patients to rooms so that

- each patient $p \in \mathcal{P}$ is assigned to a room $r \in \mathcal{R}$ in each time period $t \in \{a_p, \ldots, b_p\}$,
- in each time period, no more than C_r patients are assigned to every room $r \in \mathcal{R}$,
- no gender-mixed rooms occur.

In this paper, we concentrate on the case of constant room capacities $C_r = C$ and especially $C = 2$ reflecting the common situation that only double bedrooms exist.

In order to have more flexibility in the room assignments, we allow the option of patient transfers, i.e., patients may be moved among rooms during their stay. However, since such transfers are not desirable for staff and patients, we consider the two objectives of minimizing

- f^{Σ}: the total number of patient transfers,
- $f^{\max}$: the maximum number of transfers per patient.

Many aspects of hospital-bed management have already been investigated using a variety of methodologies such as capacity dimensioning or allocation [15]. Already in 1995, a centralised bed management was suggested [18] as well as a model for computing the capacity of emergency beds considering week-day dependent demand fluctuation [13].

Simulation models for evaluating different planning strategies were proposed in [5, 10–12, 16].

The operational task of assigning patients to beds has been considered and formalized for the first time in 2010 based on the situation in Belgium hospitals [6]. There, a most suitable room for each patient is found using a rating that considers different room equipment, ward specialisation as well as patient needs and wishes. To solve this basic patient-to-room assignment problem, a tabu search algorithm [6], a multi-neighborhood local search procedure [2], a column generation approach [17], a population-based metaheuristic [9], and a matheuristic [8] were proposed.

Since 2010, the basic problem definition has been extended to a dynamic context [3], to include uncertainties in the patients' length-of-stay [19], or to include capacities of other hospital resources [4,21]. A detailed comparison of different versions can be found in [1,8]. A further common concept in PRA are patient-to-room restrictions [6], which mean that not every patient can be assigned to every room (e.g., due to different equipment requirements). In the case of patient-to-room restrictions, PRA with $T = 1$ corresponds directly to the decision version of the red-blue transportation problem [20] and is $\mathcal{NP}$-complete, even for room capacities $C = 2$ [7].

In this paper, we consider PRA without patient-to-room restrictions for elective patients forbidding gender-mixed room assignments. We focus on transfers of patients and consider the objective functions $f^{\max}$ and f^{Σ}. Our main result is that in the case of double bedrooms there is always an optimal solution with respect to f^{Σ} for which $f^{\max} \leq 1$ holds. This justifies the restriction of the search space to solutions in which every patient is transferred at most once (e.g., as it is done in the heuristic from [2]). Unfortunately, the complexity of minimizing $f^{\max}$ or f^{Σ} is still open, even for $C = 2$. However, we provide upper bounds on f^{Σ} which are helpful in designing efficient algorithms as well as in strategic and tactical decision-making regarding staff capacity and staff scheduling.

The remainder of this paper is organized as follows. In Sect. 2, we prove that for double bedrooms there is always an optimal solution with respect to f^{Σ} where every patient has to be transferred at most once. In Sect. 3 we show that there is no need to transfer patients arriving in the first time period. In Sect. 4 we use the previous results to derive upper bounds on f^{Σ}. Finally, in Sect. 5 some conclusions can be found.

2   Every Patient Has to Be Transferred at Most Once

In this section, we study how many transfers are needed in total (objective f^{Σ}) and how often an individual patient is transferred during their stay (objective $f^{\max}$). The main result is a proof that for $C = 2$ there is an optimal solution w.r.t. f^{Σ} in which every patient is transferred at most once. This means that in real life both the wishes of staff (minimizing the total number of transfers) and the patients (being moved as few as possible) can be taken into account simultaneously.

At first, we deal with the feasibility problem. If an unlimited number of patient transfers is allowed, checking feasibility is easy, even for arbitrary constant room capacities C. Let F_t, M_t be the number of female and male patients who need a room in time period $t \in \mathcal{T}$. Then, a necessary and sufficient condition for the existence of a feasible assignment is

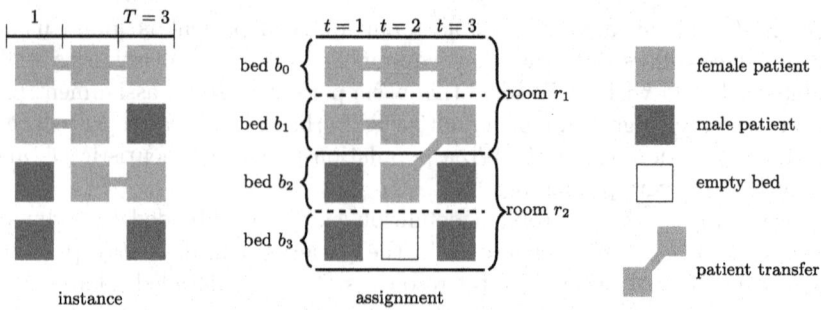

Fig. 1. Example for $T = 3$ where patient transfers are necessary for feasibility

$$\max_{t \in T} \left\{ \left\lceil \frac{F_t}{C} \right\rceil + \left\lceil \frac{M_t}{C} \right\rceil \right\} \le R. \tag{1}$$

However, in a corresponding solution several patient transfers may be necessary to achieve a feasible solution. In the following, we assume that (1) is satisfied, i.e., always a feasible solution exists. We study in more detail the question how many transfers (per patient and in total) are necessary.

Lemma 1. *For PRA with constant room capacities, in any optimal solution with respect to f^{Σ} (or $f^{\max}$) no transfer between the first two time periods is performed.*

Proof. Assume to the contrary that in an optimal solution a patient is transferred between the first two time periods from one room to another. Then there must be another patient of the same gender in the other room who leaves after the first time period. If these two patients simply switch beds in the first time period, this transfer is avoided and feasibility of the solution is not affected. Therefore, the original solution cannot be optimal with respect to f^{Σ} or $f^{\max}$.   □

If the room capacities are not constant, this result is no longer true. Suppose that we have one single and one double bedroom. If in the first time period two female patients and one male patient are in the hospital, one female patient leaves in the second period and another man comes, then the two patients who stay (one man, one woman) must switch their rooms in order to get a feasible solution.

Lemma 1 implies that no patient transfer is necessary for a planning horizon of $T = 2$. For $T = 3$, this statement is no longer valid which is shown by the example in Fig. 1. In the first time period, two female patients arrive, one stays for three periods, the other for two. In the second time period, another female patient arrives staying for two periods as well. Both in the first and third time period, two male patients arrive who stay for one period each. In this case, one patient transfer is necessary to achieve a feasible assignment, cf. Fig. 1.

In the following, we concentrate on the case $C = 2$. We start by proving that in an optimal solution w.r.t. f^{Σ}, every patient has to be transferred at most once, i.e., we have $f^{\max} \le 1$. Theoretically, a patient can be transferred at any point during their stay. However, a patient's transfer is only necessary if the following condition is satisfied:

Definition 1. *For $C = 2$, a patient-to-room assignment satisfies the* transfer condition *(TC) if*

(a) *a patient is only transferred if the patient is alone in a room or the corresponding room-mate leaves the hospital at the time period immediately before the transfer,*
(b) *a patient is never transferred to an empty room.*

At first, we prove that a restriction to patient-to-room assignments satisfying TC does not increase the number of transfers.

Lemma 2. *For $C = 2$, an optimal solution w.r.t. f^Σ satisfying TC exists.*

Proof. Let an optimal solution minimizing f^Σ be given which does not satisfy TC. It is easy to see that in an optimal solution, a patient who is not alone in a double bedroom does not have to be transferred and also a patient is not transferred to an empty room (in both cases another patient can go instead). Thus, it remains the case that there exists a patient who is transferred although the room-mate remains in hospital. By a (rather technical) interchange argument it can be shown that we can delay this transfer until one of those patients leaves the hospital without loosing feasibility or increasing the total number of transfers. We can repeat this procedure until the resulting patient-to-room assignment satisfies TC while preserving the optimality w.r.t. f^Σ. □

In the case $C = 2$, for any solution satisfying TC, we define the *transfer graph* as the digraph $D = (V, A)$ with nodes $V = \mathcal{P}$ corresponding to the patients and arcs

$$A = \{(p, q) \mid p, q \in \mathcal{P} \text{ and } p \text{ is transferred to the room in which } q \text{ already is}\}.$$

Since in a solution satisfying TC, no patient is transferred to an empty room, this graph represents all transfers performed. Furthermore, note that neither feasibility nor optimality is affected if a transfer is performed the other way round. In Fig. 2 we illustrate the reversal of a transfer between two patients a and c. While in Fig. 2a patient a is moved to patient c between the 3rd and the 4th time period, in Fig. 2b patient c is moved to a. Remark that in all figures we ignore isolated nodes.

For a given transfer graph $D = (V, A)$ let $\tilde{D} = (V, E)$ be the corresponding undirected graph with edges $E = \{\{p, q\} \mid (p, q) \in A\}$.

Lemma 3. *For any solution satisfying TC, the corresponding undirected graph $\tilde{D}$ is acyclic.*

Proof. Assume to the contrary that $(p_1, \ldots, p_k, p_1)$ with $k \geq 3$ is a cycle in $\tilde{D}$ and w.l.o.g. let (p_1, p_2) be the earliest transfer in the solution. Due to TC, the next transfer that involves p_1 or p_2 takes place after p_1 or p_2 has left the hospital.

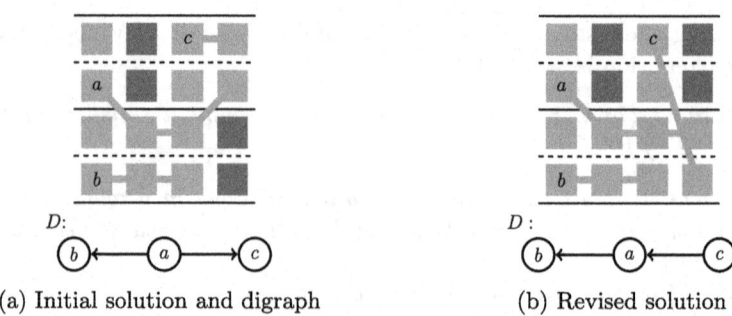

(a) Initial solution and digraph   (b) Revised solution

Fig. 2. Transfers and their representation

But then, at most one of p_1 and p_2 can be involved in another transfer, i.e., only one of the edges $\{p_k, p_1\}$ and $\{p_2, p_3\}$ can exist, which contradicts the existence of a cycle in $\tilde{D}$.   □

By using that a transfer graph does not contain cycles, we can easily transform every solution that satisfies TC into a solution in which every patient is transferred at most once, without increasing the total number of transfers:

Theorem 1. *For $C = 2$, there exists an optimal solution w.r.t. f^Σ such that every patient is transferred at most once.*

Proof. Assume that an optimal solution satisfying TC is given. We can transform it into an equivalent optimal solution in which every patient is transferred at most once if we can reorientate the arcs in D such that every node has at most one outgoing arc. According to Lemma 3, each connected component of the corresponding undirected graph $\tilde{D}$ must be a tree since $\tilde{D}$ is acyclic. Hence, such an orientation can be achieved by choosing an arbitrary node as root and orientating all edges from the leaves to the root. For example, in Fig. 2a we have chosen patient b as root and reorientated D accordingly to get the solution in Fig. 2b.   □

Theorem 1 implies that the objective functions f^Σ and $f^{\max}$ are not conflicting. Moreover, they can be optimized simultaneously by adding to the optimization problem for f^Σ the condition that every patient is transferred at most once. In the following, we call a solution which satisfies TC and in which every patient is transferred at most once, a *proper* solution. According to Theorem 1, always an optimal proper solution exists.

Finally, note that for room capacities $C > 2$, in an optimal solution w.r.t. f^Σ it may be not sufficient to transfer every patient at most once. This can be seen from the example in Fig. 3 with $C = 3$. If we restrict our considerations to solutions where every patient is transferred at most once, in total at least $f^\Sigma = 3$ transfers are necessary (cf. Fig. 3a). However, if we allow that a patient may be transferred twice, a better solution with $f^\Sigma = 2$ transfers is possible (cf. Fig. 3b).

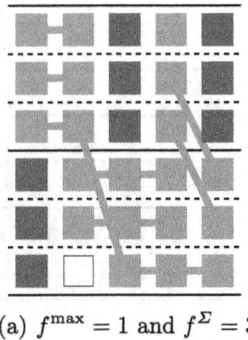

(a) $f^{\max} = 1$ and $f^\Sigma = 3$

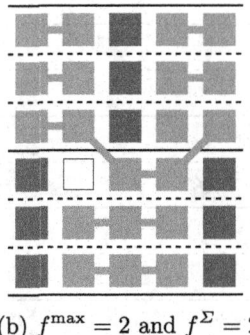

(b) $f^{\max} = 2$ and $f^\Sigma = 2$

Fig. 3. Optimal solution w.r.t. f^Σ for $C = 3$ where a patient has to be transferred twice

3 No Need to Transfer Patients Arriving in the First Period

In Sect. 2, we proved that for $C = 2$ there always exists an optimal solution with respect to f^Σ in which no patient is transferred more than once. Furthermore, $f^{\max} = 0$ if and only if $f^\Sigma = 0$. In the following we concentrate on f^Σ and assume that all considered solutions are proper, i.e., each patient is transferred at most once. We prove that an optimal solution exists in which patients arriving in the first time period do not have to be transferred. We will use this to derive upper bounds on f^Σ in Sect. 4. We call a carrives in the first time period a t_0-*patient*. However, we first need some more insights about the time dependencies between transfers in a proper solution as well as the corresponding transfer graph.

Lemma 4. *Let $D = (V, A)$ be the transfer graph corresponding to a proper solution for $C = 2$. Then, for every path in D the earliest transfer is represented by its first or last arc.*

Proof. To prove this statement, it suffices to show that among every three consecutive arcs, the middle one cannot correspond to the earliest transfer of these three. Let (a, b, c, d) be a directed path and assume the transfer (b, c) is the first of them. In order to perform the transfer (a, b) afterwards, patient c is required to leave the hospital according to TC. But then, the transfer (c, d) cannot be performed anymore, which implies that the transfer (b, c) cannot be the earliest. As the orientation of the arcs/transfers does affect neither feasibility nor optimality, the statement holds for all directed and nondirected paths of length three. $\square$

Remark that Lemma 4 especially holds for every subpath. Therefore, every path can be divided into two chronologically sorted sequences of transfers. Let $D = (V, A)$ be the transfer graph corresponding to a proper solution and $W = (p_1, p_2, \ldots, p_k)$ a directed path in D. Without loss of generality, let (p_1, p_2)

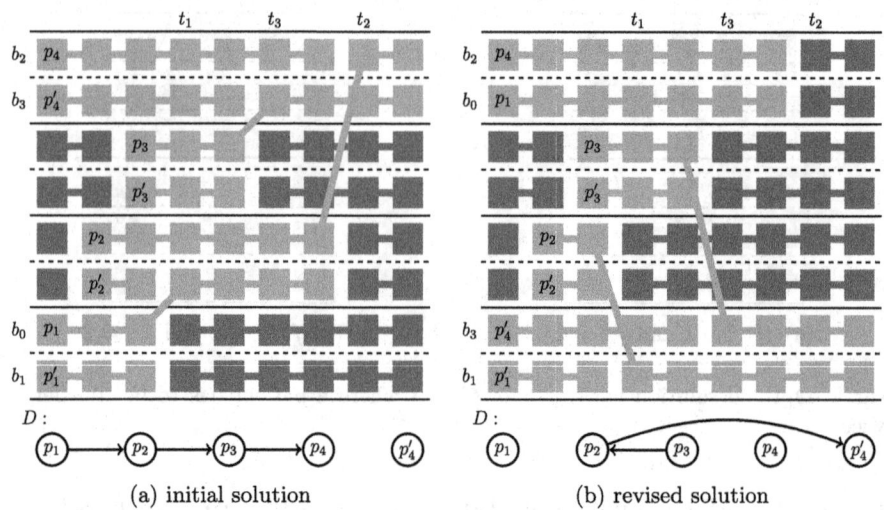

Fig. 4. Example for an assignment revision such that no t_0-patients are transferred, with $t_1 < t_3 < t_2$ and beds b_0, b_1, b_2, b_3

represent the earliest transfer in W and $(p_\tau, p_{\tau+1})$ with $\tau \in \{2, \ldots, k-1\}$ the latest transfer in W. Furthermore, we denote by t_j the time period in which the transfer (p_j, p_{j+1}) is performed, $1 \leq j < k$. Then the transfers corresponding to the subpath $(p_1, \ldots, p_\tau, p_{\tau+1})$ are sorted chronologically increasing, i.e., $t_1 < t_2 < \ldots < t_\tau$, and the transfers corresponding to the subpath $(p_\tau, p_{\tau+1}, \ldots, p_k)$ are sorted chronologically decreasing, i.e., $t_\tau > \ldots > t_{k-2} > t_{k-1}$. An example path (p_1, p_2, p_3) with $\tau = 2$ and $t_1 < t_3 < t_2$ is illustrated in Fig. 4a.

To be able to change a given solution such that it remains feasible, we do not only need information about the sequence of transfers but also about the time periods when the patients are discharged. As we assume that TC is satisfied by our solution, we can easily derive when each patient leaves the hospital:

(L1) In time period t_j patients p_{j-1} and p'_j are discharged for $1 \leq j < \tau$, where p'_j denotes the room-mate of p_j before p_j is involved in its first transfer and $p_0 := p'_1$.

(L2) In time period t_τ patients $p_{\tau-1}$ and $p_{\tau+2}$ are discharged where, in case of $\tau = k - 1$ we define $p_{k+1} := p'_k$.

(L3) In time period t_j patients p_{j+2} and p'_j are discharged for $\tau < j < k$, where p'_j denotes the room-mate of p_j before p_j is involved in its first transfer and $p_{k+1} := p'_k$.

Using the information about the sequence of transfers and about the discharge of patients, we can alter a given solution such that no t_0-patients are transferred without increasing the total number of transfers. First, we prove that no two t_0-patients are connected via a path in D.

Lemma 5. *Let $D = (V, A)$ be the transfer graph corresponding to an optimal proper solution for $C = 2$. Then all t_0-patients are in different connected components of the corresponding undirected graph $\tilde{D}$.*

Proof. We prove this lemma by contradiction. Let us assume that U is a connected component of $\tilde{D}$ which contains two t_0-patients p and q. Without loss of generality, the transfer involving p takes place earlier than the transfer involving q. Choose q as root of U and orientate U accordingly from leaves to the root. Further, let $W = (p = p_1, p_2, \ldots, p_{k-1}, p_k = q)$ be the unique directed path connecting p and q in the revised transfer graph. Let b_0 denote the bed to which p is assigned in the first time period, and b_1 that of its room-mate p_1'. Analogously, let b_2 denote the bed to which q is assigned in the first time period and b_3 that of its room-mate p_k'. Remark that in the original solution b_0 and b_1 are located in the same room as well as b_2 and b_3. For our revised solution we may change this pairing later.

We construct a revised solution by assigning all t_0-patients to their beds for their complete stay. The other patients p_j, for $2 \leq j \leq k - 1$, start also in their original beds, and we now define their transfers. We denote again with t_τ the time period of the latest transfer in W. For $1 \leq j \leq \tau - 1$, we transfer p_{j+1} in time period t_j to bed $b_{j \bmod 2}$. For $\tau < j < k$, we transfer p_j in time period t_j to bed $b_{2+(k-j) \bmod 2}$. All other bed assignments stay as before. Lastly, pair the two beds to which p_τ and $p_{\tau+1}$ are assigned to the same room and the remaining two to another room. An example of such a revision for t_0-patients p_1 and p_4 and $W = (p_1, p_2, p_3, p_4)$, with $t_1 < t_3 < t_2$, is illustrated in Fig. 4. Remark that in Fig. 4a patients p_1 and p_1' are room-mates in the first time period whereas in Fig. 4b patients p_1 and p_4 are room-mates.

The revised solution is feasible according to (L1)–(L3), and needs one transfer less than the initial solution, which is a contradiction to its optimality. Thus, no connected component of $\tilde{D}$ corresponding to an optimal proper solution contains two t_0-patients. □

Finally, we prove that there always exists an optimal proper solution in which no t_0-patients are transferred.

Theorem 2. *Let $D = (V, A)$ be the transfer graph corresponding to an optimal proper solution for $C = 2$. Then there exists an equivalent optimal proper solution in which patients arriving in the first time period are never transferred.*

Proof. For every patient p who arrives in the first time period and is transferred, we perform the following revision. Let U denote the connected component of the corresponding undirected transfer graph $\tilde{D}$ that contains p. According to Lemma 5, patient p is the only t_0-patient in U. We choose p as new root and orientate U accordingly from leaves to the root. As no additional transfers are needed, the revised transfer graph corresponds to an optimal proper solution in which p is not transferred and also no other t_0-patient is transferred instead. □

We may combine these insights with a simple greedy procedure for computing a schedule satisfying TC, to calculate a feasible proper solution in which patients arriving in the first time period are never transferred.

At first, we may construct a feasible schedule satisfying TC in polynomial time $\mathcal{O}(PR)$ by the following greedy procedure.

1. Sort the patients according to their arrival periods.
2. Assign each patient p in the list to a room as follows:
   - If there exists a room r with only one patient of the same gender in period a_p, assign p to r as well.
   - If no such room but an empty room in time period a_p exists, assign p to it.
   - Otherwise, select two rooms r_1, r_2 with one patient of the opposite gender in time period a_p each, transfer the patient from r_2 to r_1, and assign p to r_2.

Using the procedure described in the proof of Theorem 1, we can transform such a schedule into a proper solution without increasing the total number of transfers in $\mathcal{O}(P)$. Finally, we use the construction described in the proof of Theorem 2 to change the schedule in $\mathcal{O}(P)$ so that no t_0-patients are transferred.

Note that the total number of transfers in a solution computed in this way is not necessarily optimal, because the initial solution computed by the greedy procedure is not necessarily optimal. However, all modifications of the initial schedule do not increase the total number of transfers. Furthermore, as mentioned in the introduction, the complexity of computing an optimal solution w.r.t. f^Σ is still open.

4   Upper Bounds on the Number of Patient Transfers

In Sect. 2, we proved that for $C = 2$ the objective function $f^{\max}$ is bounded from above by 1. In this section, we will use this insight to determine upper bounds on the optimal value of f^Σ. As parameters, we use the number of patients P, the number of rooms R and the length of the planning horizon T. Bounds depending on P and R are helpful in operational decision-making as they take the actual situation into account. They are also helpful in tactical decision-making regarding one medical speciality department, as they reflect some of the specialities characteristics, e.g., the patients' length of stay. Bounds depending only on the length of the planning horizon T are helpful in general tactical or strategic decision-making as they hold for all wards regardless of their speciality.

First, we use the structural insights from Sect. 3 to determine an upper bound that uses the number of patients P and rooms R as parameters.

Theorem 3. *For R rooms with $C = 2$ and P patients, an optimal solution with at most $P - (2R - 1)$ patient transfers exists and this bound is tight for $R \geq 3$.*

Proof. Since we are interested in an upper bound on the total number of transfers among all instances with P patients, w.l.o.g., we may assume that in each considered instance no patient's stay can be extended to the left (i.e., the patient arrives earlier) without increasing the number of transfers.

For each instance, we consider an optimal proper solution. Remark that according to Lemma 1, no patient is transferred between the first two time periods. Let us now assume that a patient p is transferred between time periods 2 and 3. Since we may assume that patients arriving in the first time period are never transferred due to Theorem 2, p is not in hospital in time period 1. W.l.o.g., let p be female. Then there has to be a male patient assigned to the same room as p in time period 1 since otherwise the stay of p could be extended to the left, contradicting our assumption. Let r be the room to which p is assigned in time period 2. Note that all (male) patients assigned to r in time period 1 leave the hospital subsequently.

First, we prove that, w.l.o.g., we can assume that in order to make a transfer between time periods two and three necessary, at least all but one beds need to be occupied in the first time period, i.e., at least $2R - 1$ patients arrive in the first time period. Let us assume to the contrary that only $2R - 2$ beds are occupied in time period 1. If there is an empty room in period 1 or both free beds are in rooms with a male patient different from room r, we can move all patients from r to these beds in time period 1. But then the stay of p can be extended to the left, contradicting our assumption. Thus, in period 1 there are either i) a free bed in r as well as a free bed in a room with a female patient, or ii) two free beds in rooms with a female patient.

Case i): Let q be the female patient who occupies the bed to which p is assigned in time period 3. Then, q stays in hospital exactly for time periods 1 and 2. Further, let r' be a room with exactly one female patient in time period 1. Since the stay of the next patients in this room cannot be extended to the left, they have to be male. If those patients arrive in hospital after time period 2, we could assign q to r' in time periods 1 and 2, and hence the transfer of p would not be necessary, contradicting the optimality of our solution. On the other hand, if those patients arrive in hospital in time period 2, the female patient in r' in time period 1 leaves the hospital afterwards. This means that we can switch the patients in rooms r and r' in period 1, and then extend the stay of p to the left, contradicting our assumption.

Case ii): If two free beds in rooms with a female patient exist in time period 1, there must be two male patients in room r. By repeating the argumentation of Case i) twice, we again get contradictions. Thus, our assumption must be wrong and at least $2R - 1$ beds must be occupied in time period 1.

Finally, we conclude that no more than $P - (2R - 1)$ transfers are necessary since, according to Theorem 2, patients who arrive in the first time period do not have to be transferred.

The example in Fig. 5a shows that this bound is tight for $R = 3$. This example can be extended to any larger number of rooms by adding the required amount of rooms and patients who stay for the complete length of the planning horizon. These patients are obviously never transferred but, as they arrive in the first time period, the bound of $P - (2R - 1)$ is still valid. $\qquad\square$

Although we know that the bound from Theorem 3 is tight for $R \geq 3$, we do not know if this is also true for $R = 2$. We suspect, however, that a tight upper

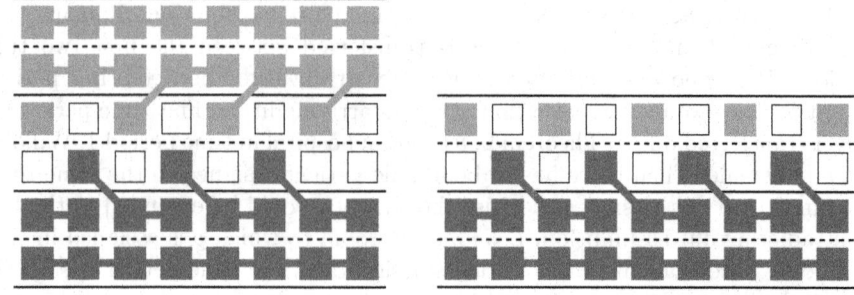

(a) $R = 3$ and $P - (2R-1)$ patient transfers, cf. Theorem 3

(b) $R = 2$ and $\frac{P-3}{2}$ patient transfers, cf. Conjecture 1

Fig. 5. Examples for tightness of bounds with R rooms and P patients

bound for f^Σ in the case $R = 2$ is much smaller. The worst example that we were able to construct, needs one transfer every two days resulting in a total of $\frac{P-3}{2}$ transfers, cf. Fig. 5b.

Conjecture 1. For $R = 2$ and $C = 2$ at most $\lfloor \frac{P-3}{2} \rfloor$ patient transfers are necessary.

Second, we use insights about the practical occurrence of patient transfers to determine tight upper bounds for f^Σ depending only on the length of the planning horizon T for $R \in \{2,3\}$.

We start by considering the situation of $R = 2$ rooms. In every room either no patient is present, one or two female, or one or two male patients are present. We call such rooms empty, female or male rooms according to the present patients' gender and mark them with -, F, and M, respectively.

Hence, in each time period $t \in \mathcal{T}$, there are six possible states of the two rooms: There are either two empty rooms (- -), one room with female patients and the other room empty (F-), one room with male patients and the other room empty (M-), two rooms with female patients (FF), two rooms with male patients (MM), or one room with female patients and one room with male patients (FM).

We introduce a directed state graph having the possible states as nodes and arcs denoting a change from one state to another in consecutive time periods, cf. Fig. 6a. Each arc is weighted with the maximal number of transfers which may be necessary to get from one state to the other. It is easy to see that these weights are 0 or 1, and that weight 1 only occurs on the arcs (FF, FM) and (MM, FM).

Lemma 6. *For $R = 2$ rooms with $C = 2$ and time horizon T, an optimal solution with at most $\lfloor \frac{T-1}{2} \rfloor$ patient transfers exists. Furthermore, there are instances achieving this bound.*

Proof. Since among the states - -, F-, and M- obviously no patient transfers occur, it is sufficient to consider the subgraph containing states FF, MM, and

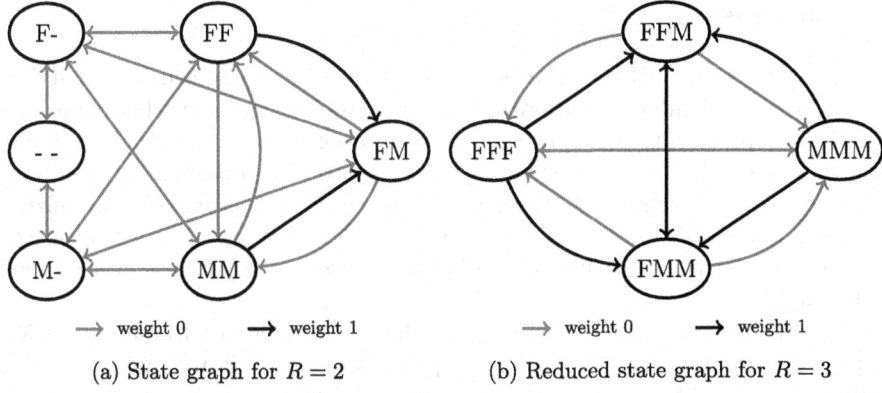

(a) State graph for $R = 2$ (b) Reduced state graph for $R = 3$

Fig. 6. State graphs

FM. Each solution of PRA with T time periods induces a directed path with $T - 1$ arcs in the state graph. If we are interested in a solution with a maximum number of patient transfers, we have to look for a path with $T - 1$ arcs having total maximum weight. Since, according to Lemma 1, always an optimal solution exists where no transfers occur between the first two time periods, and since in the state graph for any directed path each arc with weight 1 must be followed by an arc with weight 0, each solution needs at most $\lfloor \frac{T-1}{2} \rfloor$ transfers. For example, the cycle (FF, FM, FF, FM,...) has this weight, cf. Fig. 6a. That this bound is tight can be seen from the example in Fig. 5b, where $\lfloor \frac{T-1}{2} \rfloor$ patient transfers are necessary to obtain a feasible solution. $\qquad \square$

For $R = 3$, we define the states as triples of -, F and M, analogously to the case $R = 2$. We consider the corresponding reduced state graph limited to nodes where all rooms are occupied, as shown in Fig. 6b.

Lemma 7. *For $R = 3$ rooms with $C = 2$ and time horizon T, an optimal solution with at most $T - 2$ patient transfers exists. Furthermore, there are instances achieving this bound.*

Proof. As, in the case $R = 2$, each solution of PRA with T time periods induces a directed path with $T - 1$ arcs in the state graph. Since, according to Lemma 1, always an optimal solution exists where no transfers occur between the first two time periods, each solution needs at most $T - 2$ transfers. This corresponds to a path where each arc has weight 1. For example, the cycle alternating between the states FFM and FMM has this weight, cf. Fig. 6b. That this bound is tight, can be seen from the example in Fig. 5a, where $T - 2$ patient transfers are necessary to obtain a feasible solution. $\qquad \square$

Unfortunately, for $R \geq 4$ the state graphs and especially the computation of the arc weights get much more complex. Up to now, we were not able to derive upper bounds parametrized by T for a larger number of rooms.

5 Conclusion

Assigning patients to beds is an every-day task in hospital wards, but it has gained more and more attention in literature only recently. In this paper, we defined the underlying basic problem without tailoring it to the situation in one specific ward or hospital. We showed that if all rooms contain exactly two beds and gender-mixed rooms are forbidden, there exists an assignment minimizing the total number of transfers where at the same time, every patient is transferred at most once. Hence, no compromises between nurses' and patients' wishes need to be made.

In literature, e.g. [2], heuristics were already developed without discussion of potential limitations of the solution space. Our results prove that for double bedrooms optimality is not affected if we limit the solution space to assignments where every patient is transferred at most once. However, for three or more beds per room optimality with respect to the total number of transfers is lost. Apart from providing structural insights about PRA, we provided a constructive algorithm that converts any feasible assignment into one in which every patient is transferred at most once without increasing the total number of transfers.

In strategic or tactical capacity planning in hospitals, the number of transfers determines one important factor, e.g., on the workload of nurses. Our bounds provide estimates on this value and, hence, improve long term decision-making. In literature such bounds were missing so far or were just provided for a weighted sum approach combining work force needs and patient satisfaction. These are in general not suitable as input in the latter case.

Although our definition already fits the real-life task in most wards, sometimes needs and wishes of patients are considered additionally in the determination of room-mates. In future work, our aim is to use the structural insights presented in this paper to develop and test solution approaches that are applicable in real life and adaptable to different situations. Further, interesting aspects are Conjecture 1, finding an upper bound on f^Σ depending on T for $R \geq 4$, and the complexity of minimizing f^Σ and $f^{\max}$.

References

1. Abdalkareem, Z.A., Amir, A., Al-Betar, M.A., Ekhan, P., Hammouri, A.I.: Healthcare scheduling in optimization context: a review. Health Technol. **11**(3), 445–469 (2021). https://doi.org/10.1007/s12553-021-00547-5
2. Ceschia, S., Schaerf, A.: Local search and lower bounds for the patient admission scheduling problem. Comput. Oper. Res. **38**(10), 1452–1463 (2011)
3. Ceschia, S., Schaerf, A.: Modeling and solving the dynamic patient admission scheduling problem under uncertainty. Artif. Intell. Med. **56**(3), 199–205 (2012)
4. Ceschia, S., Schaerf, A.: Dynamic patient admission scheduling with operating room constraints, flexible horizons, and patient delays. J. Sched. **19**(4), 377–389 (2016). https://doi.org/10.1007/s10951-014-0407-8
5. Cochran, J.K., Bharti, A.: Stochastic bed balancing of an obstetrics hospital. Health Care Manag. Sci. **9**(1), 31–45 (2006). https://doi.org/10.1007/s10729-006-6278-6

6. Demeester, P., Souffriau, W., De Causmaecker, P., Vanden Berghe, G.: A hybrid tabu search algorithm for automatically assigning patients to beds. Artif. Intell. Med. **48**(1), 61–70 (2010)
7. Ficker, A.M.C., Spieksma, F.C.R., Woeginger, G.J.: The transportation problem with conflicts. Ann. Oper. Res. **298**(1), 207–227 (2021). https://doi.org/10.1007/s10479-018-3004-y
8. Guido, R., Groccia, M.C., Conforti, D.: An efficient matheuristic for offline patient-to-bed assignment problems. Eur. J. Oper. Res. **268**(2), 486–503 (2018)
9. Hammouri, A.I., Alrifai, B.: Investigating biogeography-based optimisation for patient admission scheduling problems. J. Theor. Appl. Inf. Technol. **70**(3), 413–421 (2014)
10. Harper, P.R.: A framework for operational modelling of hospital resources. Health Care Manag. Sci. **5**(3), 165–173 (2002). https://doi.org/10.1023/A:1019767900627
11. Harper, P.R., Shahani, A.K., Gallagher, J.E., Bowie, C.: Planning health services with explicit geographical considerations: a stochastic location-allocation approach. Omega **33**(2), 141–152 (2005)
12. Harrison, G.W., Shafer, A., Mackay, M.: Modelling variability in hospital bed occupancy. Health Care Manag. Sci. **8**(4), 325–334 (2005). https://doi.org/10.1007/s10729-005-4142-8
13. Huang, X.M.: A planning model for requirement of emergency beds. Math. Med. Biol. J. IMA **12**(3–4), 345–353 (1995)
14. Hübner, A., Kuhn, H., Walther, M.: Combining clinical departments and wards in maximum-care hospitals. OR Spectr. **40**(3), 679–709 (2018). https://doi.org/10.1007/s00291-018-0522-6
15. Hulshof, P.J.H., Kortbeek, N., Boucherie, R.J., Hans, E.W., Bakker, P.J.M.: Taxonomic classification of planning decisions in health care: a structured review of the state of the art in OR/MS. Health Syst. **1**(2), 129–175 (2012). https://doi.org/10.1057/hs.2012.18
16. Masterson, B.J., Mihara, T.G., Miller, G., Randolph, S.C., Forkner, M.E., Crouter, A.L.: Using models and data to support optimization of the military health system: a case study in an intensive care unit. Health Care Man Sci. **7**(3), 217–224 (2004). https://doi.org/10.1023/B:HCMS.0000039384.92373.c4
17. Range, T.M., Lusby, R.M., Larsen, J.: A column generation approach for solving the patient admission scheduling problem. Eur. J. Oper. Res. **235**(1), 252–264 (2014)
18. Roth, A.V., van Dierdonck, R.: Hospital resource planning: concepts, feasibility, and framework. Prod. Oper. Manag. **4**(1), 2–29 (1995)
19. Schmidt, R., Geisler, S., Spreckelsen, C.: Decision support for hospital bed management using adaptable individual length of stay estimations and shared resources. BMC Med. Inform. Decis. Making **13**(1), 1–19 (2013)
20. Vancroonenburg, W., Croce, F.D., Goossens, D., Spieksma, F.C.R.: The red-blue transportation problem. Eur. J. Oper. Res. **237**(3), 814–823 (2014)
21. Vancroonenburg, W., De Causmaecker, P., Spieksma, F.C.R., Berghe, G.V.: Scheduling elective patient admissions considering room assignment and operating theatre capacity constraints. In: Proceedings of the 5th International Conference on Applied Operational Research. Lecture Notes in Management Science vol. 5, pp. 153–158 (2013)

Tool Switching Problems in the Context of Overlay Printing with Multiple Colours

Manuel Iori[1] , Alberto Locatelli[1]([✉]) , Marco Locatelli[2] ,
and Juan-José Salazar-González[3]

[1] Department of Sciences and Methods for Engineering, University of Modena
and Reggio Emilia, Reggio Emilia, Italy
{manuel.iori,alberto.locatelli}@unimore.it
[2] Department of Engineering and Architecture, University of Parma, Parma, Italy
marco.locatelli@unipr.it
[3] Departament of Mathematics, Statistics and Operational Research,
University of La Laguna, Santa Cruz de Tenerife, Spain
jjsalaza@ull.edu.es

Abstract. This paper addresses problems arising in the context of overlay printing with multiple colours, where a finite set of jobs must be sequentially performed by a printing machine which can simultaneously accommodate a limited number of colours. Each job is associated with a subset of colours that the machine must have stored in its magazine before starting the execution. Thus, some colour switches may be required between the execution of two consecutive jobs. Since colour switches imply a reduction of productivity, minimizing them is desirable. In this regard, we address three distinct problems of increasing difficulty. All these problems can be seen as variants of the Tool Switching Problem, where each colour is treated as a tool. For each problem we discuss its complexity and propose a mathematical programming model. We evaluate the effectiveness of the models on several instances that have been generated with the aim of covering different scenarios of interest.

Keywords: Tool switching problem · Overlay printing · Complexity · Arc flow models

1 Introduction

We introduce a general class of combinatorial optimization problems arising in the context of overlay printing with multiple colours. We are given a set $J = \{1, \ldots, n\}$ of jobs to be processed sequentially on a single printing machine, a set $C = \{1, \ldots, m\}$ of colours, and a set $K = \{1, \ldots, k\}$ of slots available on the printing machine and used to allocate the colours. Each job $j \in J$ requires a subset of colours $C_j \subseteq C$ to be loaded in the magazine K in a job-dependent order before starting the execution of the job. We assume that $|C_j| \leq |K|$ holds for all $j \in J$. In practical situations, the magazine cannot usually hold all colours at

I. Ljubić et al. (Eds.): ISCO 2022, LNCS 13526, pp. 260–271, 2022.
https://doi.org/10.1007/978-3-031-18530-4_19

once (i.e., $|C| > |K|$), so that some colour switches may be necessary when performing two consecutive jobs. A colour switch implies washing the colour residue of the previous colour from the inner surface of the involved slot. The washing of a slot takes a long time, with the effect of significantly raising the setup time. After finishing the execution of a job and before starting the execution of the next job, one must decide for each slot whether to preserve the current colour or to switch to a different one. Preserving the current colour requires to fill the slot with the colour, which takes a negligible time. Instead, switching to a different colour has a significant impact on the setup time due to the cleaning process.

During the processing of a job, the printer sequentially applies the involved colours, stored in the magazine, on the printed material in accordance with its slot sequence. In many industrial printing technologies, the overlapping order of colours plays a crucial role in achieving print quality (see, e.g., [1]). Thus, the colours must be sorted along slots in the magazine so that the printer will apply them in the required order.

The overall problem is related to the so-called Tool Switching Problem (ToSP in what follows) where, for each job $j \in J$, one disregards the order of the colours and just decides into which slots its $|C_j|$ colours are loaded, in such a way that the overall setup times between consecutive jobs are minimized. We address three problem variants of increasing difficulty. For each variant we discuss its complexity and propose a mathematical programming model. In the second and third variants, we introduce the colour order constraint into the ToSP for preserving, during the processing of each job, the required overlapping order of colours.

To distinguish between the problem variants, we use different letters. More precisely, we use letter C to denote instances with *constant* setup times, and letter G for instances with *general* non-negative setup times. We use letter U for instances where for each job $j \in J$, the set C_j is *unordered*, i.e., the colours of job j can be placed in the slots in any order, and letter O for instances where the set C_j is *ordered*, i.e., the colours of job j are to be placed in the slots in the order specified in C_j, although not necessarily in consecutive slots. Finally, we use letter F for the instances where the sequence of jobs is *fixed*, and letter V for those where the sequences of jobs is *variable*.

The first and simplest problem is CUF-ToSP, where we assume uniform setup time for colour switching operations and fixed job sequence (i.e., job $j \in J$ is executed in position j of the schedule). The second addressed problem is GOF-ToSP, in which we assume non-uniform setup times for colour switching operations and fixed job sequence. In addition, in GOF-ToSP we assume that the colours required by a job must be applied by the printer in a given order, which means that they must be sorted along the slot sequence so that the printer will apply them in the required order. The third and most difficult problem is GOV-ToSP where the job sequence is not fixed. We also mention that a further variant (CUV-ToSP) is known as Job Sequencing and Tool Switching Problem (SSP) in the literature. Such problem has been addressed in different papers, including [3,10], and [2]. [3] provided a formal proof that SSP (aka CUV-ToSP) is NP-hard.

On the other hand, the first problem addressed in this paper, CUF-ToSP, can be solved in polynomial-time by means of a Keep Tool Needed Soonest policy (see [12]). For this problem, in Sect. 2 we show that the binary linear programming formulation of SSP proposed in [12] is, when specialized to CUF-ToSP, a perfect formulation, so that we also provide a different proof that the problem can be solved in polynomial time.

The second problem addressed in this paper, GOF-ToSP, is also the main motivation of this paper. Indeed, the problem originates from a real-world application arisen in a food packaging company located in the city of Reggio Emilia [7]. The specific printing technology used in this application is the flexographic print (see, e.g., [8]), which requires non-uniform setup times for colour switching operations and that the colours required by a job must be applied by the printer in a given order. Indeed, in flexographic printing, a violation of the chromatic order occurred during the printing of a job may lead to color deviations, color mixing, and reverse printing failures in the final products. In addition, non-uniform setup times for colour switching operations are caused by the fact that the process to wash a slot depends directly on the specific colours involved. For instance, if the process requires to switch a black ink cartridge with a white one, a complete cleaning of the corresponding slot is necessary to preserve the white purity, with the effect of a significant raise in the setup time. On the other hand, if the process requires to change a slot from a pale green ink to a dark green ink, then the setup time is lower because the dark green ink is not easily altered by the residuals of the pale green ink and, therefore, the relative slot needs just a partial washing. The real-world problem faced by the company generalizes the GOF-ToSP to a great extent (as it includes parallel heterogeneous machines, release and due dates, and several additional complicating features) and was solved by Iori et al. [7] by means of a constructive heuristic. For the GOF-ToSP, in Sect. 3 we state its $\mathcal{NP}$-hardness and present an arc flow model. Moreover, we provide a preprocessing method that allows us to potentially reduce the number of jobs.

The third problem addressed in this paper is GOV-ToSP, the generalization of GOF-ToSP in which the sequence of jobs is not given in advance. In Sect. 4, we state that GOV-ToSP is strongly $\mathcal{NP}$-hard and provide a descriptive arc flow model (see [11]).

In Sect. 5 we perform some preliminary experiments to evaluate the effectiveness of the proposed models for GOF-ToSP and GOV-ToSP. Such experiments have been carried out on a number of instances that have been generated starting from realistic data features and with the aim of covering different scenarios of interest. Finally, in Sect. 6 we draw some concluding remarks and provide some hints for future research.

2   CUF-ToSP

In CUF-ToSP, the order of the colours in the magazine is irrelevant, the job sequence is fixed in advance (thus, job j is executed in position j), and all colour switches have constant setup time, which, w.l.o.g., can be assumed equal to one (i.e., $c_{uv} = 1$ if $u \neq v$ and 0 if $u = v$). Thus, CUF-ToSP aims at determining the subsets of colours loaded in the magazine during the processing of each job $j \in J$, in order to minimize the total number of colour switches. For the notation it is convenient to extend J with a dummy job 0 ($C_0 = \emptyset$) obtaining $J = \{0, 1, \ldots, n\}$.

2.1 Two-Index Formulation for CUF-ToSP

The formulation that we propose for CUF-ToSP requires two sets of decision variables: a binary decision variable y_j^u taking the value 1 if during the printing of job j colour u is loaded on the magazine, and 0 otherwise; and a binary decision variable z_j^u taking the value 1 if, before starting job $j \in J$, colour u is switched with another colour loaded on the magazine. In other words, $z_j^u = 1$ corresponds to the occurrence of a colour switch. The formulation is then:

$$\min \sum_{u \in C} \sum_{j \in J \setminus \{0\}} z_j^u \tag{1}$$

$$\sum_{u \in C} y_j^u \leq |K| \qquad j \in J \tag{2}$$

$$y_j^u = 1 \qquad j \in J, u \in C_j \tag{3}$$

$$y_j^u - y_{j-1}^u \leq z_j^u \qquad j \in J \setminus \{0\}, u \in C \tag{4}$$

$$y_j^u \in \{0, 1\} \qquad j \in J, u \in C \tag{5}$$

$$z_j^u \in \{0, 1\} \qquad j \in J \setminus \{0\}, u \in C. \tag{6}$$

The objective function (1) minimizes the total number of colour switches. Constraints (2) guarantee that the magazine capacity is never exceeded, while constraints (3) guarantee that, during the printing of job $j \in J$, each colour $u \in C_j$ takes exactly a slot. Constraints (4) impose a switch operation when colour u is loaded during the printing of job j but is not loaded during the printing of job $j - 1$. We remark that the model above is the same that can be obtained from the model for SSP proposed in [12] after fixing in that model the values of the binary variables that define the sequence of jobs. Recalling that a perfect formulation of a set $S \subseteq \mathbb{R}^n$ is a linear system of inequalities $Ax \leq b$ such that the convex hull of S coincides with $\{x \in \mathbb{R}^n : Ax \leq b\}$, we state the following theorem, which is also an alternative proof of the fact that CUF-ToSP is solvable in polynomial time.

Theorem 1. *A perfect formulation of the set defined by constraints (2)–(6) is obtained by relaxing the binary requirements $y_j^u, z_j^u \in \{0, 1\}$ into $y_j^u, z_j^u \in [0, 1]$.*

Sketch of Proof. Due to space limitations, we do not give a full proof of the result. The proof is based on the observation that the matrix associated with constraints (2)–(4) is Totally Unimodular (TU), which can be seen by exploiting the Ghouila-Houri's characterization of TU matrices (see [6]).

3   GOF-ToSP

As already mentioned in the introduction, the chromatic order imposed by the flexographic printing technology is a crucial aspect in practice. To model this aspect, we introduce, in each set $C_j \subset C$, a precedence relation $\prec^j$. We read $u \prec^j v$ as "colour u must be in the slot sequence before colour v when the printer executes job j", for each $u, v \in C_j$. Thus, each job $j \in J$ is associated with a strict partially ordered set $(C_j, \prec^j)$ of colours. In GOF-ToSP, we are also given a setup time c_{uv} required to switch from colour u to colour v. More specifically, for all $u, v \in C$, the setup time c_{uv} is always non-negative and it is negligible if the two colours u and v coincide (i.e., $c_{uu} = 0$ for all $u \in C$). Moreover, we assume the triangular inequality holds for the setup times (i.e., $c_{uw} \leq c_{uv} + c_{vw}$ for all $u, v, w \in C$). In our problem, we assume that the job sequence is fixed in advance. GOF-ToSP consists of determining the assignment of the colours to the slots of the magazine during the processing of each job $j \in J$, with the aim of minimizing the total setup time. The following theorem states the complexity of GOF-ToSP in the special case of binary setup times.

Theorem 2. *GOF-ToSP is NP-hard even in case of binary setup times (i.e., $c_{uv} \in \{0, 1\}$ for all $u, v \in C$).*

Sketch of Proof. Again due to space limitations, we do not report the lengthy proof of this result, but we only mention the polynomial-time reduction on which it is based, namely that of the Disjoint Matching (DM) problem into GOF-ToSP with binary setup times. In DM, given two sets P, Q with equal cardinality and given $A_1, A_2 \in P \times Q$, we would like to establish whether there exist two disjoint matchings $M_1 \subseteq A_1$ and $M_2 \subseteq A_2$ or not. The DM problem has been proved to be NP-complete in [5] (see also [4] for a simpler proof). It is possible to reduce each DM instance into an instance of GOF-ToSP with binary setup times whose optimal value is equal to 0 if and only if DM admits a **yes** answer.

It is interesting to remark that while in the case of fixed setup times and unordered C_j a fixed sequence of jobs simplifies the problem (CUF-ToSP is solvable in polynomial-time, while CUV-ToSP is NP-hard), a fixed sequence of colours in the slots (as in GOF-ToSP with binary setup times) makes the problem hard even when the sequence of jobs is fixed.

3.1   Five-Index Arc Flow Formulation for GOF-ToSP

Given a job sequence J, a solution of GOF-ToSP can be represented as $|K|$ paths from a source to a destination visiting disjoint nodes of a specific network $\mathcal{N}$. Let $J^* = \{0, 1, \ldots, n+1\}$, where the two additional jobs 0 and $n+1$ are copies of an empty job ($C_0 = C_{n+1} = \emptyset$).

We consider the network $\mathcal{N}$ to be composed of:

- a directed graph $G = (\mathcal{V}; \mathcal{A})$, where $\mathcal{V}$ is composed of all couples (u, j) with $j \in J^*$ and $u \in C_j$, and $\mathcal{A}$ includes all arcs from node $(u, i) \in \mathcal{V}$ to node $(v, j) \in \mathcal{V}$ such that $i \in J^* \setminus \{n+1\}$, $j \in J^* \setminus \{0\}$, and $j > i$;

– a cost function $w : \mathcal{A} \to \mathbb{R}^+$ that maps each arc $((u,i),(v,j)) \in \mathcal{A}$ to a cost c_{uv} (i.e., the setup time required for switching colour u of job i with colour v of job j).

The formulation considers a copy of graph G for each slot $k \in K$ and requires a binary variable $\phi^k_{((u,i),(v,j))}$ taking the value 1 if colour $u \in C_i$ is loaded in the k-th slot and is switched with colour $v \in C_j$, 0 otherwise. We obtain:

$$\min \sum_{k \in K} \sum_{((u,i),(v,j)) \in \mathcal{A}} c_{uv} \phi^k_{((u,i),(v,j))} \tag{7}$$

$$
\sum_{\substack{(v,j): \\ ((v,j),(u,i)) \in \mathcal{A}}} \phi^k_{((v,j),(u,i))} - \sum_{\substack{(v,j): \\ ((u,i),(v,j)) \in \mathcal{A}}} \phi^k_{((u,i),(v,j))}
$$
$$
= \begin{cases} -1, & \text{if } i = 0, \\ 1, & \text{if } i = n+1, \\ 0, & \text{otherwise,} \end{cases} \qquad k \in K, (u,i) \in \mathcal{V} \tag{8}
$$

$$\sum_{k \in K} \sum_{\substack{(v,j): \\ ((u,i),(v,j)) \in \mathcal{A}}} \phi^k_{((u,i),(v,j))} = 1 \qquad (u,i) \in \mathcal{V} : i \notin \{0, n+1\} \tag{9}$$

$$
\sum_{h \geq k} \sum_{\substack{(u,i): \\ ((u,i),(v,j)) \in \mathcal{A}}} \phi^h_{((u,i),(v,j))}
$$
$$
+ \sum_{h < k} \sum_{\substack{(u,i): \\ ((u,i),(w,j)) \in \mathcal{A}}} \phi^h_{((u,i),(w,j))} \leq 1 \qquad \begin{aligned} & k \in K, \ (v,j),(w,j) \in \mathcal{V} : \\ & v \prec^j w \end{aligned} \tag{10}
$$

$$\phi^k_{((u,i),(v,j))} \in \{0,1\} \qquad ((u,i),(v,j)) \in \mathcal{A}, k \in K. \tag{11}$$

The objective function (7) minimizes the total setup times; flow conservation is imposed by constraints (8); constraints (9) guarantee that, during the printing of job $j \in J$, each colour required by j takes exactly a slot; constraints (10) guarantee that the precedence relation $\prec^j$ at each job j is fulfilled.

3.2 Preprocessing

We introduce a method that relies on the triangular inequality assumption for the setup times and allows to potentially reduce the number of jobs in an instance. Let us introduce the concept of *mergeable* job. A job i is said to be *mergeable* in job j if $C_i \subseteq C_j$ and, for each $u, v \in C_i$ such that $u \prec^i v$, it also holds that $u \prec^j v$. In such a case, if job i is next to job j, it is possible to *merge* job i into job j, i.e., consider them as a unique job, having the characteristics of job j.

Proposition 1 *Let us consider an instance of GOF-ToSP for which the triangular inequality holds. Assume job $j + 1$ is mergeable in job j. Then, there exists an optimal solution without any switching operation between jobs j and $j + 1$, i.e., we can merge job $j + 1$ into j. Similarly, if job j is mergeable in job $j + 1$, then, there exists an optimal solution without any switching operation between jobs j and $j + 1$, i.e., we can merge job j into $j + 1$.*

Sketch of Proof. Assume job $j + 1$ is mergeable in job j. An optimal solution S defines, for each job $i \in J$ and each slot $k \in K$, a colour u_i^k which lies in the k-th slot during the printing of job i. Now, we consider the solution $\overline{S}$ obtained from S imposing that $u_j^k = u_{j+1}^k$ for each $k \in K$. In view of the triangular inequality assumption, objective function value of the feasible solution $\overline{S}$ is at least as good as the objective function value of the optimal solution S. Thus, $\overline{S}$ is also an optimal solution, for which no switching operation occurs between jobs j and $j + 1$. The proof for the other case in which j is mergeable in $j + 1$ is analogous.

The operation of merging jobs can be iteratively repeated. Thus, given a job j, we merge jobs $j + 1, \ldots, j + h$ if they are all mergeable into j. Similarly, we merge jobs $j - 1, \ldots, j - r$ if they are all mergeable into j.

4   GOV-ToSP

GOV-ToSP is the generalization of GOF-ToSP in which the job sequence is not fixed. Concerning its complexity we can prove the following result.

Theorem 3. *GOV-ToSP is strongly NP-hard even when $|K| = 1$.*

Sketch of Proof. This result easily follows from the observation that there exists a polynomial-time reduction of the Traveling Salesman Problem (see, e.g., [9]) into GOV-ToSP with a single slot.

Note that the above result also states the NP-hardness of GUV-ToSP, since for $|K| = 1$ the order of the colours is irrelevant.

4.1   Six-Index Arc Flow Formulation for GOV-ToSP

A solution of GOV-ToSP can be represented as $|K|$ paths from a source to a destination visiting disjoint nodes of a specific network $\mathcal{N}$. In this section, we present an arc flow formulation that contains an additional time-index with respect to the previous formulation of Sect. 3.1. In doing so, the planning horizon is discretized into $|T|$ time periods (where $T = \{1, \ldots, n\}$), each corresponding to a time period in which a job can be processed. Let T^* be the set of time periods in T with two additional times 0 and $n + 1$, i.e., the processing times of dummy jobs 0 and $n + 1$, respectively.

Since in GOV-ToSP the job sequence is not fixed, the directed graph $G = (\mathcal{V}; \mathcal{A})$ of $\mathcal{N}$ is different from the one defined for the GOF-ToSP model. In this case, $\mathcal{V} = \{(u, i, t) : u \in C, i \in J, t \in T\} \cup \{(0, 0, 0), (0, n + 1, n + 1)\}$, where additional vertices $(0, 0, 0)$ and $(0, n + 1, n + 1)$ correspond to the source and the sink of all paths, respectively. Set $\mathcal{A}$ includes all arcs from nodes $(u, i, t) \in \mathcal{V}$ to nodes $(v, j, t + 1) \in \mathcal{V}$ such that $i \in J^* \setminus \{n + 1\}$, $j \in J^* \setminus \{0, i\}$, and $t \in T^* \setminus \{n + 1\}$. Similarly to the GOF-ToSP model, a cost function $w : \mathcal{A} \rightarrow \mathbb{R}^+$ maps each arc $((u, i, t), (v, j, t + 1)) \in \mathcal{A}$ to a cost c_{uv}.

The following formulation considers a copy of graph G for each slot $k \in K$ and requires a binary variable $\phi_{((u,i,t),(v,j,t+1))}^k$ taking the value 1 if between time

t and $t + 1$ we switch from job i to job j and from colour u to colour v in the k-th color group.

$$\min \sum_{((u,i,t),(v,j,t+1))\in\mathcal{A}} c_{uv} \sum_{k\in K} \phi^k_{((u,i,t),(v,j,t+1))} \tag{12}$$

$$\sum_{\substack{(v,j,t-1):\\((v,j,t-1),(u,i,t))\in\mathcal{A}}} \phi^k_{((v,j,t-1),(u,i,t))} - \sum_{\substack{(v,j,t+1):\\((u,i,t),(v,j,t+1))\in\mathcal{A}}} \phi^k_{((u,i,t),(v,j,t+1))}$$

$$= \begin{cases} -1, & \text{if } i = 0, \\ 1, & \text{if } i = n+1, \\ 0, & \text{otherwise,} \end{cases} \qquad \begin{array}{l} k \in K, \\ (u,i,t) \in \mathcal{V} \end{array} \tag{13}$$

$$\sum_{k\in K} \sum_{(u,i,t)\in\mathcal{V}} \sum_{\substack{(v,j,t+1):\\((u,i,t),(v,j,t+1))\in\mathcal{A}}} \phi^k_{((u,i,t),(v,j,t+1))} = 1 \qquad i \in J, u \in C_i \tag{14}$$

$$\sum_{h\geq k} \sum_{\substack{(u,i,t-1):\\((u,i,t-1),(v,j,t))\in\mathcal{A}}} \phi^h_{((u,i,t-1),(v,j,t))}$$

$$+ \sum_{h<k} \sum_{\substack{(u,i,t-1):\\((u,i,t-1),(w,j,t))\in\mathcal{A}}} \phi^h_{((u,i,t-1),(w,j,t))} \leq 1 \qquad \begin{array}{l} k \in K, \\ (v,j,t),(w,j,t)\in\mathcal{V}: \\ v \prec^j w \end{array} \tag{15}$$

$$\sum_{k\in K} \sum_{\substack{(v,j,t+1):\\((u,i,t),(v,j,t+1))\in\mathcal{A}}} \phi^k_{((u,i,t),(v,j,t+1))}$$

$$+ \sum_{k\in K} \sum_{\substack{(v,j,t'+1):\\((u',i,t'),(v,j,t'+1))\in\mathcal{A}}} \phi^k_{((u',i,t'),(v,j,t'+1))} \leq 1 \qquad \begin{array}{l} (u,i,t)\in\mathcal{V}, \\ (u',i,t')\in\mathcal{V}: t \neq t' \end{array} \tag{16}$$

$$\phi^k_{((u,i,t),(v,j,t+1))} \in \{0,1\} \qquad \begin{array}{l} k \in K, \\ ((u,i,t),(v,j,t+1))\in\mathcal{A}. \end{array} \tag{17}$$

The objective function (12) minimizes the total setup times; constraints (13) impose flow conservation; constraints (14) guarantee that, during the printing of job $j \in J$, each colour required by the job takes exactly a slot; constraints (15) guarantee that the precedence relation between colours in C_j is respected for all jobs $j \in J$; constraints (16) guarantee a consistent sequencing of jobs, i.e., among all nodes associated to a given job i, only those corresponding to a single time period can have a positive out-flow.

5 Computational Results

This section illustrates preliminary computational experiments designed and conducted to assess the effectiveness of the proposed models, and analyze their behavior. The formulations of Sects. 2.1, 3.1, and 4.1 have been implemented in Phyton by using Gurobi 9.1.2 as integer linear programming solver. The resulting algorithms are called CUF-Alg, GOF-Alg, and GOV-Alg, respectively. The tests have been executed on a 1.8 GHz Intel Core i7-10510U with 16 GB of memory. We let Gurobi use all the four available threads. The CPU time limit was set to 600 s.

5.1   Test Instances

To evaluate the effectiveness of the models, a number of instances have been generated with the aim of covering different scenarios of interest. It has been necessary to do this as the instances for CUV-ToSP available in the literature (see [2]) are difficult to adapt and extend to our problems.

The generation of a set of instances has the following input parameters: the number $|J|$ of jobs, the number $|C|$ of available colours, and the number $|K|$ of available slots. For each instance, $|J|$ jobs are created as follows: the number l of colours required by a job j is an integer value uniformly distributed in $[1; |K|]$; and the set C_j is generated by randomly selecting l different colours from C and randomly ordering them. A job is added to J if it is not mergeable in any job previously added to J. Thus, the preprocessing method described in Sect. 3.2 is used to generate instances that cannot be reduced in terms of number of jobs. Computational tests were performed on a set of 50 instances generated with various combinations of $|J| \in \{5, 10\}$, $|C| \in \{5, 20, 30\}$, and $|K| \in \{2, 5, 7, 8, 9\}$. The generated instances are divided into 5 groups of 10 instances each having the same number of jobs, colours, and slots.

5.2   Results

It follows from Theorem 1 that the CUF-ToSP model can be solved as a Linear Program (LP) and therefore large instances can be solved in very short run-time. For the sake of completeness, we report that all the proposed instances are solved in less than one second by means of CUF-Alg. Tables 1 and 2 highlight the results obtained by GOF-Alg and GOV-Alg, respectively. Entries in columns z^* and LB exhibit the integer optimal solution value (or the value of the best solution) and the final lower bound obtained by the search tree, respectively. The percentage gap, showed in column GAP, is computed as $100(z^* - LB)/z^*$, while column sec exhibits the run-time required to solve the instances (expressed in seconds). We also provide some information about the continuous relaxations of the models: z_{LP} is the value of the optimal solution at the root node of the tree, i.e., the linear relaxation optimal solution value (which we obtained by turning off the Gurobi presolve functionality); GAP_{LP} is computed as $100(z^* - z_{LP})/z^*$; and sec_{LP} exhibits the run-time required to solve the continuous relaxation (expressed in seconds). Each row in the table gives the results obtained on an instance, in addition, for each group, entries in row AVG exhibit the average values of each column. Symbol "–" indicates that no feasible solution was produced by the algorithm. Table 1 shows that GOF-Alg is able to find a solution within the time limit for almost all instances. It is worth observing that the size of proposed instances is compatible with real data (see [7]). Indeed, a flexographic printer is usually capable of holding at most 9 colours at a time, 10 jobs corresponds approximately to a week of production, and no more than 30 colours are usually required for printing 10 different jobs. As expected, the run-time increases with the increase of the instance size.

Table 1. Computational results for GOF-ToSP

Inst.	$|J|$	$|C|$	$|K|$	z^*	LB	GAP	sec	z_{LP}	GAP_{LP}	sec_{LP}
Ins01_5_5_2	5	5	2	53.0	53.0	0%	0.02	53.0	0%	0.02
Ins02_5_5_2	5	5	2	69.0	69.0	0%	0.02	69.0	0%	0.01
Ins03_5_5_2	5	5	2	57.0	57.0	0%	0.02	57.0	0%	0.01
Ins04_5_5_2	5	5	2	114.0	114.0	0%	0.02	114.0	0%	0.02
Ins05_5_5_2	5	5	2	109.0	109.0	0%	0.02	109.0	0%	0.01
Ins06_5_5_2	5	5	2	93.0	93.0	0%	0.01	93.0	0%	0.01
Ins07_5_5_2	5	5	2	54.0	54.0	0%	0.01	54.0	0%	0.01
Ins08_5_5_2	5	5	2	63.0	63.0	0%	0.01	63.0	0%	0.01
Ins09_5_5_2	5	5	2	97.0	97.0	0%	0.01	97.0	0%	0.01
Ins10_5_5_2	5	5	2	86.0	86.0	0%	0.01	86.0	0%	0.01
AVG				79.5	79.5	0%	0.02	79.5	0%	0.01
Ins01_10_20_5	10	20	5	285.0	285.0	0%	6.44	129.0	55%	0.17
Ins02_10_20_5	10	20	5	95.0	95.0	0%	0.8	51.0	46%	0.04
Ins03_10_20_5	10	20	5	207.0	207.0	0%	1.49	149.0	28%	0.13
Ins04_10_20_5	10	20	5	361.0	361.0	0%	6.24	208.0	42%	0.31
Ins05_10_20_5	10	20	5	74.0	74.0	0%	0.81	53.0	28%	0.03
Ins06_10_20_5	10	20	5	166.0	166.0	0%	1.2	110.0	34%	0.09
Ins07_10_20_5	10	20	5	230.0	230.0	0%	1.6	159.0	31%	0.18
Ins08_10_20_5	10	20	5	158.0	158.0	0%	1.26	100.0	37%	0.10
Ins09_10_20_5	10	20	5	406.0	406.0	0%	5.21	170.0	58%	0.31
Ins10_10_20_5	10	20	5	325.0	325.0	0%	2.95	202.0	38%	0.26
AVG				230.7	230.7	0%	2.8	133.1	40%	0.16
Ins01_10_30_7	10	30	7	230.0	230.0	0%	5.19	117.0	49%	0.25
Ins02_10_30_7	10	30	7	201.0	201.0	0%	23.76	100.0	50%	0.20
Ins03_10_30_7	10	30	7	350.0	350.0	0%	43.19	136.0	61%	0.45
Ins04_10_30_7	10	30	7	240.0	240.0	0%	57.96	111.0	54%	0.25
Ins05_10_30_7	10	30	7	524.0	524.0	0%	135.73	165.0	69%	0.47
Ins06_10_30_7	10	30	7	319.0	319.0	0%	32.02	147.0	54%	0.36
Ins07_10_30_7	10	30	7	308.0	308.0	0%	99.38	144.0	53%	0.41
Ins08_10_30_7	10	30	7	248.0	248.0	0%	18.78	103.0	58%	0.19
Ins09_10_30_7	10	30	7	157.0	157.0	0%	9.39	86.0	45%	0.10
Ins10_10_30_7	10	30	7	238.0	238.0	0%	3.74	130.0	45%	0.31
AVG				281.5	281.5	0%	42.91	123.9	54%	0.3
Ins01_10_30_8	10	30	8	157.0	157.0	0%	143.44	85.0	46%	0.10
Ins02_10_30_8	10	30	8	57.0	57.0	0%	6.6	39.0	32%	0.08
Ins03_10_30_8	10	30	8	235.0	235.0	0%	34.29	102.0	57%	0.23
Ins04_10_30_8	10	30	8	235.0	235.0	0%	73.76	114.0	51%	0.24
Ins05_10_30_8	10	30	8	251.0	251.0	0%	314.8	122.0	51%	0.65
Ins06_10_30_8	10	30	8	379.0	379.0	0%	23.1	186.0	51%	0.86
Ins07_10_30_8	10	30	8	200.0	200.0	0%	247.44	79.0	60%	0.33
Ins08_10_30_8	10	30	8	521.0	521.0	0%	454.34	169.0	68%	1.29
Ins09_10_30_8	10	30	8	489.0	489.0	0%	381.67	219.0	55%	0.79
Ins10_10_30_8	10	30	8	196.0	196.0	0%	186.01	80.0	59%	0.15
AVG				272.0	272.0	0%	186.54	119.5	53%	0.47
Ins01_10_30_9	10	30	9	256.0	256.0	0%	202.67	103.0	60%	0.58
Ins02_10_30_9	10	30	9	−	196.2	-	600.04	158.0	-	2.52
Ins03_10_30_9	10	30	9	340.0	340.0	0%	283.62	135.0	60%	0.68
Ins04_10_30_9	10	30	9	473.0	473.0	0%	285.96	190.0	60%	1.95
Ins05_10_30_9	10	30	9	275.0	275.0	0%	314.18	87.0	68%	0.49
Ins06_10_30_9	10	30	9	−	194.5	-	600.04	146.0	-	0.95
Ins07_10_30_9	10	30	9	474.0	354.5	34%	600.06	175.0	63%	1.44
Ins08_10_30_9	10	30	9	203.0	154.3	32%	600.02	97.0	52%	0.64
Ins09_10_30_9	10	30	9	55.0	55.0	0%	13.69	34.0	38%	0.15
Ins10_10_30_9	10	30	9	217.0	217.0	0%	62.61	105.0	52%	0.73
AVG				286.6	251.7	8%	356.29	123.0	57%	1.01

It follows from Theorem 3 that GOV-ToSP is the most difficult problem in theory. This result is also confirmed in practice by the computational experiments. GOV-Alg is able to find the minimum solution within the time limit just for the first and smaller group of instances (i.e., the instances reported in Table 2). For all the other instances, GOV-Alg failed to find even a feasible solution within the time limit, mainly because of the large number of variables and constraints of the model.

Table 2. Computational results for GOV-ToSP

Inst.	$|J|$	$|C|$	$|K|$	z^*	LB	GAP	sec	z_{LP}	GAP_{LP}	sec_{LP}
Ins01_5_5_2	5	5	2	14.0	14.0	0%	1.47	7.0	50%	0.27
Ins02_5_5_2	5	5	2	20.0	20.0	0%	1.37	10.0	50%	0.28
Ins03_5_5_2	5	5	2	23.0	23.0	0%	0.79	12.0	48%	0.26
Ins04_5_5_2	5	5	2	34.0	34.0	0%	1.96	0.0	100%	0.24
Ins05_5_5_2	5	5	2	24.0	24.0	0%	1.37	4.0	83%	0.35
Ins06_5_5_2	5	5	2	30.0	30.0	0%	1.69	0.0	100%	0.29
Ins07_5_5_2	5	5	2	19.0	19.0	0%	2.37	2.0	89%	0.23
Ins08_5_5_2	5	5	2	32.0	32.0	0%	2.11	7.0	78%	0.38
Ins09_5_5_2	5	5	2	13.0	13.0	0%	0.65	10.0	23%	0.35
Ins10_5_5_2	5	5	2	22.0	22.0	0%	0.61	3.0	86%	0.30
AVG				23.1	23.0	0%	1.44	5.5	64%	0.29

6   Conclusions and Future Research

In this paper, we addressed three distinct combinatorial optimization problems of increasing difficulty: CUF-ToSP, GOF-ToSP, and GOV-ToSP. The three problems arise in the context of overlay printing with multiple colours. In particular, GOF-ToSP is motivated by a real-world application in a company which operates in the field of packaging industry by producing and printing packaging materials for food products. For each problem we discussed its complexity and proposed a mathematical programming model. Experiments over realistic instances confirmed the theoretical results: GOV-ToSP is the most difficult problem and the proposed method is able to find the minimum solution within time limit just for the first and smaller group of instances. On the other hand, the proposed approach to solve GOF-ToSP is able to find optimal solutions in a short time and thus can provide a quick support to the company on its weekly decisions. It is worthwhile to remark that the computing times are currently not a major issue for the company. Indeed, taking into account that the planned activities of the proposed instances cover about one week, larger computing times are still feasible in practice. We also tried to implement alternative formulations for GOV-ToSP, having less variables and constraints: a four-index formulation and a five-index arc flow formulation. The behavior of these two models was worse than that of the GOV-Alg. Indeed, their GAP_{LP} was most of the time equal to 100%.

As a possible topic for future research, we are interested in introducing new methods for strengthening the proposed formulations for GOV-ToSP and GOF-ToSP (e.g., by means of valid inequalities).

References

1. Boora, S., Verma, S.: To study print contrast variation on art and map Litho paper due to effect of ink sequence. Int. J. Sci. Eng. Comput. Technol. **7**, 125–127 (2017)
2. Calmels, D.: The job sequencing and tool switching problem: state-of-the-art literature review, classification, and trends. Int. J. Prod. Res. **57**, 5005–5025 (2019)
3. Crama, Y., Kolen, A., Oerlemans, A., Spieksma, F.: Minimizing the number of tool switches on a flexible machine. Int. J. Flex. Manuf. Syst. **6**, 33–54 (1994). https://doi.org/10.1007/BF01324874
4. Fon-Der-Flaass, D.G.: Arrays of distinct representatives-a very simple NP-complete problem. Discrete Math. **1**, 295–298 (1997)
5. Frieze, A.M.: Complexity of a 3-dimensional assignment problem. Eur. J. Oper. Res. **13**, 161–164 (1983)
6. Ghouila-Houri, A.: Caractérisation des matrices totalement unimodulaires. Comptes Rendus Hebdomadaires des Séances de l'Académie des Sciences (Paris) **254**, 1192–1194 (1962)
7. Iori, M., Locatelli, A., Locatelli, M.: Scheduling of parallel print machines with sequence-dependent setup costs: a real-world case study. In: Dolgui, A., Bernard, A., Lemoine, D., von Cieminski, G., Romero, D. (eds.) APMS 2021. IAICT, vol. 631, pp. 637–645. Springer, Cham (2021). https://doi.org/10.1007/978-3-030-85902-2_68
8. Kipphan, H.: Handbook of Print Media: Technologies and Production Methods. Springer, Heidelberg (2001). https://doi.org/10.1007/978-3-540-29900-4
9. Laporte, G.: The traveling salesman problem: an overview of exact and approximate algorithms. Eur. J. Oper. Res. **59**, 231–247 (1992)
10. Laporte, G., Salazar-González, J., Semet, F.: Exact algorithms for the job sequencing and tool switching problem. IIE Trans. **36**, 37–45 (2004)
11. de Lima, V.L., Alves, C., Clautiaux, F., Iori, M., de Carvalho, J.M.V.: Arc flow formulations based on dynamic programming: theoretical foundations and applications. Eur. J. Oper. Res. **296**, 3–21 (2022)
12. Tang, C.S., Denardo, E.V.: Models arising from a flexible manufacturing machine, part I: minimization of the number of tool switches. Oper. Res. **36**, 767–777 (1988)

Optimal Vaccination Strategies for Multiple Dose Vaccinations

Jenny Segschneider[(✉)] and Arie M. C. A. Koster

Research Area Discrete Optimization, RWTH Aachen, Aachen, Germany
{segschneider,koster}@math2.rwth-aachen.de

Abstract. Due to the COVID-19 pandemic and the shortage of vaccinations during its roll-out, the question regarding the best strategy to achieve immunity in the population by adjusting the time between the two necessary vaccination doses was intensively discussed. This strategy has already been studied from various angles by various researches. However, the combinatorial optimization problem and its complexity has not been the focus of attention.

In this paper, we study the complexity of different versions of this problem by first proposing a simple approach using a matching algorithm. Then, we extend the approach by adding constraints and multiple manufacturers. Finally, we discuss a variation of the problem where three vaccinations are necessary, including the so-called "booster". This problem turns out to be NP-hard.

Keywords: Computational complexity · Graph and network algorithms

1 Introduction

The ongoing COVID-19 pandemic and the shortage of vaccinations at the beginning of their roll-out highlighted the importance of optimization problems considering allocation and scheduling of vaccines. Duijzer et al. [2] gave an overview of these problems in the literature and Wouters et al. in [11] with focus on the ongoing pandemic. For most available vaccines, each person has to receive two doses in a specific time frame, i.e., the second dose has to be given three to six weeks after the first dose. This leads to another problem: which doses are given to new patients as a first dose and which are used as a second dose for those who already received a first dose. We call this problem the Two-Dose Scheduling Problem. Most countries delayed the second dose as much as possible in order to achieve higher partial immunity by one dose only, for example Great-Britain and Canada. Other countries, like Israel and the United States, administered both doses as soon as possible in accordance with the phase 3 clinical trials (i.e., [7] for the Comirnaty vaccine from BioNTech and Pfizer).

The vaccination strategy has been subject of extensive research in the recent year. First, Jurgens and Lackner [3] examined the effect of delaying the second

© The Author(s), under exclusive license to Springer Nature Switzerland AG 2022
I. Ljubić et al. (Eds.): ISCO 2022, LNCS 13526, pp. 272–283, 2022.
https://doi.org/10.1007/978-3-031-18530-4_20

dose to 6, 9 or 12 weeks after the first dose. Silva et al. [10] proposed different so-called SEIR models (compartmental models using the compartments Susceptible, Exposed, Infectious and Recovered) to minimize the number of expected deaths due to the pandemic and included different subpopulations. Parino et al. [6] extended an SIR model by adding different stages and applied it to the COVID-19 pandemic in Italy. Lastly, Moghadas et al. [5] proposed an agent-based model using data from the United States. However to the best of our knowledge, there have been no publications regarding the complexity of the combinatorial optimization problem addressed in this work.

In this paper, we will study the Two-Dose scheduling problem to see how the problem can be modeled as a graph optimization problem and to determine the complexity of this and related problems. We introduce novel b-matching formulations for the problem and two of its three variants of the problem, one with relaxed capacities and one with different vaccines. In the meantime, these vaccinations also require a third dose, the so-called booster shot. This leads to the Three-Dose Scheduling Problem for which we prove NP-hardness.

This paper is organized as follows. In the next Sect. 2, we define the Two-Dose Scheduling problem and introduce a formulation using an Integer Linear Program. In Sect. 3, we model three variations of the problem using a b-matching Approach. Finally, we show NP-hardness of the related Three-Dose Scheduling problem in Sect. 4.

2   Problem Description and Formulation

The Two-Dose Problem scheduling problem is defined on a time horizon of n discrete time steps, e.g., weeks. For each time step $i = 1, \ldots, n$, we denote the number of delivered doses by b_i. The number of time steps between the first and second dose has to be between $\varepsilon_{\min}$ and $\varepsilon_{\max}$. Let $J(i) = \{i + \varepsilon_{\min}, \ldots, \min\{n, i + \varepsilon_{\max}\}\}$ be the set of possible time steps for a second dose fitting for a first dose in time step $i \in \{1, \ldots, n - \varepsilon_{\min}\}$. For $i \in \{n - \varepsilon_{\min} + 1, \ldots, n\}$, it is $J(i) = \emptyset$ For each $i = 1, \ldots, n$ and $j \in J(i)$, $s_{i,j}$ refers to the number of people who get the first dose in time step i and the second dose in time step j and we call the tuple (i, j) a (feasible) appointment. An appointment is feasible if and only if the time constraint $j \in J(i)$ is kept. The goal of the Two-Dose Scheduling Problem is to maximize the impact of the vaccines on the society as a whole. For this, we introduce a value $c_{i,j}$ to measure the efficacy of the first and second dose. The value $c_{i,j}$ can be computed on the basis of the studies mentioned in the introduction. We set $c_{i,j} = 0$ if (i, j) is not a feasible appointment.

Finally, for each time step $i = 1, \ldots, n$, we define with u_i the maximum capacity of the storage and with ν_i the maximum number of vaccinations that can be administered during the time step. These definitions lead to the following ILP formulation:

$$\max \sum_{i=1}^{n} \sum_{j \in J(i)} c_{i,j} s_{i,j} \tag{1}$$

$$\text{s.t.} \sum_{i=1}^{t} \left(2 \times \sum_{j=i+\varepsilon_{\min}}^{t} s_{i,j} + \sum_{j=t+1}^{\min\{n,t+\varepsilon_{\max}\}} s_{i,j} \right) \leq \sum_{i=1}^{t} b_i \qquad \forall t \in \{1,\ldots,n\} \tag{2}$$

$$\sum_{i=1}^{t} b_i - \sum_{i=1}^{t} \left(2 \times \sum_{j=i+\varepsilon_{\min}}^{t} s_{i,j} + \sum_{j=t}^{\min\{n, t+\varepsilon_{\max}\}} s_{i,j} \right) \leq u_t \qquad \forall t \in \{1,\ldots,n\} \tag{3}$$

$$\sum_{j \in J(t)} s_{t,j} + \sum_{i=0}^{t-\varepsilon_{\min}} s_{i,t} \leq \nu_t \qquad \forall t \in \{1,\ldots,n\} \tag{4}$$

$$s_{i,j} \in \mathbb{Z}_{+} \qquad \forall i \in \{1,\ldots,n\},\ j \in J(i) \tag{5}$$

$$s_{i,j} = 0 \qquad \forall i \in \{1,\ldots,n\},\ j \notin J(i). \tag{6}$$

Constraints (2) ensure that in each time step, the number of first and second doses given so far is not greater than the number of doses delivered, thus only available doses are used. Constraints (3) and (4) are optional and restrict the number of doses in storage for each time step and the number of vaccinations given in each time step (vaccination speed) respectively. With (5) and (6), we ensure integrality of feasible $s_{i,j}$, whereas nobody can be vaccinated at infeasible appointments.

3   The Matching Approach

Despite the relative easiness to formulate the problem as an ILP, the model does not provide additional insight. Computational experiments moreover show that the problem can be solved relatively fast. To understand this, we link the problem in this section to the theory of b-matching problems. First, we study the basic problem without storage capacities and vaccination speed constraints (3) and (4), afterwards we extend to capacities and multiple manufacturers.

3.1   Without Capacities

In this subsection, we reformulate the question: which doses will be given to the same person? Thus, we match two delivery times to get a feasible schedule instead of matching the points in time the vaccinations are given.

This naturally leads to a weighted b-matching problem on an undirected graph $G = (V, E)$ with vertices $V = \{v_1, \ldots, v_n\}$ for each time step and edges between each pair of vertices $E = \{\{v_i, v_j\}: v_i, v_j \in V\}$. The edge set E includes all self-loops (v_i, v_i) for $i = 1, \ldots, n$ denoting both shots are taken from the same delivery. For each $v_i \in V$, $b_{v_i} = b_i$ is given by the number of doses delivered at time step i. The weight w_e of each edge $e = \{v_i, v_j\}$ with $i \leq j$ is given by the maximum value of all appointments that can be made using doses from these time slots. We call these appointments "fitting" for the time slots i and j. More precisely, an appointment (i^*, j^*) is fitting for a pair of time slots $i \leq j$ if it satisfies the following three constraints:

1. the first shot cannot be given before time step i,
2. the second shot cannot be given before time step j,
3. the time constraint for the gap between first and second dose has to be kept.

Thus, a fitting appointment (i^*, j^*) has to satisfy $i \leq i^*$, $j \leq j^*$ and $j^* \in J(i^*)$.

From now on, we assume that an earlier vaccination has higher value than a later vaccination. Then, the fitting appointment is $(i^*, j^*) = (i, j)$ if it is already a feasible appointment. Else, there are two possible cases. If the time gap between i and j is too long ($j - i > \varepsilon_{\max}$), we delay the first dose to $i^* = j - \varepsilon_{\max}$. If the time gap is to short ($j - i < \varepsilon_{\min}$), we delay the second dose to $j^* = i + \varepsilon_{\min}$.

Example 1. If the gap between both doses is between $\varepsilon_{\min} = 3$ and $\varepsilon_{\max} = 6$ weeks, the edge between 1 and 3 has value $c_{1,4}$ since the time between the two slots is too short and we have to delay the second dose by one week. The edge between 1 and 8 would have value $c_{2,8}$, because the gap is too long and the first dose has to be delayed for the second dose to be on time.

Thus, we have $(i^*, j^*) = (\max\{i, j - \varepsilon_{\max}\}, \max\{j, i + \varepsilon_{\min}\})$. The weight of the edges is given by

$$w_{\{v_i, v_j\}} := c_{i^*, j^*} = c_{\max\{i, j - \varepsilon_{\max}\}, \max\{j, i + \varepsilon_{\min}\}} \tag{7}$$

using the assumption.

Theorem 1. *Each b-matching $(M_e)_{e \in E}$ in G corresponds to a vaccination schedule of the same value.*

Proof. For each edge $e = (v_i, v_j) \in E$, there are M_e appointments at time slots (i^*, j^*). Since the weight of each edge is the value of the matching appointment, the weight of each b-matching is the same as the value of the schedule.

The weighted b-matching problem can be solved in polynomial time as shown by Anstee [1] or Pulleyblank [8] (see also Schrijver [9]).

3.2 Include Upper Bound on Vaccination Speed and Storage Capacity

So far, we have not included a storage capacity or the vaccination speed. To include those, we return to matching the points in time the vaccinations are given, instead of matching delivery times. We design a gadget incorporating the concept of storing doses for later usage. Then, the two constraints are lower bounds on certain edges in a new graph G'. More formally, we start with a vertex for each time step and edges between each pair of time steps that form a feasible appointment.

$$\{v_i, v_j\} \in E' \iff j \in J(i) \tag{8}$$

Next, we introduce a gadget for every $i \in \{1, \ldots, n-1\}$, presented in Fig. 1. The new graph is then defined as

$$V' = \{v_i, v_i^1, v_i^2, v_i^3 : i = 1, \ldots, n-1\} \cup \{v_n, v_n^1, v_n^2\}, \tag{9}$$

$$E' = \{\{v_i, v_j\} : i = 1, \ldots, n - \varepsilon_{\min} \, ; j \in J(i)\}$$
$$\cup \{(v_i, v_i^1) : i = 1, \ldots, n\} \cup \{(v_i^1, v_i^2) : i = 1, \ldots, n\} \tag{10}$$
$$\cup \{(v_i^2, v_i^3) : i = 1, \ldots, n-1\} \cup \{(v_i^3, v_{i+1}^2) : i = 1, \ldots, n-1\}.$$

We set bounds b'_v on the vertices and weights w_e on the edges as follows:

$$b'_{v_i} = b'_{v_i^1} = b'_{v_i^2} = b'_{v_i^3} = \sum_{k=1}^{i} b_i, \tag{11}$$

$$w_e = \begin{cases} c_{i,j} & \text{if } e = \{v_i, v_j\} \\ 0 & \text{else.} \end{cases} \tag{12}$$

The interpretation is different to the simple approach: for a given b'-matching m', the value m'_{v_i, v_j} denotes the number of vaccination appointments with first dose given at time step i and second dose at time step j. Additionally, for each vertex v_i there is another edge $\{v_i, v_i^1\}$ symbolizing the storage. The added vertices and edges of the graph are shown in Fig. 1. The added gadget is used to model the storage by connecting the vertices v_i. Let a_i denote the number of doses stored in time step i. Thus, $b_i + a_{i-1} - a_i =: y_i$ doses are used in time step i. Let $m' \in \mathbb{N}^{|E'|}$ be a perfect b'-matching on G'. Since $y_i = \sum_{j \neq i} m'_{v_i, v_j}$, it holds $m'_{v_1, v_1^1} = b_1 - y_1$. Since v_1^1 has exactly 2 adjacent edges and the matching has to be a perfect b-matching, it holds $m'_{v_1^1, v_1^2} = b_1 - (b_1 - y_1) = y_1$. With the same reasoning, it holds $m'_{v_2^2, v_1^3} = b_1 - y_1 = a_1$ and $m'_{v_1^3, v_2^2} = y_1 = b_1 - a_1$. Using the argumentation for $i = 2$, we get $m'_{v_2^1, v_2^2} = y_2 = b_2 + a_1 - a_2$ and thus again $m'_{v_2^2, v_2^3} = b_1 + b_2 - y_2 - y_1 = b_1 + b_2 - (b_2 + a_1 - a_2) - (b_1 - a_1) = a_2$. We use these arguments recursively to show that, for each time step $i \in \{1, \ldots, n-1\}$, $m'_{v_i^2, v_i^3}$ equals the number of doses in storage a_i and $m'_{v_i^1, v_i^2}$ equals the number of doses administered y_i.

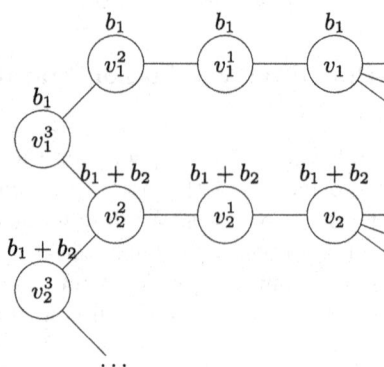

Fig. 1. Extension graph G' without the storage and vaccination speed bounds

The storage bound and maximum number of vaccinations are upper bounds on these values. We now follow $m'_{v_i,v_i^1} = \sum_{k=1}^{i} b_k - y_i = \sum_{k=1}^{i} b_k - m'_{v_i^1,v_i^2}$ and thus the upper bound on $m'_{v_i^1,v_i^2}$ is equivalent to a lower bound on (v_i, v_i^1). Similarly, the upper bound on the storage capacity $m'_{v_i^2,v_i^3}$ is equivalent to a lower bound on $m'_{v_i^3,v_{i+1}^2}$.

We include these lower bounds implicitly by adjusting the bound b' of the vertices. Let $\tilde{m} \in \mathbb{N}^{|E|}$ be such that:

$$\tilde{m}_{v_i,v_i^1} = \max\left\{0, \sum_{k=1}^{i} b_k - \nu_i\right\},$$

$$\tilde{m}_{v_i^3,v_{i+1}^2} = \max\left\{0, \sum_{k=1}^{i} b_k - u_i\right\} \quad \text{and} \tag{13}$$

$$\tilde{m}_e = 0 \qquad \text{else.}$$

Then, each feasible matching m' satisfies the two additional constraints if and only if it holds $m' \geq \tilde{m}$. Thus, instead of a b'-matching m' satisfying $m' \geq \tilde{m}$, we might also consider a β-matching $m = m' - \tilde{m}$ with

$$\beta_{v_i} = \beta_{v_i^1} = b'_{v_i} - \tilde{m}_{v_i,v_i^1} = \min\{\sum_{k=1}^{i} b_k, \nu_i\}$$

$$\beta_{v_i^2} = b'_{v_i^2} - \tilde{m}_{v_{i-1}^3,v_i^2} = \min\{\sum_{k=1}^{i} b_k, u_{i-1} + b_i\} \tag{14}$$

$$\beta_{v_i^3} = b'_{v_i^3} - \tilde{m}_{v_i^3,v_{i+1}^2} = \min\{\sum_{k=1}^{i} b_k, u_i\}.$$

Then, each (perfect) β-matching m implies a (perfect) b'-matching $m' = m + \tilde{m}$ that satisfies the two additional constraints. The adjusted β is depicted in Fig. 2.

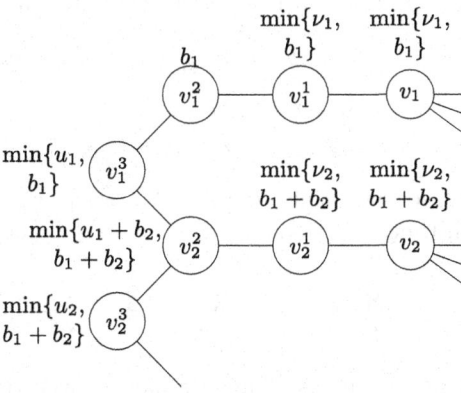

Fig. 2. Extension graph G' including the storage and vaccination speed bounds

Theorem 2. *For each perfect β-matching $(m_e)_{e \in E'}$ on G', there is a corresponding feasible schedule s with value $c^T s = (w')^T m$ satisfying the storage capacity and vaccination speed constraints and vice versa.*

Proof. Let m be a perfect β-matching in G'. Then, s with $s_{i,j} = m_{v_i, v_j}$ is the corresponding vaccination schedule. We denote by

$$y_i := \sum_{t=i+\varepsilon_{\min}}^{\min\{n, i+\varepsilon_{\max}\}} s_{i,t} + \sum_{t=\max\{0, i-\varepsilon_{\max}\}}^{i-\varepsilon_{\min}} s_{t,i} \tag{15}$$

the number of doses given in time step i and by $a_i := \sum_{t=1}^{i} b_t - \sum_{t=1}^{i} y_i = a_{i-1} + (b_i - y_i)$ the number of doses in storage. It holds $c^T s = (w')^T m$.

In order to show feasibility, we have to show the following points for each $i \in \{1, \ldots, n\}$:

(i) the vaccination speed constraint $y_i \leq \nu_i$ is satisfied,
(ii) the storage capacity constraint $a_i \leq u_i$ is satisfied, and
(iii) the delivery constraint $\sum_{t=1}^{i} y_t \leq \sum_{t=1}^{i} b_t$ is satisfied.

(i) follows from $y_i \leq b'_{v_i^1} \leq \nu_i$: it holds $y_i = b'_{v_i} - m_{v_i, v_i^1} = b'_{v_i} - (b'_{v_i^1} - m_{v_i^1, v_i^2}) = m_{v_i^1, v_i^2}$ and thus the number of doses administered in time step i is given by $m_{v_i^1, v_i^2} = y_i$.

We show by induction over $i = 1, \ldots, n$ that $m_{v_i^2, v_i^3} = a_i$ is the number of doses in storage at time step i and with $b'_{v_i^3} = u_1$ condition (ii) holds. For $i = 1$, the vertex v_1^2 has exactly two adjacent edges and $m_{v_1^1, v_1^2}$ is the number of doses used in time step 1. Thus, $m_{v_1^2, v_1^3} = b'_{v_1^2} - m_{v_1^1, v_1^2} = b_1 - m_{v_1^1, v_1^2}$ is the number of leftover doses from the first time step. Since there are no earlier time steps and the storage has to be empty, $m_{v_1^2, v_1^3}$ is the number of doses stored in time step 1 and with $b'_{v_1^3} = u_1$ it is restricted by u_1. Let the assumption hold for $i - 1 \in \{1, \ldots, n\}$ and $m_{v_{i-1}^2, v_{i-1}^3} = a_{i-1}$. Since v_{i-1}^3 has exactly two adjacent edges, it holds $m_{v_{i-1}^3, v_i^2} = u_{i-1} - a_{i-1}$ and with the same argumentation as for $i = 1$, it is $m_{v_i^1, v_i^2} = y_i$. Because m is a perfect matching, it holds

$$b'_{v_i^2} = m_{v_i^1, v_i^2} + m_{v_i^2, v_i^3} + m_{v_{i-1}^3, v_i^2}$$
$$\Leftrightarrow \qquad u_{i-1} + b_i = y_i + m_{v_i^2, v_i^3} + u_{i-1} - a_{i-1} \tag{16}$$
$$\Leftrightarrow \qquad m_{v_i^2, v_i^3} = a_{i-1} + b_i - y_i = a_i$$

and thus, $m_{v_i^2, v_i^3} = a_i$ for each time step $i \in \{1, \ldots, n\}$ and (ii) holds. This also implies (iii): for each time step, it holds $m_{v_i^2, v_i^3} = a_i \geq 0$ and thus

$$-a_i \leq 0 \;\Leftrightarrow\; \sum_{k=1}^{i} b_k - a_i \leq \sum_{k=1}^{i} b_k \;\Leftrightarrow\; \sum_{k=1}^{i} y_k \leq \sum_{k=1}^{i} b_k \tag{17}$$

by Definition of a_i and (iii) holds. Thus, we have proven (i)–(iii) and the construction is correct.

Since weighted b-matching can be solved in polynomial time (c.f. [1,8,9]), the Two-Dose Scheduling Problem with storage capacity and vaccination speed constraints can also be solved in polynomial time.

3.3 Include Multiple Vaccines and Cross-Immunization

So far, we considered a single vaccine product. In this section, we assume that there are at least two vaccines with different value c or suggested time gap between the two doses that can use the same storage space and underlie the same vaccination speed constraint. Thus, both the storage capacity constraint and the vaccination speed constraint have to hold for the sum of all available vaccines. We use a separate graph similar to G' for each vaccine and connect them via two "storage vertices" and two "speed vertices". We begin by expanding the graph G' for one vaccine and then explain how to include the other vaccines.

In Sect. 3.2, we formulated the two conditions as lower bounds on certain edges and assumed them to be part of each feasible perfect b'-matching, thus subtracting them from b' resulting in β. Now, these lower bounds depend on the other vaccines. Thus, we cannot subtract them from b', but we can add another edge to the affected vertices simulating this subtraction for every perfect b'-matching. For this, we add two edges and two vertices for each affected vertex resulting in $G'' = (V'', E'')$ with

$$
\begin{aligned}
V'' = V' &\cup \{v_i^{S_k} : i \in \{1, \dots, n-1\}, k \in \{0,1,2,3\}\} \\
&\cup \{v_i^{V_k} : i \in \{1, \dots, n\}, k \in \{0,1,2,3\}\}
\end{aligned}
\tag{18}
$$

$$
\begin{aligned}
E'' = E' &\cup \{(v_i^{S_0}, v_i^{S_2}), (v_i^{S_1}, v_i^{S_3}) : i \in \{1, \dots, n-1\}\} \\
&\cup \{(v_i^3, v_i^{S_2}), (v_{i+1}^2, v_i^{S_3}) : i \in \{1, \dots, n-1\}\} \\
&\cup \{(v_i^{V_0}, v_i^{V_2}), (v_i^{V_1}, v_i^{V_3}) : i \in \{1, \dots, n\}\} \\
&\cup \{(v_i, v_i^{V_2}), (v_i^1, v_i^{V_3}) : i \in \{1, \dots, n\}\}
\end{aligned}
\tag{19}
$$

$$
b_v'' = \begin{cases}
b_v' & \text{if } v \in V' \\
b_w' & \text{if } v \in \{v_i^{S_2}, v_i^{S_3} : i = 1, \dots, n\}, w \in V', \{v,w\} \in E'' \\
\min\{\beta_u, b_w''\} & \text{if}\{\{v,w\}, \{w,u\}\} \subseteq E'', v, w \notin V', u \in V'
\end{cases}
\tag{20}
$$

$$
w_e'' = \begin{cases}
w_e & \text{if } e \in E \\
0 & \text{else}
\end{cases}
\tag{21}
$$

The graph G'' is depicted in Fig. 3. The graph G' from the last section is depicted in gray. The additional paths attached to $v_i^{S_2}$ and $v_i^{S_3}$ are used for the storage capacity constraint; those attached to $v_i^{S_0}$ and $v_i^{S_1}$ for the vaccination speed constraint. For each of the original vertices from V', the value of b'' is the same as b'; for the new vertices $v_i^{S_2}$ and $v_i^{S_3}$ directly connected to an original vertex, the value of b'' is the same as the value of that original vertex. For the vertices $v_i^{S_0}$ and $v_i^{S_1}$, the value of b'' is the value we would get after subtraction. These

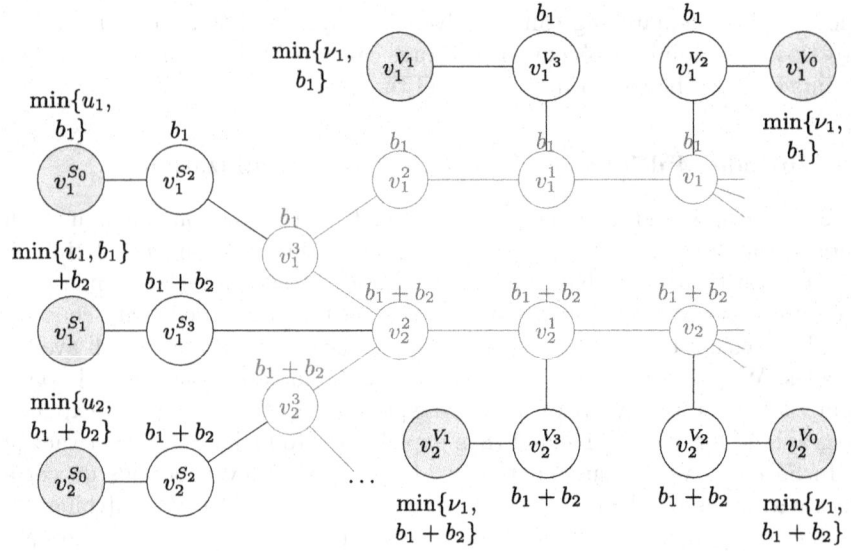

Fig. 3. Extension graph G'' including extension for multiple vaccine

vertices are the connection between different graphs for multiple vaccines that share the same storage space. In this case, the storage capacity constraint has to hold for the sum of both stored doses and they share the same vertices $v_i^{s_0}$ and $v_i^{S_1}$.

The most important difference between the two models is that in G' the two additional constraints are fixed parts of β and independent of the rest of the schedule. In G'', these constraints are realized through the value of the matching on the added vertices. For one vaccine, these are fixed, but for multiple vaccines, these values depend on the other vaccines that share the same vertices.

Theorem 3. *For each perfect b''-matching $(m_e)_{e \in E'}$ on G'' there is a corresponding feasible schedule s with value $c^T s = (w'')^T m$ satisfying storage capacity and vaccination speed constraints and vice versa.*

Proof. First, we examine the storage capacity constraint for the path $v_i^{S_0}, v_i^{S_2}, v_i^3$ with one vaccine. If $\sum_{k=1}^i b_k \leq u_i$, the storage capacity is large enough to store all delivered doses and the constraint is satisfied. Here, each perfect matching m would satisfy $m_{v_i^{S_0}, v_i^{S_2}} = \sum_{k=1}^i b_i = b''_{v_i^{S_0}}$ because $v_i^{S_0}$ has only one outgoing edge. Because of this, the other new edge on the paths has $m_{v_i^{S_2}, v_i^3} = 0$ for each perfect matching m. Because of this, nothing changes in the original part of the graph G' and the matching is equivalent to a β-matching on G'. In the case $\sum_{k=1}^i b_k > u_i$, where the storage capacity constraint is not redundant, each perfect matching satisfies $m_{v_i^{S_0}, v_i^{S_2}} = u_i$ and thus $m_{v_i^{S_2}, v_i^3} = \sum_{k=1}^i b_k - u_i$. Because v_i^3 is an adjacent vertex to this edge, the other two adjacent edges have to satisfy $m_{v_i^3, v_i^2} + m_{v_i^3, v_{i+1}^2} \leq \sum_{k=1}^{i} b_k - (\sum_{k=1}^{i} b_k - u_i) = u_i$ and thus, the matching is also a perfect β-matching on G' considering v_i^3 for each time step $i \in \{1, \ldots, n\}$. The same argumentation can be used for the other added paths. With this, it holds that each perfect β-matching on G' can be extended to a perfect b''-matching on G'' and vice versa.

4   The Three-Dose Problem

We now want to include a third dose, i.e., the so-called booster shot for COVID vaccines, into our model. Now, each person has to get three doses with the time between the first two doses in $[\varepsilon_{\min}, \varepsilon_{\max}]$ and the time between the second and third dose in $[\delta_{\min}, \delta_{\max}]$. Additionally, we expand the value c of each appointment to $c_{i,j,k}$ for each $i = 1, \ldots, n$, $j = i + \varepsilon_{\min}, \ldots, i + \varepsilon_{\max}$ and $k = j + \delta_{\min}, \ldots, j + \delta_{\max}$. However, this problem turns out to be NP-hard.

Theorem 4. *The Three-Dose Problem is NP-hard.*

Proof. We reduce from the maximum 3-dimensional matching Problem. This problem is one of Karp's 21 NP-complete Problems [4]. For given finite sets X, Y and Z with $|X| = |Y| = |Z| = m$ and $T \subseteq X \times Y \times Z$ the problem consists of finding a maximum set $M \subseteq T$ such that any $(x, y, z), (x', y', z') \in M$ it holds $x \neq x', y \neq y'$ and $z \neq z'$. This problem is known to be NP-complete.

Let $X = \{x_1, \ldots, x_m\}, Y = \{y_1, \ldots, y_m\}$ and $Z = \{z_1, \ldots, z_m\}$. For each element of $X \cup Y \cup Z$, we appoint one time step in which exactly one dose is delivered. We allocate these time slots such that the feasible appointments are exactly the triples in $X \times Y \times Z$ and appoint decreasing value to each triple to avoid storing doses if it is not necessary. Let $n = 5m$ and

$$\varepsilon_{\min} = m \quad \varepsilon_{\max} = 3m \quad \delta_{\min} = m \quad \delta_{\max} = 3m \qquad (22)$$

$$b_i = \begin{cases} 1 & i \in \{1, \ldots, m\} \\ 1 & i \in \{2m+1, \ldots, 3m\} \\ 1 & i \in \{4m+1, \ldots, 5m\} \\ 0 & \text{else} \end{cases} \qquad (23)$$

$$\hat{c}_{i,j,k} = (m - i) + (3m - j) + (5m - k) = 9m - (i + j + k) \qquad (24)$$

$$c_{i,j,k} = \begin{cases} \hat{c}_{i,j,k} & \text{if } (x_i, y_{j-2m}, z_{k-4m}) \in T \\ \hat{c}_{i,j,k} - 1 & \text{else.} \end{cases} \qquad (25)$$

The resulting Three-Dose Problem is visualized in Fig. 4. Each time slot with an available dose is marked by its corresponding element in $X \cup Y \cup Z$ and the feasible second dose for first doses given at the time slots corresponding to x_1 and x_m as well as the feasible third doses for the second doses corresponding to y_1 and y_m. We will now prove that the value of each schedule with the efficacy

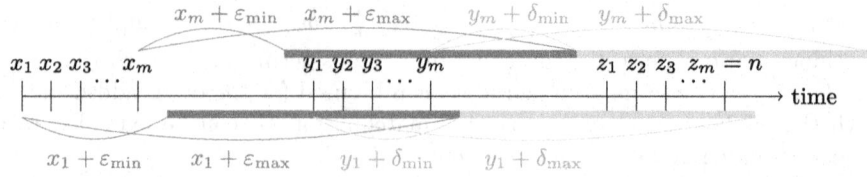

Fig. 4. Reduction, time slots for each element of $X \cup Y \cup Z$ and feasible appointments

function c is at most $S := \frac{3 \times m \times (m-1)}{2}$ and each schedule with value S defines a 3-dimensional matching

$$M = \{(x_i, y_{j-3m}, z_{k-4m}) : s_{i,j,k} > 0\} \qquad (26)$$

with $|M| = m$, thus proving the reduction.

From the definition of $\hat{c}_{i,j,k}$ follows, that $\hat{c}_{i,j,k} \geq \hat{c}_{i',j',k'}$ for all $i' \geq i$, $j' \geq j$ and $k' \geq k$. Thus, considering only efficacy $\hat{c}$, each solution delaying a dose yields a lower value than a solution that does not delay any doses with the same number of appointments. Furthermore, each solution that does not delay any doses and uses them as soon as they are available has the same value. In that case, m appointments are made and for each time slot t with $b_t \geq 0$, it holds $b_t = 1$ and there is exactly one appointment using this time slot. Thus, for each $i \in \{1, \ldots, m\}$, there is exactly one appointment (i, j, k) with $s_{i,j,k} = 1$. The same holds for each $j \in \{2m+1, \ldots, 3m\}$ and $k \in \{4m+1, \ldots, 5m\}$. Using this, we get

$$\hat{c}^T s = \sum_{i=1}^{m} \sum_{j=2m+1}^{3m} \sum_{k=4m+1}^{5m} s_{i,j,k}((m-i) + (3m-j) + (5m-k))$$

$$= \sum_{i=1}^{m}(m-i) + \sum_{j=2m+1}^{3m}(3m-j) + \sum_{k=4m+1}^{5m}(5m-k) \qquad (27)$$

$$= \sum_{i=0}^{m-1} i + \sum_{j=0}^{m-1} j + \sum_{k=0}^{m-1} k = \frac{3 \times m \times (m-1)}{2} = S$$

and each solution with m appointments and no delay has the same value. Furthermore, every solution with less appointments or delay has lower value and thus if the schedule with maximum value has value S, then there is a schedule that uses each dose as soon as it is delivered. There are three intervals of time slots with $b_i > 0$: $[1, m]$ represents the elements $X = \{x_1, \ldots, x_m\}$, $[2m+1, 3m]$ those of Y and $[4m+1, 5m]$ those of Z. Due to the choice of ε and δ and the result above, each feasible appointment (i, j, k) must satisfy $i \in [1, m], j \in [2m+1, 3m]$ and $k \in [4m+1, 5m]$. Thus, each appointment represents a triple from $X \times Y \times Z$ and vice versa. Combining the two results, we get that each schedule with value S of the Three-Dose Problem corresponds to a unique 3-dimensional matching. For the Three-Dose Problem using efficacy c, it obviously holds $c^T s \leq \hat{c}^T s \leq S$ for each solution s. If the optimal solution value is S, there is a schedule s

with $c_{i,j,k} = \hat{c}_{i,j,k}$ (and thus $(x_i, y_{j-3m}, z_{k-4m}) \in T$) for each appointment with $s_{i,j,k} > 0$. Then, for each appointment in the solution the corresponding triple is in T. Let $M = \{(x_i, y_{j-3m}, z_{k-4m}) \colon s_{i,j,k} > 0\}$, then $M \subseteq T$ and with the result above, M is a 3-dimensional matching with $|M| = m$. Therefore, there is a 3-dimensional Maximum matching for X, Y, Z, T if and only if there is a schedule for the Three-Dose Problem $(b, c, \varepsilon, \delta)$ with value S and the Three-Dose Problem is NP-hard.

This implies that for more than three doses the problem is NP-hard as well.

5 Conclusion

In this paper, we have introduced a new model for the Two-Dose Scheduling Problem using the theory of b-matching Problems. By this, the basic problem and all considered extensions can be solved in polynomial time. Finally, we have shown the NP-hardness of the Multiple-Dose Scheduling Problem for three or more doses. For future research, we are working on extending the problem to include uncertainty in the deliveries.

References

1. Anstee, R.P.: A polynomial algorithm for b-matchings: an alternative approach. Inf. Process. Lett. **24**(3), 153–157 (1987)
2. Duijzer, L.E., Van Jaarsveld, W., Dekker, R.: Literature review: the vaccine supply chain. Eur. J. Oper. Res. **268**(1), 174–192 (2018)
3. Jurgens, G., Lackner, K.: Modelled optimization of SARS-Cov-2 vaccine distribution: an evaluation of second dose deferral spacing of 6, 12, and 24 weeks. medRxiv (2021). https://doi.org/10.1101/2021.02.28.21252638
4. Karp, R.M.: Reducibility among combinatorial problems. In: Miller, R.E., Thatcher, J.W., Bohlinger, J.D. (eds.) Complexity of Computer Computations. IRSS, pp. 85–103. Springer, Boston (1972). https://doi.org/10.1007/978-1-4684-2001-2_9
5. Moghadas, S.M., et al.: Evaluation of COVID-19 vaccination strategies with a delayed second dose. PLoS Biol. **19**(4), e3001211 (2021)
6. Parino, F., Zino, L., Calafiore, G.C., Rizzo, A.: A model predictive control approach to optimally devise a two-dose vaccination rollout: a case study on COVID-19 in Italy. Int. J. Robust Nonlinear Control (2021). https://doi.org/10.1002/rnc.5728
7. Polack, F.P., et al.: Safety and efficacy of the BNT162b2 mRNA Covid-19 vaccine. New Engl. J. Med. **383**(27), 2603–2615 (2020). https://doi.org/10.1056/NEJMoa2034577. pMID: 33301246
8. Pulleyblank, W.R.: Faces of matching polyhedra (1973)
9. Schrijver, A.: Combinatorial Optimization: Polyhedra and Efficiency, vol. 24. Springer, Heidelberg (2003)
10. Silva, P.J., Sagastizábal, C., Nonato, L.G., Struchiner, C.J., Pereira, T.: Optimized delay of the second COVID-19 vaccine dose reduces ICU admissions. Proc. Natl. Acad. Sci. **118**(35), e2104640118 (2021)
11. Wouters, O.J., et al.: Challenges in ensuring global access to COVID-19 vaccines: production, affordability, allocation, and deployment. The Lancet **397**(10278), 1023–1034 (2021). https://doi.org/10.1016/S0140-6736(21)00306-8

Approximation Algorithms

Pervasive Domination

Gennaro Cordasco[1]([✉]) [ID], Luisa Gargano[2] [ID], and Adele A. Rescigno[2] [ID]

[1] Department of Psychology, University of Campania "L.Vanvitelli", Caserta, Italy
gennaro.cordasco@unicampania.it
[2] Department of Computer Science, University of Salerno, Fisciano, Italy
{lgargano,arescigno}@unisa.it

Abstract. Inspired by the implicit or explicit persuasion scenario, which characterizes social media platforms, we analyze a novel domination problem named Pervasive Partial Domination (PPD). We consider a social network modeled by a digraph $G = (V, E)$ where an arc $(u, v) \in E$ represents the capability of an individual u to persuade an individual v. We are looking for a set $S \subset V$ of social change individuals, of minimum cost, who combined enable to reach the desired behavior. The impact of S is measured by a set function $f(S)$ that is the sum of the degree of belief of all the individuals in the network and p is the desired target. We show that the natural greedy algorithm, for the PPD problem, provides an approximation guarantee, $\left(\ln \frac{p - f(\emptyset)}{\beta} + 2 \right)$ where $\beta > 0$ represents the minimum gain on the function f one can attain by bribing an additional individual when the target p is (almost) reached. The proposed solution can be generalized to the weighted partial submbodular cover problem providing a better approximation with respect to the state of the art.

Keyword: Domination

1 Introduction

Online social networks, like Facebook, Twitter, Instagram, and LinkedIn, have become the most important media for accessing a huge amount of information. Persuasion refers to the process by which individuals adjust their attitudes or opinions, revise their beliefs, or change their behaviors as a result of exposure to communication [1,4,6,21–23]. Persuasion can be implicit (social influence) or explicit (peer pressure). In both cases, it is highly pervasive in our daily life and may alter people's decisions and actions. The study of this process, due to its significant impact in different scientific fields (e.g., Economics, Sociology, Health and Political Science), has recently received extensive attention. Social influence, as well as peer pressure, can have both positive and negative effects on people's behaviors. It is true that, in a globally connected world, anyone can access a practically infinite amount of information in real-time and "knowledge is power". Social media are among the primary sources of news across the world. However, users are exposed to content of questionable accuracy (e.g., fake news, misinformation, fish tales, pseudo-science).

I. Ljubić et al. (Eds.): ISCO 2022, LNCS 13526, pp. 287–298, 2022.
https://doi.org/10.1007/978-3-031-18530-4_21

It has been shown that not only people, but also the algorithms behind social media platforms are vulnerable to manipulation. This phenomenon has been named "bias in the machine". Social media platforms, as well as search engines, employ algorithms to select or filter what people are looking for online. These algorithms are designed to select the most engaging and relevant content based on visualization history and user similarities. Doing this activity, algorithms may end up reinforcing the individual and social biases of users, thus making them even more vulnerable to misinformation (see Alexa suggesting a dangerous challenge[1]). Moreover, algorithmic flaws or biases can be exploited by virtual social agents, computer programs interacting with people on social media through text messages (chatbots), or natural language. In general, virtual agents are safe. However, in some cases, they have been used for malicious intents such as boosting misinformation. For instance, they have been used for manipulating political communication through misinformation during the U.S. midterm elections in 2010 [25]. In short, social influence affects people's and machines' behaviors as part of a complex social network.

Apart from the merit of the message, the more messages reach a person, the greater the impact these messages have on his or her behavior.

We consider a social environment where individuals are characterized by a certain degree of belief in some behavior (or opinion) and, at the same time, being part of a group, they are subject to persuasion (implicit or explicit) from their neighbors. We assume, in general, that an individual's choice may have different effects on his/her neighbors – based on the bonds of trust already existing between individuals. We study a bribery model where the structure of the network stays intact (individuals, friendships, and level of trust do not change), but there is an outside agent able to bribe some individuals (i.e., change their opinions) [17].

We would like to measure the cost of changing the social opinion through bribing/persuasion mechanisms. Measuring this cost is important because on the one hand it enables to evaluate the robustness of a network with respect to misinformation strategies (diffusion of negative behaviors) and on the other hand it allows to optimize the impact of positive strategies. E.g., we may assume that social change individuals (activists) work to make an impact (social change) in the world on a local and global level: The impact they make depends on how many people they can influence and on the degree of influence reached on each of them.

1.1   Our Model

Consider a social network represented by a digraph $G = (V, E)$ where each node $u \in V$ is associated to an individual of the social network and an arc $(u, v) \in E$ represents the capability of u of influencing v. The incoming neighborhood of an

---

individual $v \in V$ is denoted $N^{in}(v)$. Similarly, given a set $S \subseteq V$ we define the incoming neighbourhood of v restricted to S as $N_S^{in}(v) = N^{in}(v) \cap S$.

The following functions are associated to the graph G:

- A function $a : V \to [0, 1]$ indicates the degree of belief in a certain behavior, e.g., veganism or decreased consumption of meat, propensity to get vaccinated, avoid the use of drugs, etc.
- A function $b : E \to [0, 1]$ indicates, for each arc $(u, v) \in E$, the amount of influence the individual u has on v.
- A function $c : V \to \mathbb{R}_0^+$ indicates the cost to bribe an individual, that is, to make it become a social change individual.

The idea is that we want to identify a set of social change individuals (i.e., those that we choose to bribe), of minimum cost, who act to increase the group's attitude towards the desired behavior and, in this way, allow to achieve a given goal (social change). It is worth mentioning that the model focuses on changing one specific behavior, and not many at the same time.

In order to measure the impact (diffusion) made by a given set $S \subseteq V$ of bribed (social change) individuals, we define the following function:

$$f(S) = \sum_{v \in V} f_v(S), \tag{1}$$

where $f_v(S) = \begin{cases} 1 & \text{if } v \in S \\ \min \left\{ a(v) + \sum_{u \in N_S^{in}(v)} b(u, v), 1 \right\} & \text{otherwise.} \end{cases}$

In other words, the impact of S on the whole network, $f(S)$, is measured as the sum of the degree of belief of all the individuals in the network, where the degree of belief of each node v is given by the initial belief $a(v)$ to which the influence of his bribed friends $N_S^{in}(v)$ is added up to the maximum value 1.

The cost of S is the sum of the costs of each individual belonging to S,

$$c(S) = \sum_{v \in S} c(v).$$

We study the following domination problem:

PERVASIVE PARTIAL DOMINATION (PPD)
Instance: A digraph $G = (V, E)$, functions $a : V \to [0, 1]$ and $b : E \to [0, 1]$, a cost function $c : V \to \mathbb{R}_0^+$, and a bound $p \in \mathbb{R}^+$.
Goal: Find a set $S \subseteq V$ of minimum cost such that $f(S) \geq p$.

1.2   Our Results

We give a polynomial time approximation algorithm for PPD that has an approximation guarantee of $\left(\ln \frac{p - f(\emptyset)}{\beta} + 2 \right)$, where β represents the minimum gain one can attain by bribing an additional node when the target p is (almost) reached, namely $\beta = \min\{f(S \cup \{v\}) - f(S) \mid v \notin S \subset V, f(S) < p, f(S \cup \{v\}) \geq p\}$. The algorithm exploits the fact that the function f, defined in (1), is a submodular, nondecreasing set function.

2   Related Work

Bribery in Voting Systems. Our problem has been studied in terms of Social Choice theory. This line of research includes the formal study of network-based voting, as well as bribery and reputation-based systems. For instance, in [19] the authors present a rating model in which users are able to evaluate a service. The paper investigates the effect upon the network when one is able to bribe a subset of all of these users. Existing research involves the analysis of bribery in voting systems and systems of judgment aggregation [2,3]. Results show that in general the problem is not only hard, in terms of classical complexity (NP-hardness) but it is even $W[2]$-hard, with respect to its natural parameter (Parameterized complexity).

Partial Domination in Graphs. Domination problems in graphs have been extensively studied under various assumptions, both from structural and algorithmic points of view [20]. The variation of the domination problem which we study in this paper is related, but not equivalent, to the concept of vector domination problem, which, given a vector $k = (k_v \mid v \in V)$, asks to find a vector dominating set of minimum size, that is, a set S minimizing $|S|$ and such that $|N_S^{in}(v)| \geq k_v$ for all $v \in V$ [7]. Other works analyzed the effect of applying incentives to nodes to increase their susceptibility to being dominated [9,14]

We stress that all the above problems, like our PPD problem, focus on the minimization of the size (or the cost) of the set attaining the desired property. On the other hand, other problems look for solutions of bounded cost able to maximize the payoff. For instance, the budgeted version of the PPD problem can be formulated as: given a budget k, find a set $S \subseteq V$ such that $f(S)$ is maximum among all the sets of cost bounded by k. This problem can be solved using the solution provided in [26] and generalized in [5].

Inspired by the like/dislike scenario, which characterizes social networks, the work in [11] introduces a novel domination problem named Dual Domination, which, assuming that individuals are partitioned into positive and negative ones, asks for a dominating set able to maximize the set of dominated positive nodes minimizing at the same time the set of dominated negative nodes. A similar problem is also discussed in [24].

Influence Propagation in Social Networks. The study of the PPD problem is also relevant in the area of influence propagation in social networks. Indeed, maximizing or minimizing the diffusion of a certain piece of information in a social network received a lot of attention, as it has applications in several areas, including viral marketing, disease prevention, disease propagation, politics, misinformation etc. [8,10,12,15,16,18]. Usually, information diffusion is modeled as a long-term process, where information cascading occurs in several rounds. Nevertheless since time is a critical aspect, several studies model the problem as a single round of influence propagation. Our problem coincides with the case in which only one round of influence is allowed. Moreover our payoff function is not binary (influenced/not influenced) but is able to measure the degree of influence reached on each individual. Finally our approach is based on a static network while other strategies have been devised to evaluate the cost of "take over" a network by changing its structure (e.g., adding/deleting nodes and/or edges) [13].

Partial Submodular Cover. Our results can be generalized to the partial submodular cover which considers a finite set $I = \{1, 2, \ldots, n\}$, and a nondecreasing, submodular set function $f : 2^I \to \mathbb{R}_0^+$ where 2^I is the set of all the subsets of I. Given $c : I \to \mathbb{R}_0^+$ a function, which associates a cost to each element in I and a constant $p \in \mathbb{R}^+$, the problem asks for a set $S \subseteq I$ with $f(S) \geq p$ such that the value $\sum_{i \in S} c(i)$ is minimum among all sets T with $f(T) \geq p$. The special case, known as submodular cover, which can be obtained by imposing $p = f(I)$, has been studied in detail in [27]. It is worth mentioning that, as the author observed, by suitably modifying the function $f()$ (i.e., $Z() = \min\{p, f()\}$) some of the results presented for the submodular cover, also apply to the general case with $p \leq f(I)$.

3 Pervasive Partial Domination

In this section we assume that $f(\emptyset) < p$ since otherwise the empty set is the optimal solution and we have nothing to prove.

Algorithm PPD represents the natural greedy algorithm for solving the Pervasive Partial Domination problem.

Algorithm 1:   PPD $(v, f(), c(), p)$

---

1 $S = \emptyset$.

2 **while** $f(S) < p$ **do**

3 $\quad \Big|\quad v = \arg\max_{u \in V-S} \left(\frac{\min\{f(S \cup \{u\}), p\} - f(S)}{c(u)} \right)$

4 $\quad \Big\lfloor\quad S = S \cup \{v\}$

5 **return** S

Let $S^* = \{w_1, w_2, \ldots, w_s\}$ be the optimal solution for the PPD instance, and let $OPT = c(S^*)$. Let $S = \{v_1, v_2, \ldots, v_{r+1}\}$ be the solution given by the algorithm PPD, where v_k is the element added at the k-th step, for $k = 1, \ldots, r+1$. We will denote by $S_0 = \emptyset$ the starting empty set and by

$$S_k = \{v_1, v_2, \ldots, v_k\}$$

the partial solution obtained by the algorithm PPD after k steps, $k = 1, \ldots, r+1$. Notice that the last node v_{r+1} selected by the algorithm is such that $f(S_r) < p$ and $f(S_{r+1}) \geq p$.

We denote by β the following quantity

$$\beta = \min\{f(S \cup \{v\}) - f(S)\},$$

292   G. Cordasco et al.

where the minimum is over all the sets $S \subset V$ such that $f(S) < p$ and all $v \in V - S$ such that $f(S \cup \{v\}) \geq p\}$.

We will prove that

$$\frac{c(S)}{OPT} \leq \ln\left(\frac{f(S_r) - f(\emptyset)}{\beta}\right) + 2 \leq \ln\left(\frac{p - f(\emptyset)}{\beta}\right) + 2. \tag{2}$$

Remark 1. In the following, we assume $\beta > \alpha = p - f(S_r)$. Otherwise, one could apply the result in [27] (namely, (iii) of Theorem 1, with $Z() = \min\{p, f()\}$), which guarantees an approximation of $\left(\ln \frac{p-f(\emptyset)}{\alpha} + 1\right)$ and consequently the result holds.

Notice that the approximation in (2) can be much better than that in [27]. Indeed, the value $\alpha = p - f(S_r)$ can be extremely small and, since it depends on the algorithm run, it can be quite difficult to guess given the instance of the problem. On the other end, the value of β is, in general, larger than α and only depends on the instance.

3.1   Algorithm Analysis

In order to prove the desired bound, we first need the following lemma.

Lemma 1. *The diffusion function f, defined in (1), is a submodular, non-decreasing set function.*

Proof. By construction the function $f : 2^V \rightarrow \mathbb{R}_0^+$, defined in (1), is non-decreasing. Indeed, by augmenting a generic set S, for each $v \in V$, the value of any f_v is non-decreasing.

To show that function f, is submodular we have to prove that for all $S \subseteq T \subseteq V$ and for all $w \in V \setminus T$, the inequality $f(T \cup \{w\}) - f(T) \leq f(S \cup \{w\}) - f(S)$ holds. Recalling that a positive linear combination of submodular functions is a submodular function, it suffices to show that all the functions $f_v(S)$ are submodular, that is, for all $S \subseteq T \subseteq V$ and for all $w \in V \setminus T$,

$$f_v(T \cup \{w\}) - f_v(T) \leq f_v(S \cup \{w\}) - f_v(S). \tag{3}$$

Suppose first that $f_v(T) = 1$. Then $f_v(T \cup \{w\}) = 1$ and the left-hand side of inequality (3) is equal to 0. Hence inequality (3) holds since f_v is non-decreasing. From now on, we assume that $f_v(T) < 1$, which implies $f_v(T) = a(v) + \sum_{u \in N_T^{in}(v)} b(u, v)$. Since f_v is non-decreasing, we have $f_v(S) < 1$, and hence $f_v(S) = a(v) + \sum_{u \in N_S^{in}(v)} b(u, v)$. Inequality (3) simplifies to

$$f_v(T) - f_v(S) = \sum_{u \in N_{T \setminus S}^{in}(v)} b(u, v) \geq f_v(T \cup \{w\}) - f_v(S \cup \{w\}). \tag{4}$$

We may assume that $f_v(T \cup \{w\}) > f_v(S \cup \{w\})$, since otherwise the right-hand side of (4) equals 0, and inequality (4) holds.

Therefore, $f_v(S \cup \{w\}) < 1$, implying $f_v(S \cup \{w\}) = a(v) + \sum_{u \in N^{in}_{S \cup \{w\}}(v)} b(u,v)$. If also $f_v(T \cup \{w\}) < 1$ then $f_v(T \cup \{w\}) = a(v) + \sum_{u \in N^{in}_{T \cup \{w\}}(v)} b(u,v)$ and inequality (4) becomes an equality and holds

$$f_v(T) - f_v(S) = \sum_{u \in N^{in}_{T \setminus S}(v)} b(u,v) = f_v(T \cup \{w\}) - f_v(S \cup \{w\}).$$

So we may assume that $f_v(T \cup \{w\}) = 1$. Note that v does not belong to $T \cup \{w\}$ (since otherwise either $f_v(T)$ or $f_v(S \cup \{w\})$ would equal to 1). Suppose that the inequality (4) fails. Then

$$\sum_{u \in N^{in}_{T \setminus S}(v)} b(u,v) < 1 - \left(a(v) + \sum_{u \in N^{in}_{S \cup \{w\}}(v)} b(u,v) \right)$$

which implies

$$f_v(T \cup \{w\}) = a(v) + \sum_{u \in N^{in}_{T \cup \{w\}}(v)} b(u,v) < 1$$

and this contradicts the assumption that $f_v(T \cup \{w\}) = 1$.   □

Let d be the index of the first node in S such that $v_d \notin S^*$, that is $S_{d-1} \subseteq S^*$ while $v_d \notin S^*$. Note that if $d = r+1$ then v_d is the last node selected by the algorithm PPD; furthermore, since $\beta > \alpha = p - f(S_r) = p - f(S_{d-1})$ then, at the last step, the algorithm selects v_d – or another node having the same cost – in order to minimize the cost. Hence, in this case S_d would be an optimal solution and we have nothing to prove. In the following we assume that $d \leq r$.

The following results will be useful to analyze Algorithm PPD.

Lemma 2. *The following inequalities hold:*

(i) $c(v_d) \leq OPT - c(S_{d-1})$.
(ii) $\sum_{k=d+1}^{r} c(v_k) \leq (OPT - c(S_{d-1})) \ln \frac{f(S_r) - f(\emptyset)}{\beta}$.
(iii) $c(v_{r+1}) \leq OPT - c(S_{d-1})$.

Proof. Consider first (i). Let $S^* - S_{d-1} = \{w_d, \ldots, w_s\}$. Recalling that, for each $k = 1, \ldots, r$, at step k the algorithm PPD selects the element v_k such that the ratio between the marginal gain and the cost of v_k is maximum and that by Lemma 1 the function f is a non-decreasing submodular function, we have

$$f(S_{k-1} \cup \{w_d\}) - f(S_{k-1}) \leq c(w_d) \frac{f(S_k) - f(S_{k-1})}{c(v_k)}, \tag{5}$$

and for $\ell = d+1, \ldots, s$ we have

$$f(S_{k-1} \cup \{w_d, \ldots, w_\ell\}) - f(S_{k-1} \cup \{w_d, \ldots, w_{\ell-1}\}) \leq c(w_\ell) \frac{f(S_k) - f(S_{k-1})}{c(v_k)}.$$

Summing up the inequality (5) and all the above inequalities for $\ell = d+1, \ldots, s$ and recalling that $OPT - c(S_{d-1}) = \sum_{\ell=d}^{s} c(w_\ell)$ we get

$$f(S_{k-1} \cup \{w_d, \ldots, w_s\}) - f(S_{k-1}) \le \sum_{\ell=d}^{s} c(w_\ell) \frac{f(S_k) - f(S_{k-1})}{c(v_k)}$$

$$= (OPT - c(S_{d-1})) \frac{f(S_k) - f(S_{k-1})}{c(v_k)}. \quad (6)$$

When $k = d$, recalling that $f(S_{d-1} \cup \{w_d, \ldots, w_s\}) = f(S^*) \ge p$, by (6) we have

$$c(v_d) \le (OPT - c(S_{d-1})) \frac{f(S_d) - f(S_{d-1})}{p - f(S_{d-1})}.$$

Finally, by observing that $f(S_d) \le f(S_r) < p$, we obtain (i).

We prove now (ii). Fix any index k such that $d+1 \le k \le r$. We first notice that $v_d \notin S^*$ implies that

$$f(S_{k-1} \cup (S^* - S_{d-1})) \ge f(S_d \cup (S^* - S_{d-1})) = f(S^* \cup \{v_d\}) \ge p + \beta > f(S_r) + \beta.$$

From this and by using (6), we get

$$f(S_k) - f(S_{k-1}) \ge \frac{c(v_k)}{OPT - c(S_{d-1})} (f(S_r) + \beta - f(S_{k-1})). \quad (7)$$

Using (7), iteratively with $k = r, r-1, \ldots, d+1$, we have

$$\beta = f(S_r) + \beta - f(S_{r-1}) - (f(S_r) - f(S_{r-1}))$$

$$\le f(S_r) + \beta - f(S_{r-1}) - \left(\frac{c(v_r)}{OPT - c(S_{d-1})} (f(S_r) + \beta - f(S_{r-1})) \right)$$

$$= (f(S_r) + \beta - f(S_{r-1})) \left(1 - \frac{c(v_r)}{OPT - c(S_{d-1})} \right)$$

$$= (f(S_r) + \beta - f(S_{r-2}) - (f(S_{r-1}) - f(S_{r-2}))) \left(1 - \frac{c(v_r)}{OPT - c(S_{d-1})} \right)$$

$$\le \left(f(S_r) + \beta - f(S_{r-2}) - \left(\frac{c(v_{r-1})}{OPT - c(S_{d-1})} (f(S_r) + \beta - f(S_{r-2})) \right) \right) \left(1 - \frac{c(v_r)}{OPT - c(S_{d-1})} \right)$$

$$\le (f(S_r) + \beta - f(S_{r-2})) \left(1 - \frac{c(v_{r-1})}{OPT - c(S_{d-1})} \right) \left(1 - \frac{c(v_r)}{OPT - c(S_{d-1})} \right)$$

$$\le \ldots$$

$$\le (f(S_r) + \beta - f(S_{d+1})) \prod_{k=d+2}^{r} \left(1 - \frac{c(v_k)}{OPT - c(S_{d-1})} \right)$$

$$= (f(S_r) + \beta - f(S_d) - (f(S_{d+1}) - f(S_d))) \prod_{k=d+2}^{r} \left(1 - \frac{c(v_k)}{OPT - c(S_{d-1})} \right)$$

$$\le \prod_{k=d+1}^{r} \left(1 - \frac{c(v_k)}{OPT - c(S_{d-1})} \right) (f(S_r) + \beta - f(S_d))$$

$$\le \exp \left(-\frac{\sum_{k=d+1}^{r} c(v_k)}{OPT - c(S_{d-1})} \right) (f(S_r) + \beta - f(S_d)). \quad (8)$$

By (8) and recalling that $f(S_r) + \beta - f(S_d) \le f(S_r) - f(\emptyset)$, we have

$$\sum_{k=d+1}^{r} c(v_k) \le (OPT - c(S_{d-1})) \ln \frac{f(S_r) + \beta - f(S_d)}{\beta}$$

$$\le (OPT - c(S_{d-1})) \ln \frac{f(S_r) - f(\emptyset)}{\beta}.$$

We prove now (iii). Notice that at step $r + 1$ Algorithm PPD selects the element v_{r+1} such that the ratio between $\min\{f(S \cup \{u\}), p\} - f(S)$ and the cost of u is maximum. Furthermore, since $\min\{p, f(S_r)\} = f(S_r)$, we get that for each $\ell = d, \dots, s$,

$$\min\{p, f(S_r \cup \{w_d, \dots, w_\ell\})\} - \min\{p, f(S_r \cup \{w_d, \dots, w_{\ell-1}\})\} \le c(w_\ell) \frac{p - f(S_r)}{c(v_{r+1})}.$$

Summing up the above inequalities for $\ell = d, \dots, s$ and noticing that $\beta \ge f(S_{r+1}) - f(S_r) \ge p - f(S_r)$, we have $f(S_r \cup \{w_d, \dots, w_s\}) \ge p$ (recall that $s \ge d$). Hence,

$$p - f(S_r) \le (OPT - c(S_{d-1})) \frac{p - f(S_r)}{c(v_{r+1})}$$

and (iii) holds.   □

We are now ready to prove the desired approximation bound.

Theorem 1. *Algorithm PPD provides, for the Pervasive Partial Domination problem, an approximation*

$$\frac{c(S)}{OPT} \le \ln\left(\frac{f(S_r) - f(\emptyset)}{\beta}\right) + 2$$

Proof. We are going to evaluate the approximation ratio between the cost of the solution S provided by the algorithm and the cost OPT of S^*. We have

$$\frac{c(S)}{OPT} = \frac{\sum_{k=1}^{r+1} c(v_k)}{OPT}$$

$$= \frac{\sum_{k=1}^{d-1} c(v_k)}{OPT} + \frac{c(v_d)}{OPT} + \frac{\sum_{k=d+1}^{r} c(v_k)}{OPT} + \frac{c_{v_{r+1}}}{OPT}$$

$$\le \frac{c(S_{d-1})}{OPT}$$

$$+ \frac{(OPT - c(S_{d-1}))}{OPT} \qquad \text{by (i) of Lemma 2}$$

$$+ \frac{(OPT - c(S_{d-1})) \ln \frac{f(S_r) - f(\emptyset)}{\beta}}{OPT} \qquad \text{by (ii) of Lemma 2}$$

$$+ \frac{(OPT - c(S_{d-1}))}{OPT} \qquad \text{by (iii) of Lemma 2}$$

$$= \ln \frac{f(S_r) - f(\emptyset)}{\beta} - \frac{c(S_{d-1})\left(1 + \ln \frac{f(S_r)-f(\emptyset)}{\beta}\right)}{OPT} + 2$$

$$\leq \ln \frac{f(S_r) - f(\emptyset)}{\beta} + 2.$$

$\square$

4 Conclusion

In this paper we introduced and studied the Pervasive Partial Domination problem: Given a social network modeled by a digraph G, where every edge represents the influence an individual may have to the another, which minimum-cost set of nodes should be bribed so as to reach a desired behaviour? We showed a polynomial time algorithm which provides a $\left(\ln \frac{p-f(\emptyset)}{\beta} + 2\right)$ approximation on the cost of the selected set where p is the targeted impact, $f()$ is a submodular non-decreasing set function used to measure the impact of a set of bribed individuals and $\beta > 0$ represents the minimum gain on the function f one can attain by bribing an additional individual when the target p is (almost) reached. The proposed solution can be generalized to the weighted partial sumbmodular cover problem providing a better approximation with respect to the state of the art, considering also that the value of β depends only on the problem instance.

References

1. Asch, S.E.: Studies of independence and conformity: a minority of one against a unanimous majority. Psychol. Monogr. **70**, 1 (1956)
2. Baumeister, D., Erdélyi, G., Rothe, J.: How hard is it to bribe the judges? A study of the complexity of bribery in judgment aggregation. In: Brafman, R.I., Roberts, F.S., Tsoukiàs, A. (eds.) ADT 2011. LNCS (LNAI), vol. 6992, pp. 1–15. Springer, Heidelberg (2011). https://doi.org/10.1007/978-3-642-24873-3_1
3. Bartholdi, J., Tovey, C., Trick, M.: The computational difficulty of manipulating an election. Soc. Choice Welf. **6**(3), 227–241 (1989). https://doi.org/10.1007/BF00295861
4. Cascio, C.N., Scholz, C., Falk, E.B.: Social influence and the brain: persuasion, susceptibility to influence and retransmission. Curr. Opin. Behav. Sci. **3**, 51–57 (2015)
5. Cellinese, F., D'Angelo, G., Monaco, G., Velaj, Y.: Generalized budgeted submodular set function maximization. In: 43rd International Symposium on Mathematical Foundations of Computer Science (MFCS 2018), LIPIcs, vol. 117, pp. 31:1–31:14 (2018)
6. Christakis, N.A., Fowler, J.H.: Connected: The Surprising Power of Our Social Networks and How They Shape Our Lives. Brown and Co. Little, Boston (2011)
7. Cicalese, F., Milanic, M., Vaccaro, U.: On the approximability and exact algorithms for vector domination and related problems in graphs. Discret. Appl. Math. **161**, 750–767 (2013)

8. Cordasco, G., Gargano, L., Rescigno, A.A.: On finding small sets that influence large networks. Soc. Netw. Anal. Min. **6**, 94 (2016)

9. Cordasco, G., Gargano, L., Rescigno, A.A.: Threshold-bounded dominating set with incentives. In: Proceeding of ICTCS 2018, CEUR Workshop Proceedings, pp. 65–76 (2018)

10. Cordasco, G., Gargano, L., Rescigno, A.A.: Evangelism in social networks: algorithms and complexity. Networks **71**(4), 346–357 (2018)

11. Cordasco, G., Gargano, L., Rescigno, A.A.: Dual domination. In: Colbourn, C.J., Grossi, R., Pisanti, N. (eds.) IWOCA 2019. LNCS, vol. 11638, pp. 160–174. Springer, Cham (2019). https://doi.org/10.1007/978-3-030-25005-8_14

12. Cordasco, G., Gargano, L., Rescigno, A.A.: Active influence spreading in social networks. Theoret. Comput. Sci. **764**, 15–29 (2019)

13. Cordasco, G., et al.: Whom to befriend to influence people. Theoret. Comput. Sci. **810**, 26–42 (2020)

14. Cordasco, G., Gargano, L., Peters, J.G., Rescigno, A.A., Vaccaro, U.: Fast and frugal targeting with incentives. Theoret. Comput. Sci. **812**, 62–79 (2020)

15. Cordasco, G., Gargano, L., Mecchia, M., Rescigno, A.A., Vaccaro, U.: A fast and effective heuristic for discovering small target sets in social networks. In: Lu, Z., Kim, D., Wu, W., Li, W., Du, D.-Z. (eds.) COCOA 2015. LNCS, vol. 9486, pp. 193–208. Springer, Cham (2015). https://doi.org/10.1007/978-3-319-26626-8_15

16. Easley, D., Kleinberg, J.: Networks, Crowds, and Markets: Reasoning About a Highly Connected World. Cambridge University Press, Cambridge (2010)

17. Faliszewski, P., Rothe, J.: Control and bribery in voting. In: Moulin, H., Brandt, F., Conitzer, V., Endriss, U., Lang, J., Procaccia, A. (eds.) Handbook of Computational Social Choice, pp. 146–168. Cambridge University Press, Cambridge (2016)

18. Gargano, L., Hell, P., Peters, J., Vaccaro, U.: Influence diffusion in social networks under time window constraints. In: Moscibroda, T., Rescigno, A.A. (eds.) SIROCCO 2013. LNCS, vol. 8179, pp. 141–152. Springer, Cham (2013). https://doi.org/10.1007/978-3-319-03578-9_12

19. Grandi, U., Turrini, P.: A network-based rating system and its resistance to bribery. In: Proceedings of the 25th International Joint Conference on Artificial Intelligence, pp. 301–307 (2016)

20. Haynes, T.W., Hedetniemi, S., Slater, P.: Fundamentals of Domination in Graphs. Pure and Applied Mathematics (Marcel Dekker), CRC, Boca Raton (1998)

21. Hui, P., Buchegger, S.: Groupthink and peer pressure: social influence in online social network groups. In: 2009 International Conference on Advances in Social Network Analysis and Mining, pp. 53–59. IEEE (2009)

22. Huffaker, D.: Dimensions of leadership and social influence in online communities. Hum. Commun. Res. **36**(4), 593–617 (2010)

23. Pruksachatkun, Y., Pendse, S.R., Sharma, A.: Moments of change: analyzing peer based cognitive support in online mental health forums. In: Proceedings of the 2019 CHI Conference on Human Factors in Computing Systems (CHI2019), Glasgow (2019)

24. Ran, Y., Zhang, Z., Du, H., Zhu, Y.: Approximation algorithm for partial positive influence problem in social network. J. Comb. Optim. **33**, 791–802 (2017). https://doi.org/10.1007/s10878-016-0005-0

25. Ratkiewicz, J., et al.: Truthy: mapping the spread of astroturf in microblog streams. In: Proceedings of the 20th International Conference Companion on World Wide Web, pp. 249–252. ACM (2011)

26. Sviridenko, M.: A note on maximizing a submodular set function subject to a knapsack constraint. Oper. Res. Lett. **32**(1), 41–43 (2004)
27. Wolsey, L.A.: An analysis of the greedy algorithm for the submodular set covering problem. Combinatorica **2**, 385–393 (1982). https://doi.org/10.1007/BF02579435

Unified Greedy Approximability Beyond Submodular Maximization

Yann Disser$^{(\boxtimes)}$ 🄳 and David Weckbecker 🄳

TU Darmstadt, Darmstadt, Germany
{disser,weckbecker}@mathematik.tu-darmstadt.de

Abstract. We consider classes of objective functions of cardinality-constrained maximization problems for which the greedy algorithm guarantees a constant approximation. We propose the new class of γ-α-augmentable functions and prove that it encompasses several important subclasses, such as functions of bounded submodularity ratio, α-augmentable functions, and weighted rank functions of an independence system of bounded rank quotient – as well as additional objective functions for which the greedy algorithm yields an approximation. For this general class of functions, we show a tight bound of $\frac{\alpha}{\gamma} \cdot \frac{e^\alpha}{e^\alpha - 1}$ on the approximation ratio of the greedy algorithm that tightly interpolates between bounds from the literature for functions of bounded submodularity ratio and for α-augmentable functions. In particular, as a by-product, we close a gap in [Math.Prog., 2020] by obtaining a tight lower bound for α-augmentable functions for all $\alpha \geq 1$. For weighted rank functions of independence systems, our tight bound becomes $\frac{\alpha}{\gamma}$, which recovers the known bound of $1/q$ for independence systems of rank quotient at least q.

Keywords: Greedy algorithm · Cardinality-constrained maximization · Approximation ratio · Independence system · Submodularity ratio · Augmentability

1 Introduction

We consider cardinality-constrained maximization problems of the form

$$\max f(X)$$
$$\text{s.t. } |X| \leq k$$
$$X \subseteq U,$$

with a *monotone* objective function $f \colon 2^U \to \mathbb{R}_{\geq 0}$ over a finite ground set U. Additional constraints of the form $X \in \mathcal{X}$ can be modeled by the monotone objective $f'(X) := \max\{f(Y) | Y \in 2^X \cap \mathcal{X}\}$. In this way, every combinatorial, cardinality-constrained maximization problem with monotone objective function

Supported by DFG grant DI 2041/2.

300   Y. Disser and D. Weckbecker

can be captured, and we adopt this framework throughout the paper.[1] For example, the maximum weighted matching problem on a graph $G = (V, E)$ with edge weights $w \colon E \to \mathbb{R}_{\geq 0}$ yields the objective function

$$f(X \subseteq E) = \max\{\sum_{e \in M} w(e) | M \subseteq X, M \text{ is a matching in } G\}.$$

We focus on the performance of the *greedy algorithm* that iteratively produces a solution $S^{\mathrm{G}}_{f,k} = \{x_1, \ldots, x_k\}$ with

$$x_i \in \arg\max_{x \in U \setminus \{x_1, \ldots, x_{i-1}\}} f(\{x_1, \ldots, x_{i-1}\} \cup \{x\}),$$

for all $i \in [k] := \{1, \ldots, k\}$, i.e., it adds elements such that the increase in objective value is maximized in each step. The greedy algorithm is inherently incremental and may be regarded as the most natural approach for incrementally building up infrastructures that support changing active solutions (in the sense of the definition $f'(X)$ above). While this algorithm is widely used in practical applications, greedy solutions can be arbitrarily far away from optimal (e.g., for the knapsack problem). A natural question in this context is, for which objective functions f the greedy algorithm gives a good solution. We are interested in characterizing these objective functions.

Note that we consider the *adaptive* greedy solution $S^{\mathrm{G}}_{f,k}$ as opposed to the *non-adaptive* greedy solution $\tilde{S}^{\mathrm{G}}_{f,k} := S^{\mathrm{G}}_{f,\min\{k,\bar{k}\}}$, where $\bar{k} \in [|U|]$ is the smallest cardinality such that $f(S^{\mathrm{G}}_{\bar{k}} \cup \{x\}) = f(S^{\mathrm{G}}_{\bar{k}})$ for all $x \in U \setminus S^{\mathrm{G}}_{\bar{k}}$. In other words, the non-adaptive greedy algorithm terminates as soon as it cannot improve the solution further. This non-adaptive variant of the greedy algorithm has often been considered in the early literature (e.g., [15,16,22,23]). Note, that for submodular functions, i.e., functions with $f(X \cup Y) + f(X \cap Y) \leq f(X) + f(Y)$ for all $X, Y \subseteq U$, there is no difference between these two variants, and for our purposes both variants are interchangeable in the following sense.

Formally, we measure the quality of the greedy algorithm on a set of objectives $\mathcal{F}$ by the approximation ratio $\sup_{f \in \mathcal{F}} \max_{k \in [|U_f|]} f(S^*_{f,k})/f(X_{f,k})$, where U_f is the ground set of the function $f \in \mathcal{F}$, $S^*_{f,k} \in \arg\max_{X \subseteq U: |X| \leq k} f(X)$ denotes an optimum solution of cardinality at most k, and $X_{f,k} \in \{S^{\mathrm{G}}_{f,k}, \tilde{S}^{\mathrm{G}}_{f,k}\}$ refers to the (non-)adaptive greedy solution of cardinality k. We claim that the approximation ratios of both variants of the greedy algorithm coincide. To see this, observe that the non-adaptive setting is more restrictive, and that every lower bound instance in the non-adaptive setting can be made adaptive by introducing additional elements that add a vanishingly small but positive objective value when added to every solution. This implies that all our bounds on the approximation ratio of the (adaptive) greedy algorithm immediately apply to both variants.

From now on, we write $S^{\mathrm{G}}_k := S^{\mathrm{G}}_{f,k}$ and $S^*_k := S^*_{f,k}$, whenever f is clear from the context. In these terms, we are interested in characterizing the set of objectives for which the greedy algorithm has a bounded approximation ratio. Known

---

[1] Note that the objective function f may be computationally hard to evaluate. If we assume that the greedy algorithm has oracle access to f, it requires $O(|U|k)$ queries to the oracle.

examples include the objectives of maximum (weighted) (b-)matching, maximum (weighted) coverage, and many more [2,3,8,18,30], and we additionally introduce a multi-commodity flow problem (Sect. 2), where the greedy algorithm yields an approximation.

A well-known class of functions for which the greedy algorithm has a bounded approximation ratio of (exactly) $\frac{e}{e-1}$ are monotone, submodular functions [22]. This class includes the maximum coverage problem, but fails to capture many other greedily approximable settings. See Fig. 1 along with the following.

Das and Kempe [8] introduced the class of functions of bounded *submodularity ratio* as a generalization of submodular functions. Importantly, its definition depends on the greedy solutions for different cardinalities. We adapt and weaken the definition from [8] for consistency, by restricting ourselves to greedy solutions and by minimizing over all cardinalities.

Definition 1 ([8]). *The* weak submodularity ratio *of* $f \colon 2^U \to \mathbb{R}_{\geq 0}$ *is (using* $\frac{0}{0} := 1$*)*

$$\gamma(f) := \min_{X \in \{S_0^G,\ldots,S_k^G\}, Y \subseteq U \setminus X} \frac{\sum_{y \in Y}(f(X \cup \{y\}) - f(X))}{f(X \cup Y) - f(X)} \in [0,1].$$

Das and Kempe [8] showed an upper bound of $\frac{e^\gamma}{e^\gamma-1}$ on the approximation ratio of the greedy algorithm for the set of all monotone functions with submodularity ratio at least $\gamma > 0$, and Bian et al. [3] extended this to a tight bound that is additionally parameterized by the curvature of the objective. Since submodular functions have submodularity ratio 1, this bound generalizes the submodular bound. Crucially, it is easy to verify that these results carry over to the set $\tilde{\mathcal{F}}_\gamma$ of all monotone functions with *weak* submodularity ratio at least $\gamma > 0$.[2]

Another generalization of submodularity was proposed by Bernstein et al. [2]. We extend the definition by a weakened variant in order to bring it more in line with Definition 1.

Definition 2 ([2]). *The function* $f \colon 2^U \to \mathbb{R}_{\geq 0}$ *is* (weakly) α-augmentable *for* $\alpha \geq 1$, *if, for every* $X \subseteq U$ ($X \in \{S_0^G,\ldots,S_k^G\}$) *and* $Y \subseteq U$ *with* $Y \not\subseteq X$, *there exists an element* $y \in Y \setminus X$ *with*

$$f(X \cup \{y\}) - f(X) \geq \frac{f(X \cup Y) - \alpha f(X)}{|Y|}.$$

Bernstein et al. showed that the greedy algorithm has an approximation ratio of at most $\alpha \cdot \frac{e^\alpha}{e^\alpha-1}$ on the set $\mathcal{F}_\alpha$ of monotone, α-augmentable functions, for $\alpha \geq 1$, and that this bound is tight for $\alpha \in \{1,2\}$ and in the limit $\alpha \to \infty$. Since submodular functions are 1-augmentable, this bound again generalizes the submodular bound. The class of α-augmentable problems captures the objective of the maximum (weighted) α-dimensional matching problem, which is not submodular. In this paper, we introduce a natural α-commodity flow variant that

---

[2] Here and throughout we use the notation $\tilde{\mathcal{F}}$ as opposed to $\mathcal{F}$ to refer to a function class based on a *weak* definition.

is α-augmentable, and we prove a tight lower bound on the approximation ratio for all $\alpha \geq 1$.

Another well-known setting, besides submodularity, where the greedy algorithm has a bounded approximation ratio, are weighted rank functions of independence systems of bounded rank quotient [17]. An *independence system* is a tuple $(U, \mathcal{I} \subseteq 2^U)$, where $\mathcal{I}$ is closed under taking subsets and $\emptyset \in \mathcal{I}$. For a given weight function $w \colon U \to \mathbb{R}_{\geq 0}$, the *weighted rank function* of $(U, \mathcal{I})$ is given by $f(X) = \max\{\sum_{x \in Y} w(x) | Y \in \mathcal{I} \cap 2^X\}$. The *rank quotient* of an independence system $(U, \mathcal{I})$ is $q(U, \mathcal{I}) := \min_{X \subseteq U} \min_{B, B' \in \mathcal{B}(X)} |B|/|B'|$, where $\frac{0}{0} := 1$, and the set $\mathcal{B}(X)$ of all *bases* of some set $X \subseteq U$ is the set of inclusion-wise maximal subsets of $\mathcal{I} \cap 2^X$, i.e., $\mathcal{B}(X) := \{B \in \mathcal{I} \cap 2^X | \forall x \in X \setminus B \colon B \cup \{x\} \notin \mathcal{I}\}$. Jenkyns [15] and Korte and Hausmann [16] showed that the greedy algorithm has an approximation ratio of exactly $1/q$ on the set $\mathcal{F}_q$ of all weighted rank functions of independence systems with rank quotient at least $q > 0$.[3]

Our Results. Our goal is to unify and to generalize the above classes of functions on which the greedy algorithm has a bounded approximation ratio. To this end, we first observe that each one of the classes $\tilde{\mathcal{F}}_\gamma$, $\mathcal{F}_\alpha$, and $\mathcal{F}_q$ uniquely captures greedily approximable objectives (cf. Fig. 1). In particular, we construct a natural α-augmentable variant of multi-commodity flow that does not have bounded (weak) submodularity ratio (for $\alpha \in \mathbb{N} \setminus \{1\}$) and cannot be expressed as the maximization of a weighted rank function. Besides the α-dimensional matching problem, to our knowledge, the problem introduced in Sect. 2 is the only other natural α-augmentable problem to date.

Proposition 1. *For every $\gamma, q \in (0, 1)$ and $\alpha \geq 1$, it holds that*

$$\tilde{\mathcal{F}}_\gamma \nsubseteq (\mathcal{F}_\alpha \cup \mathcal{F}_q) \quad \text{and} \quad \mathcal{F}_\alpha \nsubseteq (\tilde{\mathcal{F}}_\gamma \cup \mathcal{F}_q) \quad \text{and} \quad \mathcal{F}_q \nsubseteq (\tilde{\mathcal{F}}_\gamma \cup \mathcal{F}_\alpha).$$

This motivates the following definition to consolidate all three classes.

Definition 3. *The function $f \colon 2^U \to \mathbb{R}_{\geq 0}$ is (weakly) γ-α-augmentable for $\gamma \in (0, 1]$ and $\alpha \geq \gamma$ if, for all sets $X \subseteq U$ ($X \in \{S_0^G, ..., S_k^G\}$) and all $Y \subseteq U$ with $Y \nsubseteq X$, there exists $y \in Y$ with*

$$f(X \cup \{y\}) - f(X) \geq \frac{\gamma f(X \cup Y) - \alpha f(X)}{|Y|}.$$

Note that we need to consider the weak variant of this definition if we hope to encompass the class $\tilde{\mathcal{F}}_\gamma$, which enforces its defining property only for "greedy sets", however, any upper bound on the approximation ratio immediately carries over to the same bound in the stronger definition. Also note that γ-α-augmentability only requires $\alpha \geq \gamma$, unlike α-augmentability where $\alpha \geq 1$. This is in line with the definitions of α-augmentability where $\gamma = 1$ and of the submodularity ratio where $\alpha = \gamma$. We let $\tilde{\mathcal{F}}_{\gamma, \alpha}$ denote the set of all weakly

---

[3] Note that we abuse notation, since, e.g., $\mathcal{F}_\alpha \neq \mathcal{F}_q$ for $\alpha = q = 1$. However, the set of functions we are referring to will always be clear by the naming of the indices.

γ-α-augmentable functions. The first part of our main result is that this set encompasses all functions in $\tilde{\mathcal{F}}_\gamma \cup \mathcal{F}_\alpha \cup \mathcal{F}_q$ and captures additional functions (cf. Fig. 1). Formally, we show the following.

Theorem 1. *For every $\gamma, q \in (0,1]$, every $\gamma' \in (0,1)$, every $\alpha \geq 1$, and every $\alpha' \geq \gamma'$, it holds that*

$$\tilde{\mathcal{F}}_{\gamma, \max\{\alpha, 1/q\}} \supseteq \tilde{\mathcal{F}}_\gamma \cup \mathcal{F}_\alpha \cup \mathcal{F}_q \qquad \text{and} \qquad \tilde{\mathcal{F}}_{\gamma', \alpha'} \not\subseteq \tilde{\mathcal{F}}_\gamma \cup \mathcal{F}_\alpha \cup \mathcal{F}_q.$$

Note that α' and γ' in Theorem 1 do not depend on α, γ and q. The second part of our main result is a tight bound on the approximation ratio of the greedy algorithm on $\tilde{\mathcal{F}}_{\gamma, \alpha}$ (cf. Theorem 6 and Proposition 4).

Theorem 2. *The approximation ratio of the greedy algorithm on the class $\tilde{\mathcal{F}}_{\gamma, \alpha}$ of monotone, weakly γ-α-augmentable functions, with $\gamma \in (0,1]$ and $\alpha \geq \gamma$, is exactly*

$$\frac{\alpha}{\gamma} \cdot \frac{e^\alpha}{e^\alpha - 1}.$$

Importantly, this bound recovers exactly the known bound for functions of bounded submodularity ratio, since $\tilde{\mathcal{F}}_\gamma \subseteq \tilde{\mathcal{F}}_{\gamma, \gamma}$, as well as the known bound for α-augmentable functions, since $\mathcal{F}_\alpha \subseteq \tilde{\mathcal{F}}_{1, \alpha}$. In that sense, our new bound interpolates tightly between these two bounds and generalizes them. In addition, our tight lower bound for $\tilde{\mathcal{F}}_{1, \alpha}$ is obtained with an α-augmentable function. This means that, in particular, we are able to close the gap left in [2], by showing a tight lower bound for α-augmentable objectives, for all $\alpha \geq 1$.

Corollary 1. *The approximation ratio of the greedy algorithm on the class $\mathcal{F}_\alpha$ of monotone, α-augmentable functions is exactly $\alpha \cdot \frac{e^\alpha}{e^\alpha - 1}$ for all $\alpha \geq 1$.*

Finally, we are also able to show a tight bound of α/γ for γ-α-augmentable, weighted rank functions on independence systems. Since $\mathcal{F}_q \subseteq \tilde{\mathcal{F}}_{1, 1/q}$ (by Theorem 1), our bound recovers exactly the known bound of $1/q$ for the approximation ratio of the greedy algorithm when the rank quotient is bounded from below by $q > 0$. This means that the class of monotone, weakly γ-α-augmentable functions truly unifies and generalizes the three classes $\tilde{\mathcal{F}}_\gamma$, $\mathcal{F}_\alpha$, and $\mathcal{F}_q$ of greedily approximable functions (cf. Fig. 1). Note that, in particular, the lower bound is tight already for α-augmentable functions, which implies a tight bound of α for the approximation ratio of the greedy algorithm on α-augmentable weighted rank functions.

Theorem 3. *Let $\mathcal{F}_{\mathrm{IS}} := \bigcup_{q \in (0,1]} \mathcal{F}_q$ be the set of weighted rank functions on some independence system. The approximation ratio of the greedy algorithm on the class $\tilde{\mathcal{F}}_{\gamma, \alpha} \cap \mathcal{F}_{\mathrm{IS}}$, with $\gamma \in (0,1]$ and $\alpha \geq \gamma$, is exactly $\frac{\alpha}{\gamma}$.*

The proofs of all results can be found in the full version [9].

Related Work. We can view our cardinality-constrained maximization framework as a special case of maximization over an independence system. In particular, the cardinality-constraint can be expressed as a uniform matroid contraint [17].

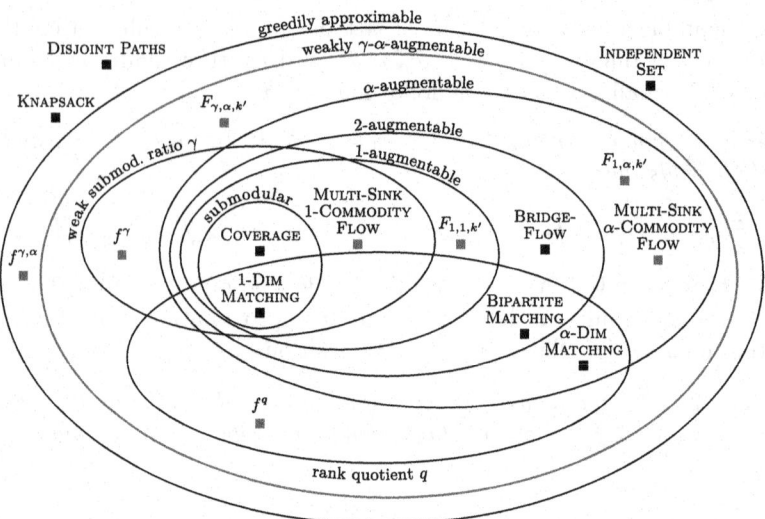

Fig. 1. Relation of the different problem classes. Newly introduced classes and problems are marked in red. The parameter k' is chosen sufficiently large, depending on γ and α. (Color figure online)

From that perspective, the most basic, non-trivial setting is the maximization of a linear (i.e., modular) objective over an independence system. Regarding the approximation ratio of the greedy algorithm, this classic setting is equivalent to the maximization of a weighted rank function, as considered in Theorem 3. This is easy to see by considering the non-adaptive variant of the greedy algorithm, and by observing that the greedy solution is guaranteed to remain feasible while the algorithm makes progress.

In that sense, the perfomance of the greedy algorithm for weighted rank function maximization has extensively been studied in the past. Rado [25] showed that the greedy algorithm is optimal for all weight functions if the underlying independence system is a matroid, and Edmonds [10] established the reverse implication. Jenkyns [15] extended this result by showing an upper bound of $1/q$ for the approximation ratio of the greedy algorithm on independence systems with rank quotient q, and Korte and Hausmann [16] gave a tight lower bound. Years later, Mestre [21] independently proved this tight bound for the subclass of k-extendible independence systems. Bouchet [4] gave a different generalization of the result by Rado and Edmonds by showing that the greedy algorithm remains optimal on symmetrical matroids.

Another prominent setting is the maximization of a submodular function over an independence system. Again, this includes cardinality-constrained maximization of a submodular objective, which is equivalent to submodular maximization over a uniform matroid. Nemhauser, Wolsey, and Fisher [23] showed that the greedy algorithm has a tight approximation ratio of $\frac{e}{e-1}$ for maximizing a monotone, submodular function under a cardinality-constraint. Krause et al. [19] observed that the approximation ratio is unbounded when maximizing the minimum of two monotone, submodular functions. Non-monotone submodular maximization over a cardinality-constraint (and knapsack constraints) was considered by Lee et al. [20]. Feldman et al. [14] analyzed a variant of the continuous greedy algorithm [28] and showed an upper bound on its approximation ratio of $(1/e-o(1))^{-1}$. This bound for the non-monotone case with cardinality-constraint was later improved by Buchbinder et al. [5] and Ene and Nguyen [11] by further adapting the (continuous) greedy algorithm. For maximizing a submodular function subject to k-extentible system and k-systems constraints, Feldman et al. [12,13] considered three variants of the greedy algorithm, a repeated greedy, a sample greedy and a simultaneous greedy. They were able to show approximation ratios of $k + O(1)$ for k-extendible system constraints and $k + O(\sqrt{k})$ for k-system constraints.

Maximization of a monotone, submodular function over a matroid was considered by Vondrák [28] and by Calinescu et al. [6], who showed that the continuous greedy algorithm has an approximation ratio of $\frac{e}{e-1}$ in this setting. Nemhauser, Wolsey, and Fisher [23], showed an upper bound of $p + 1$ for the regular greedy algorithm when maximizing over the intersection of p matroids. A generalization of this upper bound to the setting o maximizing subject to a p-system constraint was later proven by Calinescu et al. [6]. Conforti and Cornuejols [7] gave an upper bound of $p + c$ depending on the curvature c of the monotone submodular function – this interpolates between the submodular bound of [23] ($c = 1$) and the linear bound of [16] ($c = 0$). Vondrák [29] showed that the continuous greedy algorithm has an approximation ratio of at most $c\frac{e^c}{e^c-1}$ over an arbitrary matroid, and Sviridenko, Vondrák, and Ward [27] showed an improved upper bound of $\frac{e}{e-c}$ for the approximation ratio of a modified continuous greedy algorithm over a uniform matroid (i.e., a cardinality-constraint).

Other variants of the problem setting include the maximization of a monotone, submodular function over a knapsack constraint [26], and robust submodular maximization [1,24].

2   Weak Submodularity Ratio, α-Augmentability, and Independence Systems

In this section, we give an idea how to prove Proposition 1, i.e., how to separate the function classes $\tilde{\mathcal{F}}_\gamma$, $\mathcal{F}_\alpha$, and $\mathcal{F}_q$.

We start by introducing a natural α-commodity flow problem that models, e.g., production processes where output is limited by availability of all components. The objective of this problem is (exactly) α-augmentable, but, for $\alpha \in \mathbb{N} \setminus \{1\}$, does not have a bounded (weak) submodularity ratio and cannot be expressed as a weighted rank function over an independence system. This problem also gives a tight lower bound for the approximation ratio of the greedy algorithm on α-augmentable functions, for $\alpha \in \mathbb{N}$. We will extend this lower bound to all $\alpha \geq 1$ in Sect. 3.1, and thus close a gap left by [2].

Definition 4. *Let $G = (V, E)$ be a directed graph with source $s \in V$, sinks $T \subseteq V$, and arc capacities $\mu \colon E \to \mathbb{R}_{\geq 0}$. We define an s-T-flow to be a function $\vartheta \colon E \to \mathbb{R}_{\geq 0}$ that satisfies*

$$
\begin{aligned}
\vartheta(e) \leq \mu(e) && \forall e \in E && \textit{(capacity constraint)}, \\
\mathrm{ex}_\vartheta(v) = 0 && \forall v \in V \setminus (\{s\} \cup T) && \textit{(flow conservation)}, \\
\mathrm{ex}_\vartheta(t) \geq 0 && \forall t \in T && \textit{(T are sinks)},
\end{aligned}
$$

where (using $\delta^+(v) := (\{v\} \times V) \cap E$, $\delta^-(v) := (V \times \{v\}) \cap E$) the excess of a vertex $v \in V$ is defined as $\mathrm{ex}_\vartheta(v) := \sum_{e \in \delta^-(v)} \vartheta(e) - \sum_{e \in \delta^+(v)} \vartheta(e)$.

We extend this notion to multi-commodity flows, where each commodity has an independent capacity function.

Definition 5. *Let $\alpha \in \mathbb{N}$ and $G = (V, E)$ be a graph, let $s \in V$ and $T \subseteq V$, and let $\vec{\mu} = (\mu_i \colon E \to \mathbb{R}_{\geq 0})_{i \in [\alpha]}$ be capacity functions. A multicommodity-flow in G w.r.t. $\vec{\mu}$ is a tuple $\vec{\vartheta} = (\vartheta_1, ..., \vartheta_\alpha)$, where ϑ_i is an $s - T - flow$ in G with respect to capacities μ_i. The minimum-excess of the sink vertex $t \in T$ in $\vec{\vartheta}$ is*

$$
\mathrm{minex}_{\vec{\vartheta}}(t) := \min_{i \in [\alpha]} \mathrm{ex}_{\vartheta_i}(t).
$$

For convenience, we let $\mu(u, v) := \mu((u, v))$, $\vartheta(u, v) := \vartheta((u, v))$, and we let $\mathrm{ex}_\vartheta(V') := \sum_{v \in V'} \mathrm{ex}_\vartheta(v)$ for $V' \subseteq V$, and $\mathrm{minex}_{\vec{\vartheta}}(T') := \sum_{t \in T'} \mathrm{minex}_{\vec{\vartheta}}(t)$ for $T' \subseteq T$ in the following.

An instance of the problem MULTI-SINK α-COMMODITY FLOW, for $\alpha \in \mathbb{N}$, is given by a tuple $(G, s, T, \vec{\mu})$, where $G = (V, E)$ is a directed graph, $s \in V$ is a source vertex, $T \subseteq V$ contains sink vertices, and $\vec{\mu} = (\mu_i \colon E \to \mathbb{R}_{\geq 0})_{i \in [\alpha]}$ are capacity functions. The problem is to find a subset of sinks $X \subseteq T$ with $|X| = k$ that maximizes the objective function

$$
f(X) = \max_{\vec{\vartheta} \in \mathcal{M}_{G,\mu}} \mathrm{minex}_{\vec{\vartheta}}(X),
$$

where $\mathcal{M}_{G,\mu}$ denotes the set of all multicommodity-flows in G w.r.t. capacities $\vec{\mu}$.

Theorem 4. *For every $\alpha \in \mathbb{N}$, the objective of MULTI-SINK α-COMMODITY FLOW is monotone and α-augmentable.*

For $\alpha = 2$, MULTI-SINK α-COMMODITY FLOW problem is equivalent to the BRIDGEFLOW problem considered in [2]. We generalize the tight lower bound construction for BRIDGEFLOW to arbitrary $\alpha \in \mathbb{N}$ (see full version [9] for details).

With this, we obtain a lower bound for the approximation ratio of the greedy algorithm on $\mathcal{F}_\alpha$ for $\alpha \in \mathbb{N}$ that tightly matches the upper bound of [2], i.e., we obtain Corollary 1 for $\alpha \in \mathbb{N}$. In particular, it follows that the objective of MULTI-SINK α-COMMODITY FLOW is not β-augmentable for any $\beta < \alpha$. We will generalize the lower bound to all $\alpha \geq 1$ in Sect. 3.1.

Theorem 5. *For $\alpha \in \mathbb{N}$, the greedy algorithm has an approximation ratio of at least $\alpha \frac{e^\alpha}{e^\alpha - 1}$ for MULTI-SINK α-COMMODITY FLOW.*

2.1   Separating Function Classes

We are now ready to show Proposition 1 for $\alpha \in \mathbb{N} \setminus \{1\}$. The case $\alpha \geq 1$ will be addressed in Sect. 3.1.

In order to seperate $\mathcal{F}_\alpha$ for $\alpha \in \mathbb{N} \setminus \{1\}$, we have shown that the objective of MULTI-SINK α-COMMODITY FLOW does not have a (weak) submodularity ratio bounded away from zero, and cannot be represented as the weighted rank function of some independence system. Separating $\tilde{\mathcal{F}}_\gamma$ and $\mathcal{F}_q$ for every $\gamma, q \in (0,1)$ is done by defining a simple according function.

3 γ-α-Augmentability

In this section, we argue that the class $\tilde{\mathcal{F}}_{\gamma,\alpha}$ of weakly γ-α-augmentable functions unifies and generalizes the classes $\tilde{\mathcal{F}}_\gamma$, $\mathcal{F}_\alpha$, and $\mathcal{F}_q$. We start by proving the first half of Theorem 1. The second half will be shown in Sect. 3.1, together with lower bounds for the approximation ratio of the greedy algorithm.

Since (weak) γ-α-augmentability implies (weak) γ'-α'-augmentability for all $\gamma \geq \gamma'$ and $\alpha \leq \alpha'$, the following proposition implies the first part of Theorem 1.

Proposition 2. *For every* $\gamma, q \in (0,1]$, *and every* $\alpha \geq 1$, *it holds that*

$$\tilde{\mathcal{F}}_{1,\alpha} \supseteq \mathcal{F}_\alpha \quad \text{and} \quad \tilde{\mathcal{F}}_{\gamma,\gamma} \supseteq \tilde{\mathcal{F}}_\gamma \quad \text{and} \quad \tilde{\mathcal{F}}_{\gamma,\gamma/q} \supseteq \mathcal{F}_q.$$

Having shown that $\tilde{\mathcal{F}}_{\gamma,\alpha}$ subsumes the other three classes of functions, we now prove the upper bound of Theorem 2 for this class. Observe that the upper bound trivially carries over to the class of monotone, γ-α-augmentable (not weakly) functions.

Theorem 6. *The approximation ratio of the greedy algorithm on the class* $\tilde{\mathcal{F}}_{\gamma,\alpha}$ *of monotone, weakly* γ-α-*augmentable functions, with* $\gamma \in (0,1]$ *and* $\alpha \geq \gamma$, *is at most*

$$\frac{\alpha}{\gamma} \cdot \frac{e^\alpha}{e^\alpha - 1}.$$

3.1   A Critical Function

To obtain the tight lower bound of Theorem 2 for weakly γ-α-augmentable problems and to separate this class from $\tilde{\mathcal{F}}_\gamma \cup \mathcal{F}_\alpha \cup \mathcal{F}_q$, we introduce a function that is inspired by a construction in [3] for the submodularity ratio.

We fix $\gamma \in (0,1]$ and $\alpha \geq \gamma$. Let $k \in \mathbb{N}$ with $k > \alpha$, and let $A = \{a_1, ..., a_k\}$ and $B = \{b_1, ..., b_k\}$ be disjoint sets. We set $U = A \cup B$, define $\xi_i := \frac{1}{k}(\frac{k-\alpha}{k})^{i-1}$ and let $h(x) := \frac{\gamma^{-1}-1}{k-1}x^2 + \frac{k-\gamma^{-1}}{k-1}x$. For our purpose, the important facts about h are $h(0) = 0$, $h(1) = 1$, $h(k) = \frac{k}{\gamma}$ and that h is convex. With this in mind, we define the function $F_{\gamma,\alpha,k} : 2^U \to \mathbb{R}_{\geq 0}$ by

$$F_{\gamma,\alpha,k}(X) = \max_{X' \subseteq X} \left\{ \frac{h(|\{b_1\} \cap X'| \cdot |B \cap X'|)}{k} \left(1 - \alpha \sum_{\substack{i \in [k]: \\ a_i \in A \cap X'}} \xi_i \right) + \sum_{\substack{i \in [k]: \\ a_i \in A \cap X'}} \xi_i \right\}.$$

We show that our modification of the function introduced in [3] retains the same structure in regard to greedy solutions.

Proposition 3. *For $i \in [k]$, the greedy algorithm picks the element a_i in iteration i, and, for $i \in [2k] \setminus [k]$, the greedy algorithm picks the element b_{i-k} in iteration i.*

With this, we can show that $F_{\gamma,\alpha,k}$ is weakly γ-α-augmentable.

Lemma 1. *For every $\gamma \in (0,1]$, every $\alpha \geq \gamma$, and every $k \in \mathbb{N}$ with $k > \alpha$, it holds that $F_{\gamma,\alpha,k} \in \tilde{\mathcal{F}}_{\gamma,\alpha}$.*

It is straightforward to bound the approximation ratio of the greedy algorithm for $F_{\gamma,\alpha,k}$.

Proposition 4. *The approximation ratio of the greedy algorithm for maximizing the function $F_{\gamma,\alpha,k}$, with $\gamma \in (0,1]$, $\alpha \geq \gamma$ and $k \in \mathbb{N}$ with $k > \alpha$, is at least*

$$\frac{\alpha}{\gamma} \frac{1}{1 - (1 - \frac{\alpha}{k})^k}.$$

Now, the tight lower bound of Theorem 2 follows in the limit $k \to \infty$.

It turns out that, for $\gamma = 1$, the function $F_{\gamma,\alpha,k}$ is α-augmentable. Together with Proposition 4, this extends the lower bound of Theorem 5 to all $\alpha \geq 1$ and thus proves Corollary 1.

For k large enough, it can even be shown that $F_{1,\alpha,k} \notin (\tilde{\mathcal{F}}_\gamma \cup \mathcal{F}_q)$ for fixed $\gamma, q \in (0,1]$, which separates $\mathcal{F}_\alpha$ for $\alpha \geq 1$, and thus closes the gap left in Sect. 2.1.

3.2 γ-α-Augmentability on Independence Systems

To tightly capture the class $\mathcal{F}_q$ of weighted rank functions on independence systems, we show a stronger bound for the approximation ratio of the greedy algorithm on monotone, (weakly) γ-α-augmentable functions. In particular, it was already shown in [2] that the objective function of α-DIMENSIONAL MATCHING is (exactly) α-augmentable, while the greedy algorithm yields an approximation ratio of α, which beats the upper bound of $\alpha \cdot \frac{e^\alpha}{e^\alpha - 1}$ for this case. We show that this can be explained by the fact that α-DIMENSIONAL MATCHING can be represented via a weighted rank function over an indepencence system. We first show the upper bound of Theorem 3.

Proposition 5. *Let $\mathcal{F}_{IS} := \bigcup_{q \in (0,1]} \mathcal{F}_q$ be the set of weighed rank functions on some independence system. The approximation ratio of the greedy algorithm on the class $\tilde{\mathcal{F}}_{\gamma,\alpha} \cap \mathcal{F}_{IS}$ is at most $\frac{\alpha}{\gamma}$, for every $\gamma \in (0,1]$ and $\alpha \geq \gamma$.*

The lower bound of Theorem 3 follows directly from the well-known tight bound of $1/q$ for $\mathcal{F}_q$ [15] and the fact that every weighted rank function over an independence system with rank quotient q is γ-$\frac{\gamma}{q}$-augmentable, by Proposition 2.

4 Outlook

The vision guiding our work is to precisely characterize the set of cardinality-constrained maximization problems for which the greedy algorithm yields an approximation, and to tightly bound the corresponding approximation ratio.

In this paper, we have made progress towards this goal by unifying and generalizing important classes of greedily approximable maximization problems, and by providing tight bounds on the approximation ratio for the resulting generalized class of problems. While this brings us closer to a full characterization, there are still settings that are not captured by (weak) γ-α-augmentability.

Proposition 6. *For $\gamma \in (0, 1]$ and $\alpha \geq \gamma$, there exists a monotone function $f^{\gamma,\alpha}$ that is not weakly γ-α-augmentalbe, and for which the greedy algorithm computes an optimum solution.*

Proof (Sketch). Let U be any ground set of size $|U| > \frac{1}{\gamma}$ and consider the objective function $f^{\gamma,\alpha} \colon 2^U \to \mathbb{R}_{\geq 0}$ with

$$f^{\gamma,\alpha}(X) = |X|^2,$$

and show that it is not weakly γ-α-augmentable. Yet, picking elements in any order is obviously optimal. □

We leave it as an open problem to find a natural generalization of weak γ-α-augmentability that captures a larger set of greedily approximable objectives. The challenge is to find a meaningful generalization in terms of a natural definition that does not directly depend on the behavior of the greedy algorithm, but rather enforces some structural property of the objective function. In that sense, the dependency of weak γ-α-augmentability on the greedy solutions $S_0^G, \ldots, S_k^G$ is a significant flaw. Note that we needed to introduce this dependency in order to encompass settings with bounded (weak) submodularity ratios, since the definition of the latter depends on the greedy solutions as well. Importantly, our upper bound on the approximation ratio of the greedy algorithm carries over to the stronger notion of γ-α-augmentability that requires the defining property to hold for *all* sets X, and not just the greedy solutions. Our tight lower bound does not immediately translate to this, more restrictive, definition, and it remains an open problem to construct a tight lower bound in this setting as well.

References

1. Anari, N., Haghtalab, N., Naor, S., Pokutta, S., Singh, M., Torrico, A.: Structured robust submodular maximization: offline and online algorithms. In: Proceedings of the 22nd International Conference on Artificial Intelligence and Statistics (AISTATS), pp. 3128–3137 (2019)
2. Bernstein, A., Disser, Y., Groß, M., Himburg, S.: General bounds for incremental maximization. Math. Program. **191**, 953–979 (2020). https://doi.org/10.1007/s10107-020-01576-0

3. Bian, A., Buhmann, J., Krause, A., Tschiatschek, S.: Guarantees for greedy maximization of non-submodular functions with applications. In: Proceedings of the 34th International Conference on Machine Learning (ICML), pp. 498–507 (2017)
4. Bouchet, A.: Greedy algorithm and symmetric matroids. Math. Program. **38**(2), 147–159 (1987). https://doi.org/10.1007/BF02604639
5. Buchbinder, N., Feldman, M., Naor, J., Schwartz, R.: Submodular maximization with cardinality constraints. In: Proceedings of the Annual ACM-SIAM Symposium on Discrete Algorithms (SODA), pp. 1433–1452 (2014). https://doi.org/10.1137/1.9781611973402.106
6. Calinescu, G., Chekuri, C., Pál, M., Vondrák, J.: Maximizing a monotone submodular function subject to a matroid constraint. SIAM J. Comput. **40**(6), 1740–1766 (2011). https://doi.org/10.1137/080733991
7. Conforti, M., Cornuéjols, G.: Submodular set functions, matroids and the greedy algorithm: tight worst-case bounds and some generalizations of the Rado-Edmonds theorem. Discret. Appl. Math. **7**, 251–274 (1984)
8. Das, A., Kempe, D.: Approximate submodularity and its applications: subset selection, sparse approximation and dictionary selection. J. Mach. Learn. Res. **19**, 1–34 (2018)
9. Disser, Y., Weckbecker, D.: Unified greedy approximability beyond submodular maximization. arXiv:2011.00962 (2020)
10. Edmonds, J.: Matroids and the greedy algorithm. Math. Program. **1**, 127–136 (1971). https://doi.org/10.1007/BF01584082
11. Ene, A., Nguyen, H.L.: Constrained submodular maximization: beyond 1/e. In: 2016 IEEE 57th Annual Symposium on Foundations of Computer Science (FOCS), pp. 248–257 (2016). https://doi.org/10.1109/FOCS.2016.34
12. Feldman, M., Harshaw, C., Karbasi, A.: Greed is good: near-optimal submodular maximization via greedy optimization. In: Proceedings of the 30th Conference on Learning Theory (COLT), pp. 758–784 (2017)
13. Feldman, M., Harshaw, C., Karbasi, A.: Simultaneous greedys: a swiss army knife for constrained submodular maximization. arXiv:2009.13998 (2020)
14. Feldman, M., Naor, J., Schwartz, R.: A unified continuous greedy algorithm for submodular maximization. In: 2011 IEEE 52nd Annual Symposium on Foundations of Computer Science (FOCS), pp. 570–579 (2011). https://doi.org/10.1109/FOCS.2011.46
15. Jenkyns, T.: The efficacy of the "greedy" algorithm. In: Proceedings of the 7th Southeastern Conference on Combinatorics, Graph Theory and Computing, pp. 341–350 (1976)
16. Korte, B., Hausmann, D.: An analysis of the greedy heuristic for independence systems. Ann. Discrete Math. **2**, 65–74 (1978)
17. Korte, B., Vygen, J.: Combinatorial Optimization: Theory and Algorithms. Springer, Heidelberg (2012). https://doi.org/10.1007/978-3-642-24488-9
18. Krause, A., Golovin, D.: Submodular function maximization. Tractability Pract. Approach. Hard Prob. **3**, 71–104 (2011). https://doi.org/10.1017/CBO9781139177801.004
19. Krause, A., McMahan, H., Guestrin, C., Gupta, A.: Robust submodular observation selection. J. Mach. Learn. Res. **9**, 2761–2801 (2008)
20. Lee, J., Mirrokni, V., Nagarajan, V., Sviridenko, M.: Non-monotone submodular maximization under matroid and knapsack constraints. In: Proceedings of the 41st Annual ACM Symposium on Theory of Computing (STOC), pp. 323–332 (2009). https://doi.org/10.1145/1536414.1536459

21. Mestre, J.: Greedy in approximation algorithms. In: Azar, Y., Erlebach, T. (eds.) ESA 2006. LNCS, vol. 4168, pp. 528–539. Springer, Heidelberg (2006). https://doi.org/10.1007/11841036_48

22. Nemhauser, G., Wolsey, L., Fisher, M.: An analysis of approximations for maximizing submodular set functions - I. Math. Program. **14**, 265–294 (1978). https://doi.org/10.1007/BF01588971

23. Nemhauser, G., Wolsey, L., Fisher, M.: An analysis of approximations for maximizing submodular set functions - II. Math. Program. Stud. **8**, 73–87 (1978)

24. Orlin, J., Schulz, A., Udwani, R.: Robust monotone submodular function maximization. Math. Program. **172**, 505–537 (2015). https://doi.org/10.1007/s10107-018-1320-2

25. Rado, R.: A theorem on independence relations. Q. J. Math. **13**, 83–89 (1942). https://doi.org/10.1093/qmath/os-13.1.83

26. Sviridenko, M.: A note on maximizing a submodular set function subject to a knapsack constraint. Oper. Res. Lett. **32**(1), 41–43 (2004). https://doi.org/10.1016/S0167-6377(03)00062-2

27. Sviridenko, M., Vondrák, J., Ward, J.: Optimal approximation for submodular and supermodular optimization with bounded curvature. In: Proceedings of the 26th Annual ACM-SIAM Symposium on Discrete Algorithms (SODA), pp. 1134–1148 (2015)

28. Vondrák, J.: Optimal approximation for the submodular welfare problem in the value oracle model. In: Proceedings of the 40th Annual ACM Symposium on Theory of Computing (STOC), pp. 67–74 (2008). https://doi.org/10.1145/1374376.1374389

29. Vondrák, J.: Submodularity and curvature: the optimal algorithm. RIMS Kôkyûroku Bessatsu **B23**, 253–266 (2010)

30. Williamson, D., Shmoys, D.: The Design of Approximation Algorithms. Cambridge University Press, Cambridge (2011)

Neighborhood Persistency of the Linear Optimization Relaxation of Integer Linear Optimization

Kei Kimura[1]([⊠])[iD] and Kotaro Nakayama[2]

[1] Faculty of Information Science and Electrical Engineering, Kyushu University,
Fukuoka, Japan
kkimura@inf.kyushu-u.ac.jp
[2] Oplan Incorporated, Tokyo, Japan

Abstract. For an integer linear optimization (ILO) problem, persistency of its linear optimization (LO) relaxation is a property that for every optimal solution of the relaxation that assigns integer values to some variables, there exists an optimal solution of the ILO problem in which these variables retain the same values. Although persistency has been used to develop heuristic, approximation, and fixed-parameter algorithms for special cases of ILO, its applicability remains unknown in the literature. In this paper, we propose a stronger property called *neighborhood persistency* and show that the LO relaxation of ILO on unit-two-variable-per-inequality (UTVPI) systems is a maximal class of ILO such that its LO relaxation has (neighborhood) persistency. Our result on neighborhood persistency generalizes the previous results of Nemhauser and Trotter, Hochbaum et al., and Fiorini et al., and implies fixed-parameter tractability and two-approximability for ILO on UTVPI systems where the objective function and the variables are non-negative.

Keywords: Integer linear optimization · Linear optimization · Unit-two-variable-per-inequality system · Persistency

1 Introduction

In this paper, we mainly investigate the ILO problem on a unit-two-variable-per-inequality (UTVPI) system. In this problem, we are given a matrix $A \in \{-1, 0, 1\}^{m \times n}$ with at most two nonzero elements per row, an integer vector $b \in \mathbb{Z}^m$, and a rational vector $w \in \mathbb{Q}^n$, and our task is to compute the optimal value of the following ILO problem:

$$
\begin{aligned}
\text{minimize} \quad & w^{\mathrm{T}} x \\
\text{subject to} \quad & Ax \geq b, \\
& x \in \mathbb{Z}^n.
\end{aligned}
\tag{1}
$$

The first author is partially supported by JST, ACT-X Grant Number JPMJAX200C, Japan, and JSPS KAKENHI Grant Numbers JP19K22841, JP21K17700.

I. Ljubić et al. (Eds.): ISCO 2022, LNCS 13526, pp. 312–323, 2022.
https://doi.org/10.1007/978-3-031-18530-4_23

ILO on UTVPI systems has many applications in practice and theory. Practical applications include map labeling [2] and scheduling [17], and theoretical applications include various problems in graph theory and combinatorial optimization such as the vertex cover problem, the maximum independent set problem, a disjoint path problem [13], and the minimum clique cover problem [3]. ILO on UTVPI systems is strongly NP-hard, since it includes the vertex cover problem, which is NP-hard. When UTVPI systems are *monotone* (i.e., when each constraint is of the form $x_p - x_q \geq c$) they are sometimes called *difference constraint systems* (DCSs), and ILO on DCSs is solvable in polynomial time by minimum cost flow algorithms (see, e.g., [1]). ILO on DCSs includes the dual linear optimization problem of the shortest path problem, a fundamental problem in combinatorial optimization. The feasibility problem of UTVPI systems has also been extensively studied. The feasibility problem is solvable in polynomial time, and many algorithms have been proposed ([7,8,13,15,16]). The feasibility problem also appears in practice, e.g., abstract interpretation [10].

It is quite common to solve an ILO problem by first solving its *linear optimization (LO) relaxation* and rounding up or down the obtained LO solution. Here the linear optimization relaxation of an ILO problem is a problem where the integrality condition of the variables (i.e., $x \in \mathbb{Z}^n$) is dropped (or changed to $x \in \mathbb{R}^n$). The optimal solution of the LO relaxation is sometimes called the optimal *fractional* solution of the ILO problem, whose optimal solution is called the optimal *integer* solution.

The LO relaxation of certain ILO subclasses has *persistency*. Persistency of the LO relaxation of an ILO problem is a property that for every optimal fractional solution that assigns integer values to some variables, an optimal *integer* solution exists in which these variables retain the same values. If the LO relaxation of an ILO problem has persistency, one can solve the problem by obtaining an optimal fractional solution (in polynomial time) by solving linear optimization, substituting the integer values of the fractional solution to the corresponding variables, and solving the resulting problem with fewer variables. This algorithmic framework gives not only a fast heuristic algorithm but also a theoretically fast one. Indeed, persistency was first shown in the LO relaxation of an ILO formulation of the vertex cover problem [12], which is ILO on special UTVPI systems, and used to obtain a fixed-parameter algorithm for the vertex cover problem [9]. The persistency result in [12] is generalized to special cases of ILO on UTVPI systems [4,5].

Our Contribution
In this paper, we show that the LO relaxation of ILO on UTVPI systems in general has persistency. More strongly, we propose *neighborhood persistency*, which is stronger than persistency, and show that the LO relaxation of ILO on UTVPI systems in general has neighborhood persistency.[1] Neighborhood per-

---

[1] Neighborhood persistency resembles the generalization of persistency [4]. However, that work assumed that a solution of the relaxation problem is extreme, although it is not in the neighborhood persistency.

sistency of the LO relaxation is a property that for every optimal fractional solution x^*, there exists an optimal integer solution in the *integer neighborhood* of x^*, where for vector $x \in \mathbb{R}^n$ its integer neighborhood $N(x)$ is defined as $N(x) = \{z \in \mathbb{Z}^n \mid |z_j - x_j| < 1 \ (j = 1, \ldots, n)\}$. To obtain our result, we use the celebrated strong duality theorem on LO, which was not used in the proofs of the previous results on persistency mentioned above.

We also show that ILO on UTVPI systems is a maximal subclass of ILO with its LO relaxation having persistency in the sense that if we allow (i) an inequality with *three* variables whose coefficients are all one or (ii) an inequality with two variables whose coefficients are in $\{1, 2\}$, then not even persistency holds for the LO relaxation of such ILO.

From our neighborhood-persistency result, we can solve ILO on UTVPI systems by (i) solving the LO relaxation and then (ii) solving a *binary* ILO problem (i.e., each variable takes value zero or one), since an optimal integer solution exists that rounds up or down the fractional values of the optimal LO solution. Using this two-step algorithm we show fixed-parameter tractability (in terms of the solution size) of the ILO on UTVPI systems with non-negative objective functions and non-negative variables. We also obtain another proof of two-approximability of such ILO problems, which was first shown by Hochbaum et al. [5]. Note that a (half-integral) optimal solution of the LO relaxation of an ILO problem on a UTVPI system can be efficiently computed by first transforming it to an ILO problem on a DCS by a previously proposed method [5] and solving the transformed ILO problem by a minimum cost flow algorithm.

Previous and Related Work

The vertex cover problem is, given an undirected graph $G = (V, E)$, to compute the minimum size of vertex subset $C \subseteq V$ such that every edge in E has at least one end vertex in C. It is well known that this problem is formulated as ILO on special UTVPI systems as follows. The variable set is $\{x_i \mid v_i \in V\}$, where each variable is binary, the objective function is $\sum_{i=1}^{|V|} x_i$, and the linear system is $\{x_i + x_j \geq 1 \mid \{i, j\} \in E\}$. Nemhauser and Trotter [12] showed that the LO relaxation of this ILO formulation has persistency. Generalizing this result, Hochbaum et al. [5] showed that persistency also holds for the LO relaxation of ILO on UTVPI systems if the variables are binary and the coefficients in the objective function are non-negative. Note that persistency and neighborhood persistency are the same when the variables are binary. Fiorini et al. [4] gave another generalization that persistency holds for the LO relaxation of ILO on UTVPI systems if each inequality is of the form $x_i + x_j \leq c$ for some integer c. It should be noted that optimal solutions of the LO relaxation are assumed to be half-integral in these persistency results, while not in our (neighborhood) persistency result in this paper.

A previous work [5] showed that one can obtain a two-approximate solution of the (feasible) ILO problem (2) by rounding up or down a half-integral optimal solution of the LO relaxation if w is non-negative and the variables take non-negative values. The work [5] also showed that any ILO problem on a two-variable-per-inequality system (i.e., input matrix A is in $\mathbb{Q}^{m \times n}$ and each row of A has at most two nonzero elements) can be reduced to a binary ILO problem on a UTVPI system of pseudo-polynomial size if upper and lower bounds exists on the value of each variable. From the reduction and persistency of the LO relaxation of binary ILO on UTVPI systems (for non-negative objective functions and variables), we can obtain upper and lower bounds on the values of the variables in an optimal integer solution of an ILO problem on a two-variable-per-inequality system. However, this does not imply the neighborhood persistency of the LO relaxation of ILO on UTVPI systems.

The persistency of the LO relaxation of the ILO formulation of the vertex cover problem is generalized to the so-called k-submodular relaxation [6] where k is any positive integer. In k-submodular relaxation, a problem with values in $\{1, \ldots, k\}$ is relaxed to one with values in $\{0, 1, \ldots, k\}$. Although exactly one relaxed value (i.e., value 0) exists in k-submodular relaxation, there exists infinite relaxed values (i.e., all the values in $\mathbb{R} \setminus \mathbb{Z}$) in our relaxation of ILO to LO.

Outline

The rest of our paper is organized as follows. Section 2 formally defines our problem and (neighborhood) persistency, and provides useful results. Section 3 shows our main result, namely, neighborhood persistency of the LO relaxation of ILO on UTVPI systems. Section 4 gives examples of ILO problems on non-UTVPI systems such that their LO relaxations lack persistency, which shows the maximality of ILO on UTVPI systems among ILO with its LO relaxation having (neighborhood) persistency. Section 5 shows that ILO on UTVPI systems with non-negative objective functions and non-negative variables is fixed-parameter tractable and two-approximable by using neighborhood persistency. Section 6 concludes our paper.

Due to the page limitation, proofs of the claims marked with (*) are omitted in this paper.

2   Preliminaries

We consider an integer linear optimization (ILO) problem of the following form throughout the paper:

$$
\begin{aligned}
\text{minimize} \quad & w^{\mathrm{T}} x \\
\text{subject to} \quad & Ax \geq b, \\
& x \in \mathbb{Z}^n,
\end{aligned}
\tag{2}
$$

where $A \in \{-1, 0, 1\}^{m \times n}$ is a matrix having at most two nonzero elements per row, $b \in \mathbb{Z}^m$, $w \in \mathbb{Q}^n$, and m, n are positive integers.[2] An optimal solution of

---

[2] Our formulation of an ILO problem does not include the non-negativity constraints on the variables, i.e., the constraint $x \geq 0$. However, such constraints can be incorporated to the linear system $Ax \geq b$.

K. Kimura and K. Nakayama

the ILO problem (2) is called an *optimal integer solution*. If the ILO problem (2) has a solution, then it is called *feasible*.

The following linear optimization (LO) problem is called the *LO relaxation* of the ILO problem (2):

$$
\begin{aligned}
\text{minimize} \ \ & w^\mathrm{T} x \\
\text{subject to} \ \ & Ax \geq b, \\
& x \in \mathbb{R}^n.
\end{aligned}
\tag{3}
$$

An optimal solution of the LO problem (3) is called an *optimal fractional solution*.

Now, we provide the key notions of this paper.

Definition 1 (Persistency). *LO relaxation* (3) *is* persistent *if for every optimal fractional solution that assigns integer values to some variables, there exists an optimal integer solution of* (2) *in which these variables retain the same values. Namely, for every fractional solution* x^*, *there exists an optimal integer solution* z^* *such that* $x_j^* \in \mathbb{Z}$ *implies* $x_j^* = z_j^*$ *for each* $j = 1, \ldots, n$.

Definition 2 (Integer Neighborhood). *For vector* $x \in \mathbb{R}^n$, *its* integer neighborhood $N(x) \subseteq \mathbb{Z}^n$ *is defined as* $N(x) = \{z \in \mathbb{Z}^n \mid |z_j - x_j| < 1 \ (j = 1, \ldots, n)\}$.

We focus on the following property, which is stronger than persistency.

Definition 3 (Neighborhood Persistency). *LO relaxation* (3) *is* neighborhood persistent *if for every optimal fractional solution* x^*, *there exists an optimal integer solution of* (2) *in integer neighborhood* $N(x^*)$ *of* x^*.

Note that for vector $x \in \mathbb{R}^n$ and $j \in \{1, \ldots, n\}$ if $x_j \in \mathbb{Z}$, then $z_j = x_j$ for any $z \in N(x)$. Thus, neighborhood persistency implies (ordinary) persistency.

We will show our main result using the following form of the strong duality of LO.

Theorem 1 (Theorem 5.4 in [14]). *Let* A *be an integer matrix, let* b *be an integer vector, and let* w *be a rational vector. If at least one of* $\min\{w^\mathrm{T} x \mid Ax \geq b, x \in \mathbb{R}^n\}$ *or* $\max\{b^\mathrm{T} y \mid A^\mathrm{T} y = w, y \geq 0, y \in \mathbb{R}^m\}$ *is bounded, then* $\min\{w^\mathrm{T} x \mid Ax \geq b, x \in \mathbb{R}^n\} = \max\{b^\mathrm{T} y \mid A^\mathrm{T} y = w, y \geq 0, y \in \mathbb{R}^m\}$.

For $a \in \mathbb{R}$, define $\lceil a \rceil = \min\{n \in \mathbb{Z} \mid n \geq a\}$ and $\lfloor a \rfloor = \max\{n \in \mathbb{Z} \mid n \leq a\}$.

3 Main Results

In this section, we show the following theorem, which is the main result of this paper.

Theorem 2. *If the integer linear optimization problem* (2) *is feasible, then its linear optimization relaxation* (3) *is neighborhood persistent.*

Proof. Assume that the ILO problem (2) is feasible. If the LO relaxation (3) does not have an optimal fractional solution (i.e., it is unbounded), then the condition of the neighborhood persistency of the LO relaxation vacuously holds. Therefore, we assume that the LO relaxation (3) has an optimal fractional solution in what follows. Then the ILO problem (2) is bounded and feasible (by assumption), and thus, it has an optimal integer solution. Fix an optimal integer solution z^* and an optimal fractional solution x^* in what follows. We show that there exists an optimal integer solution of the ILO problem (2) in integer neighborhood $N(x^*)$ of x^*, which shows the theorem.

Define $z \in \mathbb{Z}^n$ as

$$
z_j = \begin{cases} x_j^* & (x_j^* \in \mathbb{Z}), \\ \lceil x_j^* \rceil & (x_j^* \notin \mathbb{Z}, x_j^* < z_j^*), \\ \lfloor x_j^* \rfloor & (x_j^* \notin \mathbb{Z}, x_j^* > z_j^*), \end{cases} \tag{4}
$$

for $j = 1, \ldots, n$. Note that $z \in N(x^*)$. We show that z is an optimal integer solution of the ILO problem (2). For this, we show that (i) z is a feasible solution of the ILO problem (2) (in Claim 1 below), and (ii) $w^T z \leq w^T z^*$ (in Claim 2 below). These imply that z is an optimal integer solution of the ILO problem (2), since so is z^*.

Claim 1. *z is a feasible solution of the ILO problem* (2).

Proof. The proof consists of case analysis with many cases. Due to the page limitation, we only deal with the case where the ith constraint of $Ax \geq b$ is of the form $x_p + x_q \geq b_i$ for some $p, q \in \{1, \ldots, n\}$ with $p \neq q$, $z_p \geq x_p^*$, and $z_q < x_q^*$. Let $\alpha = z_p - x_p^*$ and $\beta = x_q^* - z_q$. By definition, we have $0 \leq \alpha < 1$ and $0 < \beta < 1$. If $\alpha \geq \beta$, then we have

$$
z_p + z_q = x_p^* + x_q^* + (\alpha - \beta) \geq x_p^* + x_q^* \geq b_i. \tag{5}
$$

If $\alpha < \beta$, then we have $0 < \beta - \alpha < 1$, and since $x_p^* + x_q^* = z_p + z_q + (\beta - \alpha)$ and $z_p + z_q \in \mathbb{Z}$, we have $\lfloor x_p^* + x_q^* \rfloor = z_p + z_q$. Since $b_i \in \mathbb{Z}$, we have $\lfloor x_p^* + x_q^* \rfloor \geq b_i$, and thus, $z_p + z_q = \lfloor x_p^* + x_q^* \rfloor \geq b_i$. Hence, z satisfies the ith constraint of $Ax \geq b$.

Claim 2. $w^T z \leq w^T z^*$.

Proof. To show $w^T z \leq w^T z^*$, we use the duality theorem of linear optimization (see Theorem 1). The following is the dual LO problem of (3):

$$
\begin{aligned}
\text{maximize} \quad & b^T y \\
\text{subject to} \quad & A^T y = w, \\
& y \geq 0, \\
& y \in \mathbb{R}^m.
\end{aligned} \tag{6}
$$

Since we are assuming that the LO problem (3) has an optimal solution, the dual LO problem (6) also has an optimal solution and the optimal values of LO problems (3) and (6) are the same by Theorem 1. Fix an optimal solution y^* of the LO problem (6) in what follows. We have $A^T y^* = w$ from the equality in the LO problem (6). Thus, $w^T z$ can be rewritten as $w^T z = (y^*)^T A z = \sum_{i=1}^m y_i^* A_i z$, where A_i is the ith row of A for each $i = 1, \ldots, m$. Similarly, we have $w^T z^* = \sum_{i=1}^m y_i^* A_i z^*$. Therefore, to show $w^T z \leq w^T z^*$, it suffices to show that

$$y_i^* A_i z \leq y_i^* A_i z^* \tag{7}$$

for each $i = 1, \ldots, m$. We show (7) in what follows.

Recall that x^* is an optimal solution of the LO problem (3). The following condition called the *complementary slackness* condition holds (see, e.g., Sect. 5.5 in [14]): $y_i^* (A_i x^* - b_i) = 0$ for each $i = 1, \ldots, m$. For our purpose, we use the following equivalent form of the complementary slackness condition:

$$\text{If } y_i^* > 0, \text{ then } A_i x^* - b_i = 0 \tag{8}$$

for each $i = 1, \ldots, m$. Now we are ready to show (7) for each $i = 1, \ldots, m$.

Fix $i \in \{1, \ldots, m\}$. If $y_i^* = 0$, then $y_i^* A_i z \leq y_i^* A_i z^*$ since $y_i^* A_i z = y_i^* A_i z^* = 0$. Therefore, we assume that $y_i^* > 0$ in what follows. Then it suffices to show that $A_i z \leq A_i z^*$ for showing (7). Note that we have $A_i x^* - b_i = 0$ from (8).

Since $Ax \geq b$ is a UTVPI system, $A_i x \geq b_i$ is of the form $\pm x_p \pm x_q \geq b_i$ or $\pm x_p \geq b_i$ for some $p, q \in \{1, \ldots, n\}$ with $p \neq q$. We divide the proof into cases by the (non-)integrality of x_p^* and x_q^*. The proof of single-variable case is easy and omitted. Note that $A_i z^* \geq b_i$, since z^* is a feasible solution.

Case 1: $x_p^* \in \mathbb{Z}$ *and* $x_q^* \in \mathbb{Z}$. Since $z_p = x_p^*$ and $z_q = x_q^*$ by the definition of z, we have $A_i z = A_i x^* = b_i$. Since we have $A_i z^* \geq b_i$, we have $A_i z \leq A_i z^*$.

Case 2: $x_p^* \in \mathbb{Z}$ *and* $x_q^* \notin \mathbb{Z}$ *(This case includes the case of $x_p^* \notin \mathbb{Z}$ and $x_q^* \in \mathbb{Z}$ by the symmetry of the constraints).* This case does not occur, since $x_p^* \in \mathbb{Z}$ and $x_q^* \notin \mathbb{Z}$ imply that $A_i x^* \notin \mathbb{Z}$, and from $b_i \in \mathbb{Z}$ we cannot have $A_i x^* = b_i$.

Case 3: $x_p^* \notin \mathbb{Z}$ *and* $x_q^* \notin \mathbb{Z}$. We divide into cases where $A_i x \geq b_i$ is $x_p + x_q \geq b_i$, $x_p - x_q \geq b_i$, $-x_p + x_q \geq b_i$, or $-x_p - x_q \geq b_i$. Further, we divide into cases by the small and large comparison of the values x_p^*, x_q^* and z_p^*, z_q^*. Since $x_p^*, x_q^* \notin \mathbb{Z}$ and $z_p^*, z_q^* \in \mathbb{Z}$, we have four cases: (i) $x_p^* < z_p^*$ and $x_q^* < z_q^*$, (ii) $x_p^* < z_p^*$ and $x_q^* > z_q^*$, (iii) $x_p^* > z_p^*$ and $x_q^* < z_q^*$, or (iv) $x_p^* > z_p^*$ and $x_q^* > z_q^*$. Since these cases can be proven in similar ways, we only show the case of (i) $x_p^* < z_p^*$ and $x_q^* < z_q^*$, and omit the proof of the remaining cases.

In the following, we assume that $x_p^* < z_p^*$ and $x_q^* < z_q^*$. Then we have $z_p \leq z_p^*$ and $z_q \leq z_q^*$ by the definition of z. Let $x_p^* = \alpha_p + \beta_p$ and $x_q^* = \alpha_q + \beta_q$, where $\alpha_p, \alpha_q \in \mathbb{Z}$ and $0 < \beta_p, \beta_q < 1$. Note that we have $z_p = \lceil x_p^* \rceil = x_p^* + 1 - \beta_p$ and $z_q = \lceil x_q^* \rceil = x_q^* + 1 - \beta_q$ by definition.

Case 3.1: $A_i x = x_p + x_q$. From $z_p \leq z_p^*$ and $z_q \leq z_q^*$, we have $z_p + z_q \leq z_p^* + z_q^*$. Hence, we have $A_i z \leq A_i z^*$.

Case 3.2: $A_i x = x_p - x_q$. From $A_i x^* = b_i$, we have

$$x_p^* - x_q^* = \alpha_p - \alpha_q + \beta_p - \beta_q = b_i. \tag{9}$$

Since $b_i \in \mathbb{Z}$ and $-1 < \beta_p - \beta_q < 1$, we have $\beta_p - \beta_q = 0$. Hence, we have

$$z_p - z_q = (x_p^* + 1 - \beta_p) - (x_q^* + 1 - \beta_q) \tag{10}$$

$$= x_p^* - x_q^* - (\beta_p - \beta_q) = x_p^* - x_q^* = b_i. \tag{11}$$

Since $z_p^* - z_q^* \geq b_i$, we have $z_p - z_q \leq z_p^* - z_q^*$, i.e., $A_i z \leq A_i z^*$.

Case 3.3: $A_i x = -x_p + x_q$. We can show $A_i z \leq A_i z^*$ in a similar way as in Case 3.2.

Case 3.4: $A_i x = -x_p - x_q$. Let $z_p^* = x_p^* + \gamma_p$ and $z_q^* = x_q^* + \gamma_q$, where $\gamma_p, \gamma_q > 0$. Since $-x_p^* - x_q^* = b_i$ from $A_i x^* = b_i$, we have

$$-z_p^* - z_q^* = -(x_p^* + \gamma_p) - (x_q^* + \gamma_q) \tag{12}$$

$$= -x_p^* - x_q^* - (\gamma_p + \gamma_q) < b_i. \tag{13}$$

This contradicts that z^* is feasible. Hence, this case does not occur. This completes the proof.

From Claims 1 and 2, we conclude that z is an optimal integer solution of the ILO problem (2). This completes the proof of Theorem 2.

4   Maximality of UTVPI Systems

In this section, we show that ILO on UTVPI systems is a maximal subclass of ILO with its LO relaxation having (neighborhood) persistency in the following sense: If we allow (i) an inequality with *three* variables whose coefficients are all one or (ii) an inequality with two variables whose coefficients are in $\{1, 2\}$, then even persistency does not hold for the LO relaxation of the ILO problem (2).

Example 1. Consider the following ILO problem:

$$
\begin{aligned}
\text{minimize}\ \ & 3x_1 + x_2 \\
\text{subject to}\ \ & x_1 + x_2 + x_3 \geq 2, \\
& x_1 - x_3 \geq 0, \\
& -x_2 \geq -1, \\
& x \in \mathbb{Z}^3.
\end{aligned}
\tag{14}
$$

By using an (I)LO solver, one can check that $(x_1^*, x_2^*, x_3^*) = (0.5, 1, 0.5)$ is an optimal fractional solution of the LO relaxation of (14), and $(z_1^*, z_2^*, z_3^*) = (1, 0, 1)$ is an optimal integer solution of (14), whose objective value is 3. On the other hand, if we fix $x_2 = x_2^* = 1$ in (14), then we obtain the following ILO problem:

$$
\begin{aligned}
\text{minimize}\ \ & 3x_1 + 1 \\
\text{subject to}\ \ & x_1 + x_3 \geq 1, \\
& x_1 - x_3 \geq 0, \\
& x_1, x_3 \in \mathbb{Z}.
\end{aligned}
\tag{15}
$$

By using an ILO solver, one can check that the ILO problem (15) has an optimal integer solution $(z_1^*, z_3^*) = (1, 0)$ whose objective value is 4. If the LO relaxation of the ILO problem (14) has persistency, then the optimal values of the ILO problems (14) and (15) must be equal. However, they are different and we conclude that the LO relaxation of (14) does not have persistency.

Example 2. Consider the following ILO problem:

$$
\begin{aligned}
\text{minimize } & 3x_1 + x_2 \\
\text{subject to } & 2x_1 + x_2 \geq 2, \\
& -x_2 \geq -1, \\
& x \in \mathbb{Z}^2.
\end{aligned}
\tag{16}
$$

By using an (I)LO solver, one can check that $(x_1^*, x_2^*) = (0.5, 1)$ is an optimal fractional solution of the LO relaxation of (16), and $(z_1^*, z_2^*) = (1, 0)$ is an optimal integer solution of (16), whose objective value is 3. On the other hand, if we fix $x_2 = x_2^* = 1$ in (16), then we obtain the following ILO problem:

$$
\begin{aligned}
\text{minimize } & 3x_1 + 1 \\
\text{subject to } & 2x_1 \geq 1, \\
& x_1 \in \mathbb{Z}.
\end{aligned}
\tag{17}
$$

The ILO problem (17) has the unique optimal integer solution $z_1^* = 1$ whose objective value is 4. Since the optimal values of ILO problems (16) and (17) differ, we conclude that the LO relaxation of (16) does not have persistency.

From Examples 1 and 2, together with Theorem 2, we see that ILO on UTVPI systems is a maximal subclass of ILO with its LO relaxation having (neighborhood) persistency.

5 Fixed-Parameter Tractability and Two-Approximability for Special Cases

In this section, we consider ILO on UTVPI systems with non-negative objective functions and non-negative variables and address the following ILO problem:

$$
\begin{aligned}
\text{minimize } & w^{\mathrm{T}} x \\
\text{subject to } & Ax \geq b, \\
& x \geq 0, \\
& x \in \mathbb{Z}^n,
\end{aligned}
\tag{18}
$$

where $A \in \{-1, 0, 1\}^{m \times n}$ is a matrix having at most two nonzero elements per row, $b \in \mathbb{Z}^m$, $w \in \mathbb{Q}_+^n$ (where $\mathbb{Q}_+$ denotes the set of non-negative rational numbers), and m, n are positive integers. We show that the ILO problem (18) is both fixed-parameter tractable and two-approximable in what follows.

From our main result (Theorem 2) we can reduce solving the ILO problem (18) to solving an ILO problem with *binary* variables, i.e., each variable takes value zero or one. Indeed, let x^* be an optimal fractional solution of the LO relaxation of (18) and let $\lfloor x^* \rfloor$ be a vector obtained from x^* by taking componentwise $\lfloor \cdot \rfloor$. Then from Theorem 2 the ILO problem (18) is equivalent to the following problem:

$$\begin{aligned} \text{minimize} \quad & (w')^{\mathrm{T}} x' + w^{\mathrm{T}} \lfloor x^* \rfloor \\ \text{subject to} \quad & A' x' + A \lfloor x^* \rfloor \geq b, \\ & x' \in \{0,1\}^{n - |I(x^*)|}, \end{aligned} \qquad (19)$$

where $I(x^*) = \{j \in \{1, \ldots, n\} \mid x_j^* \in \mathbb{Z}\}$, and w' (resp, A') is a restriction of w (resp., columns of A) to $\{1, \ldots, n\} \setminus I(x^*)$. In turn, the ILO problem (19) is equivalent to solving

$$\begin{aligned} \text{minimize} \quad & (w')^{\mathrm{T}} x' \\ \text{subject to} \quad & A' x' \geq b - A \lfloor x^* \rfloor, \\ & x' \in \{0,1\}^{n - |I(x^*)|}. \end{aligned} \qquad (20)$$

The ILO problem (20) with binary variables is fixed-parameter tractable [11][3] (i.e., for positive integer k there exists an algorithm that solves the problem in time $f(k)s^{O(1)}$ where s is the input size of the problem) and two-approximable [5] (i.e., there exists a polynomial time algorithm that outputs a feasible solution (if it exists) where the objective value is at most twice the optimal value). We show that we can obtain the same results for the (non-binary) ILO problem (18).

Let k be a positive integer. From Theorem 2, the ILO problem (18) has a solution whose objective value is at most k if and only if the ILO problem (20) has a solution whose objective value is at most $k - w^T \lfloor x^* \rfloor (\leq k)$. Moreover, the optimal value of the LO relaxation of (18) is the sum of the optimal value of the LO relaxation of (20) and $w^T \lfloor x^* \rfloor$. Consequently, the following results on fixed-parameter tractability holds from previous results [11]. Notation $O^*()$ hides functions that are polynomial in the input size in what follows.

Theorem 3. *Given an ILO problem* (18) *such that $w_j \geq 1$ for all $j \in \{1, \ldots, n\}$ and positive integer k, it can be verified in time $O^*(1.3788^k)$ if* (18) *has a feasible solution whose objective value is at most k. An optimal integer (if it exists) can be obtained in time $O^*(1.2377^n)$.*

When w is an all-one vector (i.e., $w_j = 1$ for all $j \in \{1, \ldots, n\}$), we obtain the following:

Theorem 4. *Let I be an ILO problem* (18) *such that $w_j = 1$ for all $j \in \{1, \ldots, n\}$ and k be a positive integer. Then*

---

[3] ILO on UTVPI systems with non-negative objective functions and binary variables is equivalent to the weighted min one 2-SAT problem in [11].

- *it can be checked in time* $O^*(1.2738^k)$ *whether* I *has a feasible solution whose objective value is at most* k. *An optimal integer solution (if it exists) can be obtained in time* $O^*(1.2114^n)$ *and polynomial space, or* $O^*(1.2108^n)$ *and exponential space;*
- *it can be checked in time* $O^*(2.3146^{k-\mathrm{OPT_{LO}}})$ *whether* I *has a feasible solution whose objective value is at most* k *where* $\mathrm{OPT_{LO}}$ *is the optimal value of the LO relaxation of* I;
- *there exists a randomized polynomial time algorithm that produces ILO problem* I' *on a UTVPI system with an all-one objective function vector and binary variables, and* k' *such that* I' *has a number of variables and inequalities polynomial in* $k - \mathrm{OPT_{LO}}$ *and if* I *has a feasible solution whose objective value is at most* k, *then* I' *has a feasible solution whose objective value is at most* k', *and if* I *has no feasible solution with the objective value at most* k, *then with the probability at least half,* I' *has no feasible solution whose objective value is at most* k'.

For approximability, we obtain the following, using the fact that the ILO problem (20) is two-approximable [5].

Theorem 5 (*). *ILO on UTVPI systems with non-negative objective functions and non-negative variables is 2-approximable.*

The two-approximability of ILO on UTVPI systems with non-binary variables is already known [5]. Therefore, we obtain another proof of that fact using neighborhood persistency.

6   Conclusion

We introduced neighborhood persistency of the linear optimization (LO) relaxation of integer linear optimization (ILO), which is a property stronger than persistency, and show that ILO on unit-two-variable-per-inequality (UTVPI) systems is a maximal subclass of ILO with its LO relaxation having (neighborhood) persistency. Our persistency result generalizes known results on special cases of ILO on UTVPI systems [4,5,12]. Using neighborhood persistency, we obtain fixed-parameter algorithms and another proof of the two-approximability for special cases of ILO on UTVPI systems. An interesting future direction will be to find a (maximal) subclass of ILO with its LO relaxation having (neighborhood) persistency that is incomparable to ILO on UTVPI systems. Future works will also include generalizations of our result to nonlinear objective functions.

Acknowledgments. We would like to thank the anonymous reviewers for many helpful comments and suggestions.

References

1. Ahuja, R.K., Magnanti, T.L., Orlin, J.B.: Network Flows: Theory, Algorithms, and Applications. Prentice Hall, Hoboken (1993)

2. Bekos, M.A., Kaufmann, M., Papadopoulos, D., Symvonis, A.: Combining traditional map labeling with boundary labeling. In: Černá, I., et al. (eds.) SOFSEM 2011. LNCS, vol. 6543, pp. 111–122. Springer, Heidelberg (2011). https://doi.org/10.1007/978-3-642-18381-2_9

3. Bonomo, F., Oriolo, G., Snels, C., Stauffer, G.: Minimum clique cover in claw-free perfect graphs and the weak Edmonds-Johnson property. In: Goemans, M., Correa, J. (eds.) IPCO 2013. LNCS, vol. 7801, pp. 86–97. Springer, Heidelberg (2013). https://doi.org/10.1007/978-3-642-36694-9_8

4. Fiorini, S., Joret, G., Weltge, S., Yuditsky, Y.: Integer programs with bounded subdeterminants and two nonzeros per row arXiv:2106.05947v2

5. Hochbaum, D.S., Megiddo, N., Naor, J.S., Tamir, A.: Tight bounds and 2-approximation algorithms for integer programs with two variables per inequality. Math. Program. **62**, 69–83 (1993). https://doi.org/10.1007/BF01585160

6. Iwata, Y., Wahlström, M., Yoshida, Y.: Half-integrality, LP-branching, and FPT algorithms. SIAM J. Comput. **45**(4), 1377–1411 (2016)

7. Jaffar, J., Maher, M.J., Stuckey, P.J., Yap, R.H.C.: Beyond finite domains. In: Borning, A. (ed.) PPCP 1994. LNCS, vol. 874, pp. 86–94. Springer, Heidelberg (1994). https://doi.org/10.1007/3-540-58601-6_92

8. Lahiri, S.K., Musuvathi, M.: An efficient decision procedure for UTVPI constraints. In: Gramlich, B. (ed.) FroCoS 2005. LNCS (LNAI), vol. 3717, pp. 168–183. Springer, Heidelberg (2005). https://doi.org/10.1007/11559306_9

9. Lokshtanov, D., Narayanaswamy, N.S., Raman, V., Ramanujan, M.S., Saurabh, S.: Faster parameterized algorithms using linear programming. ACM Trans. Algorithms **11**(2), 15:1–15:31 (2014). https://doi.org/10.1145/2566616

10. Miné, A.: The octagon abstract domain. Higher-Order Symb. Comput. **19**, 31–100 (2006). https://doi.org/10.1007/s10990-006-8609-1

11. Misra, N., Narayanaswamy, N., Raman, V., Shankar, B.S.: Solving min ones 2-SAT as fast as vertex cover. Theoret. Comput. Sci. **506**, 115–121 (2013)

12. Nemhauser, G.L., Trotter, L.E.: Vertex packings: structural properties and algorithms. Math. Program. **8**(1), 232–248 (1975)

13. Schrijver, A.: Disjoint homotopic paths and trees in a planar graph. Discret. Comput. Geom. **6**(4), 527–574 (1991). https://doi.org/10.1007/BF02574704

14. Schrijver, A.: Combinatorial Optimization. Springer, Heidelberg (2003)

15. Subramani, K.: On deciding the non-emptiness of 2SAT polytopes with respect to first order queries. Math. Log. Q. **50**, 281–292 (2004). https://doi.org/10.1002/malq.200310099

16. Subramani, K., Wojciechowski, P.: Analyzing lattice point feasibility in UTVPI constraints. In: Beck, J.C. (ed.) CP 2017. LNCS, vol. 10416, pp. 615–629. Springer, Cham (2017). https://doi.org/10.1007/978-3-319-66158-2_39

17. Upadrasta, R., Cohen, A.: Sub-polyhedral scheduling using (unit-)two-variable-per-inequality polyhedra. In: Proceedings of the 40th Annual ACM SIGPLAN-SIGACT Symposium on Principles of Programming Languages, pp. 483–496 (2013). https://doi.org/10.1145/2429069.2429127

Polynomial-Time Approximation Schemes for a Class of Integrated Network Design and Scheduling Problems with Parallel Identical Machines

Yusuke Saito and Akiyoshi Shioura[✉]

Tokyo Institute of Technology, Tokyo 152-8550, Japan
shioura.a.aa@m.titech.ac.jp

Abstract. In the integrated network design and scheduling problem (INDS-P), we are asked to repair edges in a graph by using parallel machines so that the performance of the network is recovered by a certain level, and the objective is to minimize the makespan required to finish repairing edges. The main aim of this paper is to show that polynomial-time approximation schemes exist for some class of the problem (INDS-P), including the problems associated with minimum spanning tree, shortest path, maximum flow with unit capacity, and maximum-weight matching.

Keywords: Network optimization · Parallel machine scheduling · Polynomial-time approximation scheme · Approximation algorithm

1 Introduction and Results

Network optimization problems aim at constructing networks of good performance. Let us consider a situation where an existing network is damaged due to a disaster such as an earthquake, a flood, or a hurricane. In this situation, it is required to repair the network as quickly as possible so that the performance of the network is recovered by a certain level. A mathematical model of this problem is provided in [15] (see also [14]), which is referred to as the integrated network design and scheduling problem. In this paper, we deal with a variant of the integrated network design and scheduling problem, which we denote by (INDS-P), where parallel identical machines are used to repair edges and the objective is minimizing the makespan[1].

1.1 Problem Definition

The definition of the problem (INDS-P) is explained in more detail. Let $G = (V, E)$ be a (directed or undirected) graph representing the network, where V is

---

[1] This problem is denoted as "$Pm|\beta|C_{\max}$-Threshold" in [15].

I. Ljubić et al. (Eds.): ISCO 2022, LNCS 13526, pp. 324–335, 2022.
https://doi.org/10.1007/978-3-031-18530-4_24

the vertex set and E is the edge set. We are given a set $E_0 \subseteq E$ of edges that are still alive after the disaster and the remaining edges in $E \setminus E_0$ are damaged. We select some damaged edges and repair them as quickly as possible so that the performance of the repaired graph reaches a certain desired level.

Selected damaged edges are repaired by using m parallel identical machines, where m is assumed to be a constant throughout this paper. Time required to repair an edge $e \in E \setminus E_0$ is given by a non-negative integer $p(e)$. Each edge can be processed by a single machine and is not allowed to be processed simultaneously by two or more machines. In addition, preemption is not allowed for each edge, i.e., once we start repairing an edge, then we need to continue it until the repair finishes. The objective of the problem (INDS-P) is to minimize the makespan, i.e., the time required to finish repairing edges. The makespan is represented as $\max\{p(X_i) \mid i = 1, 2, \ldots, m\}$, where X_i is a set of edges repaired by the i-th machine and $p(X_i) = \sum_{e \in X_i} p(e)$.

A set $X \subseteq E \setminus E_0$ of repaired edges is selected so that the performance of the repaired graph $(V, X \cup E_0)$ reaches a certain level. The performance level of the repaired graph is determined by the edge set $X \cup E_0$ of the repaired graph and represented by a function $\varphi : 2^E \to \mathbb{R} \cup \{\pm\infty\}$, which we call a *performance function*. The value $\varphi(X \cup E_0)$ is given by the optimal value of a certain network optimization problem on the graph $(V, X \cup E_0)$, such as minimum spanning tree, shortest path, and maximum flow (see [15]). With a performance function φ and a threshold value W, the constraint on the performance level is given as $\varphi(X \cup E_0) \leq W$ (resp., $\varphi(X \cup E_0) \geq W$) if φ is defined by a minimization problem (resp., a maximization problem).

In summary, the problem (INDS-P) is formulated as follows:

(INDS-P) Minimize $\max\{p(X_i) \mid i = 1, 2, \ldots, m\}$
subject to $\varphi(X \cup E_0) \leq W$ (or $\varphi(X \cup E_0) \geq W$),
$X = \bigcup_{i=1}^{m} X_i,$
$X_1, X_2, \ldots, X_m \subseteq E \setminus E_0$ are mutually disjoint.

It is easy to see that this problem is NP-hard since it is a generalization of parallel identical machine scheduling minimizing the makespan (see also [15]). Hence, we focus on approximability of the problem (INDS-P), especially in the cases with the following specific families of performance functions:

- Performance functions associated with the minimum spanning tree.
  Assume that $G = (V, E)$ is an undirected graph with non-negative integer edge length $\ell(e)$ ($e \in E$). The value of $\varphi_{\mathrm{MST}}(Y)$ for $Y \subseteq E$ is defined as the length of a minimum spanning tree in the edge-induced subgraph (V, Y); $\varphi_{\mathrm{MST}}(Y) = +\infty$ if the graph (V, Y) has no spanning tree. We denote by Φ_{MST} the family of performance functions associated with the minimum spanning tree.

- Performance functions associated with the single-source single-destination shortest path.
  Assume that $G = (V, E)$ is a directed graph with two distinct vertices $s, t \in V$ and non-negative integer edge length $\ell(e)$ ($e \in E$). The value of $\varphi_{\mathrm{SP}}(Y)$ for

$Y \subseteq E$ is defined as the length of a shortest path from s to t in the graph (V, Y); $\varphi_{\mathrm{SP}}(Y) = +\infty$ if the graph (V, Y) has no path from s to t. We denote by Φ_{SP} the family of performance functions associated with the single-source single-destination shortest path.

- Performance functions associated with the maximum-weight matching. Assume that $G = (V, E)$ is an undirected graph with non-negative integer edge weight $w(e)$ ($e \in E$). The value of $\varphi_{\mathrm{MM}}(Y)$ for $Y \subseteq E$ is defined as the weight of a maximum-weight matching in the graph (V, Y). We denote by Φ_{MM} the family of performance functions associated with the maximum-weight matching.

- Performances function associated with the maximum flow with unit capacity. Assume that $G = (V, E)$ is a directed graph with two distinct vertices $s, t \in V$. The value of $\varphi_{\mathrm{UMF}}(Y)$ for $Y \subseteq E$ is defined as the amount of a maximum flow from s to t in the graph (V, Y). We denote by Φ_{UMF} the family of performance functions associated with the maximum-weight matching.

1.2 Our Results

The main aim of this paper is to show that polynomial-time approximation schemes (PTASes, for short) exist for the problem (INDS-P) with some families of performance functions, including Φ_{MST}, Φ_{SP}, Φ_{UMF}, and Φ_{MM}. Recall that an approximation algorithm for (INDS-P) is a PTAS if for any feasible instance of (INDS-P) and any constant $\varepsilon > 0$, the algorithm finds in polynomial time a feasible solution $(X_1, X_2, \ldots, X_m)$ such that

$$\max\{p(X_i) \mid i = 1, 2, \ldots, m\} \leq (1 + \varepsilon)C^*,$$

where C^* is the optimal value of (INDS-P).

To state the main result, consider the special case of (INDS-P) with $m = 1$, i.e., only a single machine is available. We denote this special case as (INDS-S), which is formulated as

$$\begin{aligned} \text{(INDS-S) Minimize } & p(X) \\ \text{subject to } & \varphi(X \cup E_0) \leq W \quad (\text{or } \varphi(X \cup E_0) \geq W), \\ & X \subseteq E \setminus E_0. \end{aligned}$$

We show that (INDS-P) admits a PTAS if its special case (INDS-S) admits a PTAS. That is, the existence of a PTAS for (INDS-P) depends on the performance function φ, and irrelevant to the number m of machines.

Theorem 1. *Suppose that the problem* (INDS-S) *with some performance function* φ *admits a PTAS (or a polynomial-time exact algorithm). Then, the problem* (INDS-P) *with the same performance function* φ *also admits a PTAS.*

As shown in Sect. 3, if $\varphi \in \Phi_{\mathrm{MST}} \cup \Phi_{\mathrm{SP}} \cup \Phi_{\mathrm{MM}}$ then a PTAS exists for the problem (INDS-S). We also show that if $\varphi \in \Phi_{\mathrm{UMF}}$, then (INDS-S) can be solved in polynomial time. Hence, the following result is obtained as an immediate corollary of Theorem 1.

Corollary 1. *If $\varphi \in \Phi_{\mathrm{MST}} \cup \Phi_{\mathrm{SP}} \cup \Phi_{\mathrm{UMF}} \cup \Phi_{\mathrm{MM}}$, then the problem* (INDS-P) *admits a polynomial-time approximation scheme.*

Theorem 1 follows from the next lemma, stating that (INDS-S) and (INDS-P) are almost equivalent in terms of the approximation ratio; see Sect. 2 for a proof of the lemma.

Lemma 1. *Suppose that the problem* (INDS-S) *with some performance function φ admits a polynomial-time $(1 + \alpha)$-approximation algorithm with some $\alpha \geq 0$. Then, for any constant $\delta > 0$, the problem* (INDS-P) *with the same performance function φ admits a polynomial-time $(1 + \alpha + \delta)$-approximation algorithm.*

The problem (INDS-P) can be regarded as a bicriteria optimization problem minimizing the makespan and also minimizing (or maximizing) the performance function. Therefore, we may consider the following problem obtained by swapping the objective function and the constraint on the network performance level:

(INDS2-P) Minimize (or Maximize) $\varphi(X \cup E_0)$

$$\text{subject to} \quad p(X_i) \leq C \ (i = 1, 2, \ldots, m),$$
$$X = \bigcup_{i=1}^{m} X_i,$$
$$X_1, X_2, \ldots, X_m \subseteq E \setminus E_0 \text{ are mutually disjoint.}$$

We also consider the special case of (INDS2-P) with $m = 1$:

(INDS2-S) Minimize (or Maximize) $\varphi(X \cup E_0)$

$$\text{subject to} \quad p(X) \leq C, \ X \subseteq E \setminus E_0.$$

For $\alpha > 0$, we say that an algorithm for (INDS2-P) is a $(1, 1 + \alpha)$-approximation algorithm if it finds a feasible solution $X = \bigcup_{i=1}^{m} X_i$ satisfying

$$\varphi(X \cup E_0) \leq W^* \text{ (for minimization problem)},$$
$$\varphi(X \cup E_0) \geq W^* \text{ (for maximization problem)},$$
$$p(X_i) \leq (1 + \alpha)C \quad (i = 1, 2, \ldots, m),$$

where W^* is the optimal value of (INDS2-P). We also say that (INDS2-P) admits a bicriteria PTAS if it admits a polynomial-time $(1, 1 + \alpha)$-approximation algorithm for every constant $\alpha > 0$.

Theorem 2. *Suppose that the problem* (INDS2-S) *with some performance function φ admits a bicriteria PTAS (or a polynomial-time exact algorithm). Then, the problem* (INDS2-P) *with the same performance function φ also admits a bicriteria PTAS.*

As in the case of (INDS-P), we obtain the following corollary on the existence of bicriteria PTASes for some specific performance functions.

Corollary 2. *If $\varphi \in \Phi_{\mathrm{MST}} \cup \Phi_{\mathrm{SP}} \cup \Phi_{\mathrm{UMF}} \cup \Phi_{\mathrm{MM}}$, then the problem* (INDS2-P) *admits a bicriteria PTAS.*

Theorem 2 follows immediately from the next lemma, to be proven in Sect. 2.

Lemma 2. *Suppose that the problem* (INDS2-S) *with some specific performance function* φ *admits a polynomial-time* $(1, 1 + \alpha)$*-approximation algorithm with some* $\alpha \geq 0$. *Then, for any constant* $\delta > 0$, *the problem* (INDS2-P) *with the same performance function* φ *admits a polynomial-time* $(1, 1 + \alpha + \delta)$*-approximation algorithm.*

Remark 1. The problem (INDS2-P) includes as a special case the multiple knapsack problem with identical capacity knapsacks. The multiple knapsack problem is a natural generalization of the knapsack problem, where multiple knapsacks are given and items can be packed into one of the knapsacks. It is known that the multiple knapsack problem admits a PTAS, even in the case where capacity of knapsacks are different [6,11]. It is not clear how to apply the techniques used in [6,11] to obtain a PTAS for (INDS2-P). □

Remark 2. In the problems (INDS-P) and (INDS2-P), we may assume, without loss of generality, that $E_0 = \emptyset$. Indeed, Any problem instance with $E_0 \neq \emptyset$ can be reduced to the one with $E_0 = \emptyset$ by setting $p(e) = 0$ for all $e \in E_0$; it is easy to see that the instance after the reduction is essentially equivalent to the original one since all edges e with $p(e) = 0$ can be processed immediately when the machines start processing. □

1.3   Related Work

The integrated network design and scheduling problem is originated by Nurre et al. [14], where the network performance level is determined by the amount of a maximum flow, and the objective is to maximize the cumulative performance level over a finite time horizon. The same problem with different kinds of performance functions is considered in [15], where NP-hardness of the problems is proved and some heuristic algorithms based on dispatching rules are presented. The problem of this type is also referred to as the incremental network design problem and the incremental combinatorial optimization problem, and approximation algorithms with theoretical guarantee have been proposed [4,7,8,10].

In addition to the maximization of the cumulative performance level, Nurre and Sharkey [15] deal with the problem of minimizing the makespan under the performance level constraint, which is nothing but the problem (INDS-P) discussed in this paper. To the best of our knowledge, no approximation algorithm with nontrivial bound is known for (INDS-P) and its special cases with specific performance functions, while a PTAS exists for (INDS-S) if the performance function is given by minimum-weight spanning tree, shortest path, or maximum-weight matching (see Sect. 3 for details).

A network optimization problem similar to (INDS-P) is discussed by Averbakh, Pereira, et al. (see [1,2,18], etc.) in the name "the network construction/reconstruction problem," and various theoretical results as well as computational study have been presented. A major difference between the network construction/reconstruction problem and (INDS-P) is that in the former problem each edge to be repaired should be connected with a fixed depot vertex by existing (alive and already repaired) edges, while there is no such restriction in our problem.

The problem (INDS-P) in this paper is also related to *scheduling problems with rejection* (see, e.g., [3,17]). In scheduling problem with rejection, we do not need to process all jobs and can reject some jobs, which yields penalty cost. This scheduling problem is a bicriteria optimization problem, where the objective is to optimize some standard objective function such as makespan, total flow time, etc., and also to minimize the total rejection cost. The problem (INDS-P) can be regarded as a scheduling problem with nonlinear rejection cost; the edge set $E \setminus (X \cup E_0)$ can be regarded as the set of rejected edges, and the difference of the values $\varphi(E)$ and $\varphi(X \cup E_0)$ can be regarded as the rejection cost, which is nonlinear in general.

In the literature of scheduling with rejection, the total rejection cost is often assumed to be linear, i.e., it is given as the sum of rejection cost for rejected jobs. Recently, a nonlinear (submodular) rejection cost function is considered in some papers [12,13,19], where constant-factor approximation algorithms are proposed, while we focus on PTASes in this paper.

2   Proofs of Lemmas 1 and 2

To the end of this paper, we assume, without loss of generality, that $E_0 = \emptyset$ in the problems (INDS-P) and (INDS2-P) (see Remark 2).

2.1   Proof of Lemma 1

We give a proof of Lemma 1 in the case where the performance level constraint is $\varphi(X) \leq W$; the case of $\varphi(X) \geq W$ can be proven similarly and omitted.

Let $C^* \in \mathbb{Z}$ be the optimal value of the problem (INDS-P), which is not known in advance. Also, let $\delta' > 0$ be a real number satisfying

$$(1 + \alpha + \delta')(1 + \delta') \leq 1 + \alpha + \delta.$$

In the following, we present a polynomial-time algorithm such that for a given real number C, if $C \geq C^*$ then the algorithm finds a feasible solution $(X_1, X_2, \ldots, X_m)$ of (INDS-P) satisfying

$$p(X_i) \leq (1 + \alpha + \delta')C \ (i = 1, 2, \ldots, m), \quad \varphi(X) \leq W \text{ with } X = \bigcup_{i=1}^{m} X_i; \quad (1)$$

note that if $C < C^*$ then the algorithm may find a feasible solution satisfying (1), or stop without finding any solution. Using this algorithm and binary search with respect to C, we can find C with $C^* \leq C \leq (1+\delta')C^*$ and a feasible solution $X_1, X_2, \ldots, X_m$ of (INDS-P) satisfying (1), for which the approximation ratio is $(1 + \alpha + \delta')(1 + \delta') \leq 1 + \alpha + \delta$. Hence, $(X_1, X_2, \ldots, X_m)$ is an $(1 + \alpha + \delta)$-approximate solution of (INDS-P).

We now show that if $C \geq C^*$ then a feasible solution satisfying (1) can be computed in polynomial time, even when the value C^* is not known in advance. Let

$$E_S = \{e \in E \mid p(e) < \delta'C\}, \qquad E_L = E \setminus E_S;$$

i.e., E_S (resp., E_L) is the set of edges with small (resp., large) processing time. Since C^* is the optimal value of (INDS-P), there exists a feasible (and optimal) solution $X_1^*, X_2^*, \ldots, X_m^*$ of (INDS-P) such that

$$p(X_i^*) \leq C^* \quad (i = 1, 2, \ldots, m), \qquad \varphi(X^*) \leq W \text{ with } X^* = \bigcup_{i=1}^{m} X_i^*.$$

We first guess the assignment of "large" edges $X_1^* \cap E_L, X_2^* \cap E_L, \ldots, X_m^* \cap E_L$ to m machines in the optimal solution[2]. Since $p(e) \geq \delta'C$ for each $e \in E_L$, it holds that

$$|X_i^* \cap E_L| \leq C^*/(\delta'C) \leq C/(\delta'C) = 1/\delta' \quad (i = 1, 2, \ldots, m),$$

where the second inequality is by the assumption $C \geq C^*$. This implies that there exist at most $n^{m/\delta'}$ possible choices of the assignment $X_1^* \cap E_L, \ldots, X_m^* \cap E_L$. Note that m and δ' are assumed to be constant numbers in this paper, and therefore $n^{m/\delta'}$ is polynomial in the input size of the problem.

Suppose that the sets $X_1^* \cap E_L, X_2^* \cap E_L, \ldots, X_m^* \cap E_L$ are guessed correctly, and let $X_L = \bigcup_{i=1}^{m}(X_i^* \cap E_L)$. We then consider the following problem:

$$\text{Minimize} \quad \widehat{p}(X) \qquad \text{subject to} \quad \varphi(X) \leq W, \; X \subseteq E, \tag{2}$$

where

$$\widehat{p}(e) = \begin{cases} p(e) & (e \in E \setminus E_L), \\ 0 & (e \in X_L), \\ (1 + \alpha)mC + 1 & (e \in E_L \setminus X_L), \end{cases}$$

This is an instance of (INDS-S), and its optimal value is bounded by $\widehat{p}(X^*)$. Since the optimal value of (INDS-P) is at most C and the set X_L is guessed correctly, an upper bound of the value $\widehat{p}(X^*)$ can be obtained as follows:

$$\begin{aligned} \widehat{p}(X^*) = \widehat{p}(X^* \cap E_S) + \widehat{p}(X_L) &= \widehat{p}(X^* \cap E_S) \\ &= p(X^* \cap E_S) \\ &= p(X^*) - p(X_L) \leq mC - p(X_L). \end{aligned} \tag{3}$$

Let $X^{**} \subseteq E$ be a $(1 + \alpha)$-approximate solution of the problem (INDS-S) given above and $X_S = X^{**} \cap E_S$. Since the optimal value of (INDS-S) is at most $mC - p(X_L)$ by (3), it holds that

$$\widehat{p}(X^{**}) \leq (1 + \alpha)(mC - p(X_L)). \tag{4}$$

---

[2] By guessing we mean trying all possible assignments by enumeration.

This inequality, together with the definition of $\widehat{p}$, implies that

$$X^{**} \cap (E_L \setminus X_L) = \emptyset.$$

The inequality (4) also implies that

$$p(X_S) + p(X_L) = \widehat{p}(X_S) + p(X_L) = \widehat{p}(X^{**}) + p(X_L) \le (1 + \alpha)mC. \quad (5)$$

We finally construct a feasible solution $(X_1, X_2, \ldots, X_m)$ of (INDS-P) satisfying (1) by assigning edges in $X_S \cup X_L$ to m machines appropriately. Edges in X_L are assigned to m machines according to the guessed assignment. Then, edges in X_S are assigned to m machines in a greedy way as follows. We first assign edges in X_S to the 1st machine until its makespan exceeds $(1+\alpha)C$; then assign remaining edges in X_S to the 2nd machine until its makespan exceeds $(1 + \alpha)C$, and so on. We stop this iteration if we assign all edges in X_S. Due to the inequality (5), all edges in X_S can be assigned to some of the m machines.

Since each edge in X_S has processing time at most $\delta'C$, the makespan of each machine is at most

$$(1 + \alpha)C + \delta'C = (1 + \alpha + \delta')C.$$

Hence, $(X_1, X_2, \ldots, X_m)$ is a feasible solution of (INDS-P) satisfying (1). This concludes the proof of Lemma 1.

Remark 3. It may be possible to obtain a feasible solution of (INDS-P) satisfying (1), even in the case with $C < C^*$; in such a case we have $(1+\alpha+\delta')C \ge C^*$. □

2.2 Proof of Lemma 2

We give a proof of Lemma 2 in the case where the objective is the minimization of performance function $\varphi(X)$; the case of maximization can be proven similarly and omitted.

Let W^* be the optimal value of (INDS2-P). Then, there exists a feasible (and optimal) solution $X_1^*, X_2^*, \ldots, X_m^*$ of (INDS2-P) such that

$$\varphi(X^*) = W^* \text{ with } X^* = \bigcup_{i=1}^{m} X_i^*, \qquad p(X_i^*) \le C \ (i = 1, 2, \ldots, m).$$

Also, let

$$E_S = \{e \in E \mid p(e) < \delta C\}, \qquad E_L = E \setminus E_S;$$

i.e., E_S (resp., E_L) is the set of edges with small (resp., large) processing time. We first guess the assignment of "large" edges $X_1^* \cap E_L, X_2^* \cap E_L, \ldots, X_m^* \cap E_L$ to m machines in the optimal solution. Since $p(e) \ge \delta C$ for each $e \in E_L$, it holds that

$$|X_i^* \cap E_L| \le C/(\delta C) = 1/\delta \qquad (i = 1, 2, \ldots, m).$$

This implies that there exist at most $n^{m/\delta}$ possible choices of the assignment $X_1^* \cap E_L, \ldots, X_m^* \cap E_L$. Since m and δ are constant numbers, $n^{m/\delta}$ is polynomial in the input size of the problem.

Suppose that the sets $X_1^* \cap E_L, X_2^* \cap E_L, \ldots, X_m^* \cap E_L$ are guessed correctly, and let $X_L = \bigcup_{i=1}^{m}(X_i^* \cap E_L)$. We then consider the following problem:

$$\text{Minimize} \quad \varphi(X) \qquad \text{subject to} \quad \widehat{p}(X) \leq mC - p(X_L), \ X \subseteq E, \qquad (6)$$

where

$$\widehat{p}(e) = \begin{cases} p(e) & (e \in E \setminus E_L), \\ 0 & (e \in X_L), \\ (1+\alpha)mC + 1 & (e \in E_L \setminus X_L), \end{cases}$$

This is an instance of (INDS2-S), and its optimal value is at most W^* since X^* is a feasible solution; indeed, X^* satisfies the inequality constraint:

$$\begin{aligned} \widehat{p}(X^*) = \widehat{p}(X^* \cap E_S) + \widehat{p}(X_L) &= p(X^* \cap E_S) \\ &= p(X^*) - p(X_L) \\ &= \sum_{i=1}^{m} p(X_i^*) - p(X_L) \leq mC - p(X_L). \end{aligned}$$

Let $X^{**} \subseteq E$ be a $(1, 1+\alpha)$-approximate solution of the problem (INDS2-S) given above and $X_S = X^{**} \cap E_S$. Since the optimal value of (INDS2-S) is at most W^*, it holds that

$$\begin{aligned} \varphi(X^{**}) &\leq W^*, \\ \widehat{p}(X^*) &\leq (1+\alpha)(mC - p(X_L)). \end{aligned} \qquad (7)$$

This inequality, together with the definition of $\widehat{p}$, implies that

$$X^{**} \cap (E_L \setminus X_L) = \emptyset.$$

The inequality (7) also implies that

$$p(X_S) + p(X_L) = \widehat{p}(X_S) + p(X_L) = \widehat{p}(X^{**}) + p(X_L) \leq (1+\alpha)mC. \qquad (8)$$

We finally construct a feasible solution $(X_1, X_2, \ldots, X_m)$ of (INDS2-P) satisfying

$$\varphi(X) \leq W^* \text{ with } X = \bigcup_{i=1}^{m} X_i, \quad p(X_i) \leq (1+\alpha+\delta)C \ (i = 1, 2, \ldots, m)$$

by assigning edges in $X_S \cup X_L$ to m machines appropriately. Edges in X_L are assigned to m machines according to the guessed assignment. Then, edges in X_S are assigned to m machines in a greedy way as follows. We first assign edges in X_S to the 1st machine until its makespan exceeds $(1+\alpha)C$; then assign remaining edges in X_S to the 2nd machine until its makespan exceeds $(1+\alpha)C$, and so on. We stop this iteration if we assign all edges in X_S. Due to the inequality (8), all edges in X_S can be assigned to some of the m machines.

Since each edge in X_S has processing time at most δC, the makespan of each machine is at most

$$(1 + \alpha)C + \delta C = (1 + \alpha + \delta)C.$$

Hence, $(X_1, X_2, \ldots, X_m)$ is a feasible solution of (INDS2-P) satisfying the desired conditions. This concludes the proof of Lemma 2.

3   Proofs of Corollaries 1 and 2

Throughout this section we assume, without loss of generality, that $E_0 = \emptyset$.

3.1   Case of $\Phi \in \{\Phi_{\mathrm{MST}}, \Phi_{\mathrm{SP}}\}$.

The problem (INDS-S) with $\varphi \in \Phi_{\mathrm{MST}} \cup \Phi_{\mathrm{SP}}$ is reformulated as follows:

Minimize $p(X)$
subject to $\ell(X) \le W$,
    $X \subseteq E$ is a spanning tree (or s-t path) in the graph (V, E).

This is the minimum spanning tree problem (or the shortest s-t path problem) with a knapsack constraint. It is easy to see that (INDS2-S) has the same problem structure as (INDS-S). The minimum spanning tree problem with a knapsack constraint has a PTAS due to Ravi and Goemans [16], and the shortest s-t path problem with a knapsack constraint has a fully PTAS due to Hassin [9]. Hence, Corollary 1 holds in the case $\Phi \in \{\Phi_{\mathrm{MST}}, \Phi_{\mathrm{SP}}\}$. Note that a PTAS can be converted into a $(1, 1 + \varepsilon)$-approximation algorithm for every $\varepsilon > 0$ (see, e.g., [16, Section 1]). Hence, Corollary 2 also holds in this case.

Case of $\Phi = \Phi_{\mathrm{MM}}$. The problem (INDS2-S) with $\varphi \in \Phi_{\mathrm{MM}}$ is reformulated as follows:

Maximize $w(X)$
subject to $p(X) \le C$, $X \subseteq E$ is a matching in the graph (V, E).

This is the maximum-weight matching problem with a knapsack constraint, for which a PTAS exists (see Berger et al. [5]). By using this PTAS for (INDS2-S) and binary search, we can easily obtain a $(1, 1 + \varepsilon)$-approximation algorithm of (INDS2-S) for every $\varepsilon > 0$, and also a PTAS for (INDS-S) (see, e.g., [16, Section 1]). Hence, Corollaries 1 and 2 hold if $\Phi = \Phi_{\mathrm{MM}}$.

Case of $\Phi = \Phi_{\mathrm{UMF}}$. In the problem (INDS-S) with $\varphi \in \Phi_{\mathrm{UMF}}$, we need to find an edge set $X \subseteq E$ that minimizes the value $p(X)$ under the constraint that the amount of a maximum flow in (V, X) is at least W. Since for each edge $e \in E$ its capacity is one and $p(e) \ge 0$, we can assume that for each $e \in X$ contained in the optimal solution of the problem, we have positive flow on the edge e. Hence, we can regard this problem as a minimum-cost flow problem, where we minimize the total flow cost $\sum_{e \in E} p(e)f(e)$ under the constraint that the amount of the

s-t flow represented by the vector $f(e)$ $(e \in E)$ is at least W. This observation shows that (INDS-S) with $\varphi \in \Phi_{UMF}$ can be solved in polynomial time.

Similarly, the problem (INDS2-S) with $\varphi \in \Phi_{UMF}$ can be regarded as the problem of maximizing the amount of an s-t flow under the constraint that the total cost of the flow is at most C. Hence, we can obtain an optimal solution of (INDS2-S) by solving the problem (INDS-S) for all possible value of W. Since each edge has unit capacity, the amount of flow is at most $|E|$, the number of edges in G. That is, it suffices to solve (INDS-S) for $W = 0, 1, \ldots, |E|$, and then find a maximum s-t flow under the cost constraint, which can be done in polynomial time.

References

1. Averbakh, I., Pereira, J.: The flowtime network construction problem. IIE Trans. **44**, 681–694 (2012)
2. Averbakh, I., Pereira, J.: Network construction problems with due dates. Eur. J. Oper. Res. **244**, 715–729 (2015)
3. Bartal, Y., Leonardi, S., Marchetti-Spaccamela, A., Sgall, J., Stougie, L.: Multiprocessor scheduling with rejection. SIAM J. Discret. Math. **13**, 64–78 (2000)
4. Baxter, M., Elgindy, T., Ernst, A.T., Kalinowski, T., Savelsbergh, M.W.P.: Incremental network design with shortest paths. Eur. J. Oper. Res. **238**, 675–684 (2014)
5. Berger, A., Bonifaci, V., Grandoni, F., Schäfer, G.: Budgeted matching and budgeted matroid intersection via the gasoline puzzle. Math. Program. **128**, 355–372 (2011). https://doi.org/10.1007/s10107-009-0307-4
6. Chekuri, C., Khanna, S.: A polynomial time approximation scheme for the multiple knapsack problem. SIAM J. Comput. **35**, 713–728 (2005)
7. Engel, K., Kalinowski, T., Savelsbergh, M.W.P.: Incremental network design problem with minimum spanning trees. J. Graph Algorithms Appl. **21**, 417–432 (2017)
8. Goemans, M.X., Unda, F.: Approximating incremental combinatorial optimization problems. In: Proceedings of APPROX/RANDOM, LIPIcs, vol. 81, pp. 6:1–6:14 (2017)
9. Hassin, R.: Approximation schemes for the restricted shortest path problem. Math. Oper. Res. **17**, 36–42 (1992)
10. Kalinowski, T., Matsypura, D., Savelsbergh, M.W.P.: Incremental network design with maximum flows. Eur. J. Oper. Res. **242**, 51–62 (2015)
11. Kellerer, H.: A polynomial time approximation scheme for the multiple knapsack problem. In: Hochbaum, D.S., Jansen, K., Rolim, J.D.P., Sinclair, A. (eds.) APPROX/RANDOM -1999. LNCS, vol. 1671, pp. 51–62. Springer, Heidelberg (1999). https://doi.org/10.1007/978-3-540-48413-4_6
12. Liu, X., Li, W.: Approximation algorithm for the single machine scheduling problem with release dates and submodular rejection penalty. Mathematics **8**, 133 (2020)
13. Liu, X., Li, W.: Approximation algorithms for the multiprocessor scheduling with submodular penalties. Optim. Lett. **15**(6), 2165–2180 (2021). https://doi.org/10.1007/s11590-021-01724-1
14. Nurre, S.G., Cavdaroglu, B., Mitchell, J.E., Sharkey, T.C., Wallace, W.A.: Restoring infrastructure systems: an integrated network design and scheduling (INDS) problem. Eur. J. Oper. Res. **223**, 794–806 (2012)

15. Nurre, S.G., Sharkey, T.C.: Integrated network design and scheduling problems with parallel identical machines: complexity and dispatching rules. Networks **63**, 303–326 (2014)
16. Ravi, R., Goemans, M.X.: The constrained minimum spanning tree problem. In: Karlsson, R., Lingas, A. (eds.) SWAT 1996. LNCS, vol. 1097, pp. 66–75. Springer, Heidelberg (1996). https://doi.org/10.1007/3-540-61422-2_121
17. Shabtay, D., Gaspar, N., Kaspi, M.: A survey on offline scheduling with rejection. J. Sched. **16**, 3–28 (2013). https://doi.org/10.1007/s10951-012-0303-z
18. Wang, T., Averbakh, I.: Network construction/restoration problems: cycles and complexity. J. Comb. Optim. (2021, published online)
19. Zheng, H., Gao, S., Liu, W., Wu, W., Du, D.-Z., Hou, B.: Approximation algorithm for the parallel-machine scheduling problem with release dates and submodular rejection penalties. J. Comb. Optim. (2022, published online)

Author Index

Printed in the United States
by Baker & Taylor Publisher Services